Canonical Nonlinear Modeling

Canonical Nonlinear Modeling

S-System Approach to Understanding Complexity

Edited by

Eberhard O. Voit
Department of Biostatistics, Epidemiology, and Systems Science
Medical University of South Carolina

VNR VAN NOSTRAND REINHOLD
New York

Library of Congress Catalog Card Number 90-48603
ISBN 0-442-00479-6

Manufactured in the United States of America

Published by Van Nostrand Reinhold
115 Fifth Avenue
New York, New York 10003

Chapman and Hall
2-6 Boundary Row
London, SE1 8HN

Thomas Nelson Australia
102 Dodds Street
South Melbourne 3205
Victoria, Australia

Nelson Canada
1120 Birchmount Road
Scarborough, Ontario M1K 5G4, Canada

16 15 14 13 12 11 10 9 8 7 6 5 4 3 2 1

Library of Congress Cataloging-in-Publication Data

Canonical nonlinear modeling: S-system approach to understanding complexity / Eberhard O. Voit, editor.
p. cm.
Includes index.
ISBN 0-442-00479-6
1. Biological systems—Mathematical models—Congresses. 2. Differential equations—Congresses. 3. Differential equations. Nonlinear—Congresses. I. Voit, Eberhard O. II. Title: S-system approach to understanding complexity.
QH323.5.C364 1991
574′.01′5118—dc20 90-48603
CIP

Contributors

Ahmad M. Bin Razali, Department of Biostatistics, Epidemiology, and Systems Science, Medical University of South Carolina, Charleston, SC 29425-2503; and: Iabatan Statistik, Pusat Pengajian Kuantitatif, Universiti Kebangsaan Malaysia, 43600 Bangi, Malaysia.

Enric I. Canela, Departament de Bioquímica i Fisiologia, Facultat de Química, Universitat de Barcelona, 08071 Barcelona, Spain.

Marta Cascante, Departament de Bioquímica i Fisiologia, Facultat de Química, Universitat de Barcelona, 08071 Barcelona, Spain.

Rafael Franco, Departament de Bioquímica i Fisiologia, Facultat de Química, Universitat de Barcelona, 08071 Barcelona, Spain.

Katsuya Hayashi, Department of Agricultural Chemistry, Faculty of Agriculture, Kyushu University, Fukuoka 812, Japan.

Douglas H. Irvine, Department of Molecular, Cellular and Developmental Biology, University of Colorado, Boulder, Colorado 80309-0347.

Thomas Johnson, Department of Economics and Business, Box 8110, North Carolina State University, Raleigh, NC 27695-8110.

Donald C. Lewis, Department of Mathematics, College of Charleston, Charleston, SC 29424.

Mitchelle S. Morrison, Department of Biostatistics, Epidemiology, and Systems Science, Medical University of South Carolina, Charleston, SC 29425-2503.

Masahiro Okamoto, Department of Biochemical Engineering and Science, Faculty of Computer Science and Systems Engineering, Kyushu Institute of Technology, Iizuka, Fukuoka 820, Japan.

Philip F. Rust, Department of Biostatistics, Epidemiology, and Systems Science, Medical University of South Carolina, Charleston, SC 29425-2503.

Peter J. Sands, CSIRO Division of Forest Research, P.O. Box 4008, Queen Victoria Terrace, Canberra, ACT 2600, Australia.

Michael A. Savageau, Department of Microbiology and Immunology, 6730

Medical Science Building II, The University of Michigan, Ann Arbor, MI 48109-0620.

Albert Sorribas, Departament de Ciencies Mediques Basiques, Divisio de l'Estudi General de Lleida, Universitat de Barcelona, 25007 Lleida, Spain.

Joni A. Torsella, Department of Biostatistics, Epidemiology, and Systems Science, Medical University of South Carolina, Charleston, SC 29425-2503.

Eberhard O. Voit, Department of Biostatistics, Epidemiology, and Systems Science, Medical University of South Carolina, Charleston, SC 29425-2503.

Volker Weinberger, Department of Biostatistics, Epidemiology, and Systems Science, Medical University of South Carolina, Charleston, SC 29425-2503. Current Address: Fachbereich Medizinische Informatik, Fachhochschule Heilbronn, Max-Planck-Str. 39, D-7100 Heilbronn, West Germany.

Jun-ji Yoshii, Hitachi Software Engineering Co., Ltd., Biotechnology Project, 1-2-20, Kohrinboh, Kanazawa 920, Japan.

Contents

CONTRIBUTORS		v
PREFACE		ix
Chapter 1.	20 Years of S-Systems *Michael A. Savageau*	1
PART I	**GENERAL METHODS OF CANONICAL MODELING**	
Chapter 2.	Introduction to S-Systems *Eberhard O. Voit, Michael A. Savageau, and Douglas H. Irvine*	47
Chapter 3.	Optimal Strategies for Modeling Biochemical Systems *Albert Sorribas*	67
Chapter 4.	Sensitivity Analysis *Marta Cascante, Rafael Franco, and Eric Canela*	76
Chapter 5.	The Method of Controlled Mathematical Comparison *Douglas H. Irvine*	90
Chapter 6.	Symbolic Analysis of S-Systems with MACSYMA *Volker Weinberger*	110
Chapter 7.	Analysis of Complex Dynamic Networks with ESSYNS *Douglas H. Irvine, Michael A. Savageau, and Eberhard O. Voit*	133
PART II	**APPLICATIONS**	
Chapter 8.	Completely Uncoupled or Perfectly Coupled Circuits for Inducible Gene Regulation *Michael A. Savageau and Peter J. Sands*	145

Chapter 9. GTP-Dependent Kinetic Proofreading
Masahiro Okamoto, Jun-ji Yoshii, and Katsuya Hayashi 166

Chapter 10. An Analysis of Forestry Data
Joni A. Torsella and Ahmad M. Bin Razali 181

Chapter 11. Estimating Parameters of S-Systems
Thomas Johnson 200

PART III MATHEMATICAL PROPERTIES OF S-SYSTEMS

Chapter 12. Canonical Nonlinear Forms: Recasting
Eberhard O. Voit 213

Chapter 13. Applications of Recasting in Statistical Distributions
Philip F. Rust 238

Chapter 14. Recasting as a Modeling Tool
Mitchelle S. Morrison 258

Chapter 15. Algebraic Properties of Canonical Forms
Eberhard O. Voit 278

Chapter 16. Qualitative Analysis of S-Systems: Hopf Bifurcations
Donald C. Lewis 304

References 345

Index 361

Preface

Nature is a complex system of complex systems. Complexity is not only evident in large systems such as rain forests, but even in individual cells composed of thousands of constituents that interact in a marvellously orchestrated scheme to perform their assigned duty in life. Understanding a complex system requires insight regarding the relationships between the global behavior of the system and the properties of its components. Only if these relationships are truly understood is it possible to explain why one system design is selected over a multitude of imaginable alternative designs. Such understanding also allows reliable predictions of system responses to internal or external perturbations. Small perturbations are a fact of daily life for every organism. In addition, natural systems routinely exhibit substantial alterations in their development, for instance, when new mutants are tested against the predominant wild type or when environmental conditions change significantly. Perturbations also can be consequences of human interference with nature. They can be intentionally induced to improve suboptimal conditions, as in medicine or the breeding of new crops. They may be unintentionally induced as well, as can be seen in nearly every ecosystem examined today. Understanding complexity in nature means understanding the responses of natural systems to perturbations.

Biology has met the challenge of enormous complexity primarily with reductionistic methods. The underlying philosophy has been that sufficient characterization of the components reveals how an organism functions. In fact, there is no doubt that the reductionistic approach has been extremely successful. Over the past three decades, biochemistry, genetics, and microbiology have led to an unprecedented increase in knowledge about the molecular constituents of living matter. In relatively simple viral or bacterial systems, almost all components are biochemically identified and many of the regulatory processes that govern these systems are fairly well characterized. Nevertheless, it has become clear that the reductionistic approach alone is not sufficient to yield an understanding deep enough to allow reliable predictions of system responses under novel conditions.

What is needed, in addition to the reductionistic methodology, is a systematic approach to reconstructing the integrated system from knowledge about its component parts and representing its dynamic behavior *in vivo*. Because natural systems are typically composed of large numbers of constituents and interactions, modulations, and control mechanisms, the only language capable of simultaneously capturing all essential aspects of a complex system appears to be mathematics. Among the various candidates for a mathematical system description, linear formulations usually receive top ranking, even though it is well known that nature typically exhibits nonlinear characteristics that are difficult if not impossible to capture with linear models. The main reason for the choice of a linear representation is that powerful mathematical tools and recipes are available for most aspects of linear analysis. This is because linear representation always have the same structure. All functions involved are linear by definition, and the differences between linear models are solely caused by the numerical values of their parameters. In contrast, most current nonlinear modeling methods have no common basis and, as a consequence, one frequently encounters problems that require a novel mathematical structure. A unified approach seems to be missing. We need a canonical nonlinear formalism general enough to represent a substantial portion of the nonlinear behaviors in nature.

Twenty years ago, Dr. Michael Savageau published the first concepts of a new formalism for the analysis of complex biological systems. This formalism was called the *power-law formalism* because it is based on a particular type of ordinary differential equation in which the component processes are characterized by power-law functions. The formalism resulted from a combination of ideas in biochemistry and network theory, and was found to possess some of the most significant advantages of linear systems analysis. Among different implementations of the power-law formalism, one mathematical representation was identified as being particularly useful for the analysis of synergistic and saturable biological systems and this representation has been called an *S-system*. S-systems are highly structured nonlinear ordinary differential equations and allow simple mathematical analyses.

Given the homogeneous form of S-systems and their mathematical advantages, the obvious question in seeking a canonical form for nonlinear systems is whether S-systems are general enough to represent all essential properties of complex systems. Initially, this question could only be answered with the experience that S-systems are quite flexible and capable of generating typical nonlinear characteristics like saturation and limit cycles. However, it is now known that virtually any ordinary differential equation can be *recast* exactly in S-system form. That is, a differential equation can be transformed into an S-system that generates the same responses as the nontransformed equation. This fact is astounding. It demonstrates that the S-system structure is not just another nonlinear model with some nice mathematical features. Rather, this S-system structure can be considered a true canonical form that includes virtually all differentiable nonlinearities as special cases. This generality in scope, combined with specificity in structure, is a prerequisite for classifying, comparing, and generalizing nonlinearities, and for developing generally applicable methods of canonical nonlinear modeling.

In March 1989, an international group of researchers met at the Medical University of South Carolina in Charleston, S.C. for the First S-system Symposium. The exchange of novel concepts and applications was exciting, and long discussions between experts and newcomers showed the need for a book leading from the basic concepts of canonical nonlinear modeling with S-systems to the current frontiers.

This book is the result of the collaborative effort of the leading S-system experts. Each chapter has been thoroughly reviewed by peers and revised by the authors and the editor. All references have been collected at the end of the book to facilitate reading and access to the original literature in the area. An index of key words allows the reader quickly to find topics of particular interest.

The authors have agreed to this extra effort in order to produce a book that serves two purposes. First, the book is intended to give the reader an easy, but not oversimplified, entry into the area of canonical modeling with S-systems. Since a modest mathematical background is sufficient to understand almost all material, the book addresses a wide audience of scientists interested in complex systems. Second, the book offers a representative snapshot of most of the current activities in S-system research. Because the field of canonical nonlinear modeling is young and vigorous, the reader will find that all chapters, whether they describe the development of methods or particular applications, set directions for new research. Having understood the material of the book, the reader will be prepared to follow and contribute to future research developments.

The book begins with a synopsis of the development of canonical nonlinear modeling and the prominent role of S-systems. Chapter 2 introduces the mathematical concepts of S-systems and the underlying power-law formalism. Chapters 3, 4, and 5 extend the modeling methods to more advanced topics and show how S-systems can help us to understand the design principles of complex systems. Chapters 6 and 7 address two complementary types of computer-aided analysis. Chapter 6 outlines the potential of symbolic analyses with MACSYMA, and Chapter 7 describes ESSYNS, a comprehensive program that was developed specifically for the numerical analysis of S-systems. All methodological chapters include sufficient detail so that the interested reader can actually employ the material in new contexts. Chapters 8 through 11 deal with biological applications and demonstrate how specific problems can be analyzed quantitatively with canonical representations. Chapters 12 through 16 elucidate prominent mathematical properties of S-systems and related canonical forms and contain many seeds for new research on the canonical nonlinear forms themselves.

Many people have helped to produce this volume. I would like to thank and applaud the contributors and peer reviewers for taking the time to produce science of the highest quality, for accepting and providing critique, and for rigorously rewriting and improving their chapters. Special appreciation should be given to Dr. Philip F. Rust who co-organized the First S-system Symposium and throughout the symposium preparations and the completion of this book was always willing to help solve the scientific and logistic problems. The First S-system Symposium and this book could not have been realized without the support of Burroughs-Wellcome Corporation, the National Institutes of Health, Sigma Xi, the USDA Forestry Service, and Zenith Data Systems.

It is my hope that many readers of this book will become intrigued by the beauty and potential of canonical nonlinear modeling with S-systems and that some, having learned the basic methodology, will be eager to use it in their own research. When these readers join the growing S-system family, then this book will have served its purpose.

Eberhard O. Voit
Charleston, SC

Canonical Nonlinear Modeling

1. 20 Years of S-Systems

MICHAEL A. SAVAGEAU

It is with mixed feelings that I have accepted Dr. Voit's request to recount the development of S-systems. On the one hand, I feel some sense of satisfaction about the developments over the past twenty years. Also there is great excitement about the possibilities for the next twenty years. On the other hand, one should always have some misgivings about "recounting the early days." It sounds too much like a pending retirement and here I am still in my forties!

In the end he convinced me by saying that it didn't have to be a finely polished scientific presentation. In fact it might be more informative if I were to reminisce, to describe the origin of those ideas that worked and those that didn't, and to include some of the personal influences that may have contributed to the development of the field. As Root-Bernstein [74] has pointed out, the creative process of discovery is considered a prime component of the scientific enterprise, and yet it receives almost no attention in the training of scientists, and most practicing scientists are loathe to discuss the issue openly. I share his view.

Nevertheless, my personality (or perhaps my own scientific training) makes it difficult for me to approach the subject. I am all too aware that scientists are notorious for attempting to rewrite history for their own polemical purposes. Scientists are no different from others in this regard, but they are often more hypocritical in doing so because the practice is so at odds with the conventional image of science as the disinterested pursuit of truth. For those who are self-critical and have a sense of history, it is well known that it is all too easy to abuse history, either purposefully through fraud or inadvertently through ignorance and self-deception. Remember what has been said about Herodotus, "He is the Father of History and the Patron Saint of Liars." The answer is not to avoid all discussion of history, but rather to have more and better history.

Indeed, what is most needed is to develop a critical sense of history as part of the training of every scientist.

As it stands, scientists often exhibit a naive ahistorical attitude toward their discipline that leads them to believe that such abuses are not a part of science. Broad and Wade [54] have argued persuasively that fraud and self-deception are common in science, although there is a spectrum between these two manifestations that often makes it difficult to distinguish their relative contributions in any given instance. These authors summarize their penetrating study of the scientific process as follows: "From its earliest days, science has been an arena in which men have striven for two goals: to understand the world and to achieve recognition for their personal efforts in doing so. This duality of purpose lies at the foundation of the scientific enterprise. Only through recognition of the double goal can the motives of scientists, the behavior of the scientific community, and the process of science itself be properly understood."

The best immunity is to be on guard against rhetoric, whose purpose is persuasion and not the elucidation of truth, and above all to read critically the original sources and to make up one's own mind. With these forewarnings, let us begin.

INTRODUCTION

The topic of S-systems has become much more diverse than anyone ever dreamed possible twenty years ago. Thus, there are many different themes that one might use to organize this presentation. After some consultation with Dr. Voit, it was agreed that I would adopt a chronological account. Such an approach provides a unique perspective for those new to the field—and make no mistake, this is still a very young field. This perspective is difficult to obtain from the rigid style of the modern research paper, in which one is not allowed to describe failed experiments, the roads not taken, personal influences, and so forth.

So let me begin at the beginning, with my earliest scientific impressions having something to do with what is now called the power-law formalism, and in particular the S-system representation within this nonlinear formalism.

1954–1958

My first vivid memory in this regard is that of a fourteen-year-old in junior high school. As a part of our general science class we read a weekly science newsletter that was supposed to awaken us to the wonders of science. I cannot remember its name, but it contained some assorted facts and stories about science in the news. Most of the time students paid little attention, and for some it was a joke. But one issue took my fancy: it was a story about the principle of similarity, which goes back at least to Galileo.

Galileo noted that the proportions of an organism must necessarily change as its size increases. If the diameter of a sphere were to increase 4-fold, then its mass, which is proportional to its volume, would increase 64-fold. On the other hand, the diameter of a typical supporting structure, with strength proportional to cross-sectional area, would have to increase 8-fold. Hence, the diameter of the supporting structure increases disproportionately more than the diameter of the sphere. Similarly, if the height of a person were to increase by 4-fold, then the mass would increase by an even greater factor, and the diameter of the legs to support this mass would increase by more than 4-fold. With this type of scaling it is impossible to keep the same shape—in fact, the person would begin to take on the proportions of, say, an elephant.

This really struck me, because there was a simple beauty in this scientific principle that has to do with power-law functions and relative change, and it went all the way back to Galileo. These concepts were to reappear repeatedly in my education, and even though the context might differ—be it metabolic physiology, biochemical kinetics, growth, morphogenesis, sensory physiology, and nonbiological subjects such as economics, physics, communication theory, *etc.*—I recognized them as familiar friends and warmly remembered my first encounter with them through the eyes of Galileo. In Table 1.1, I have listed a number of investigators who have contributed to this tradition in biology. We shall come back to point out a few of these developments in more detail as we go along. I eventually learned that Robert Rosen [73] and others had written extensively on the concept of similarity and pointed out its long and very rich tradition, but for me it started with the simple newsletter and the story of Galileo. So you see that made a very strong impression on me.

There was another equally strong impression that was coupled to it. The next day, in my enthusiasm, I told a friend about this scientific insight that I had gained by reading Galileo's account. He laughed at me and told me I was crazy.

Table 1.1. Principle of Relative Change in Biology: The Power-Law Function

$$\left(\frac{dy}{dx}\right)\left(\frac{x}{y}\right) = a \text{ or } y = bx^a$$

1638	Galileo	Morphology
1838	Rameaux and Sarrus	Physiology
1847	Bergmann	Physiology
1867	Guldber and Waage	Chemical Kinetics
1902	Brown, Henri	Biochemical Kinetics
1917	D'Arcy Thompson	Growth and Form
1932	Huxley, Teissier	Growth and Development
1950	Needham	Biochemistry
1960	Bertalanffy	Biological Systems

He obviously didn't see the same sort of beauty in the idea, if indeed he saw the point at all. This was my first lesson on the human side of science!

1958–1962

The next impressions that I recall with any clarity came up in college when I was introduced to modern physics and engineering. It was then that I internalized for the first time the notion that there is an art to approximation. The real world is complex and multifaceted—indeed, this is a cliche. One can never hope to deal with it directly and entirely; everything we do in science involves approximation. But there's an art to it; not just any approximation will do.

> *We should make things as simple as possible, but not simpler.*
>
> —Einstein

One must choose approximations judiciously. They should simplify so that one might see more clearly the basic principles, and at the same time they must adequately represent the phenomenon under study. One can go too far and lose the very essence of the phenomenon under study. Thus, approximation involves the related issue of representation.

This issue came up in practically every science course that I took, and to the extent that there was a common approach, it invariably involved progressive approximation by series representations; for example, Taylor series and Fourier series were among the most common. These tools from mathematical analysis guaranteed a representation that could be as accurate as desired provided certain fairly reasonable assumptions about the process were satisfied. These are indeed remarkable theorems that still inspire awe. But perhaps only slightly less marvelous is the observation that in many practical or experimental situations one obtains an acceptable representation by using only the first few terms of such a series. Indeed just the zero- and first-order terms constitute the linear approximation on which much of classical mathematics, science, and engineering are based. Nowhere was the judicious use of approximation more beautifully illustrated than in network theory.

Network theory, in the tradition beginning with Euler and leading through Kirchhoff to Bode and modern analysis and synthesis, was to have a profound impact on me in other respects as well. It immediately gives one the impression of being a logically structured and rigorous theory. There were general theorems about the topology of networks (Euler) that allowed one to formulate the network equations in the least redundant fashion. These theorems demonstrated the equivalence of different methods for writing the fundamental network equations (Kirchhoff's laws). The duality between the flux and the node representations was proved, and this led to an identification of the minimum number of equations needed to describe a given network. This was obviously a very extensively developed theory, and it really appealed to my sense of aesthetics. In my mind, it became the most useful theory for dealing with complex systems. Most of the early portions of this course on network theory were devoted to analysis in the time domain, and of course the systems were linear. Nevertheless, when we were introduced to the frequency domain, through the work of Bode, I learned an important lesson in representing nonlinear functions.

Bode was to have a profound influence on me in several respects; initially it was his insight into the meaning of logarithmic plots and how they can be used to simplify the analysis of signal propagation in complex networks. I had encountered logarithmic plots many times before in physics and chemistry, but what I remembered was simply their utilitarian value. If you had a wide range of data and you wanted to have it all fit on the same plot, then you took the logarithm. This was never seen as having any fundamental significance. In fact, a few years later a well-known physical biochemist, reflecting the views that I had been presented with up to this point in my education, said that you get an "A" if you can represent the relationship between two variables in a rectilinear plot; you get a "B" if you have to use a semi-logarithmic plot; and you get a "C" if you have to use a log-log plot. The obvious reason is that anything can be fit with a log-log plot.

Bode saw it differently. When a linear network is stimulated by an input signal that is a pure sinusoid of a given frequency, the output signal monitored at some other point in the network will also be a pure sinusoid of the same frequency. However, the amplitude and phase of the signal typically differ from those of the input signal. The amplitude of the output signal relative to that of the input signal represents the magnitude of the "transfer function" between input and output, and the early network theorists showed that the transfer function is a special type of rational function in the frequency variable. If one plots this transfer function in relation to the frequency of the sinusoid, one obtains a complex characterization of the network. Although one could show that the physical parameters that characterize the components of the network are related to the parameters of the rational function, it was difficult to analyze a given network and readily predict the shape of the transfer function when plotted against frequency. Conversely, the problem of synthesizing a given physical network to have some desired shape for the transfer function was nearly impossible. Bode showed that both analysis and synthesis could be greatly simplified by representing the transfer function in log-log coordinates, which in network theory is called a Bode plot.

Bode showed that the numerator and denominator of the rational function could be factored into products:

$$V(S) = \frac{\epsilon_0(S + \mu_1)(S + \mu_2) \cdots (S + \mu_m)}{\phi_0(S + v_1)(S + v_2) \cdots (S + v_n)}$$

Then by taking logarithms these products could be separated into a simple sum of terms, of which there are only four distinct types:

$\log G$	G, a positive real number
$\log S^\sigma$	σ, a positive integer
$\log\left(\frac{S}{\mu} + 1\right)$	μ, a positive real number
$\log\left[\left(\frac{S}{S_0}\right)^2 + 2\zeta\left(\frac{S}{S_0}\right) + 1\right]$	S_0, a positive real number,
	ζ, a positive real number less than one

Moreover, this separation is not only of the mathematical terms in the transfer function, but also of the graphical components in the plot of the transfer function (see Fig. 1.1), and of the physical components that constitute the network. [This presentation actually depicts the modified Bode analysis that is appropriate for biochemical rate laws (see below).]

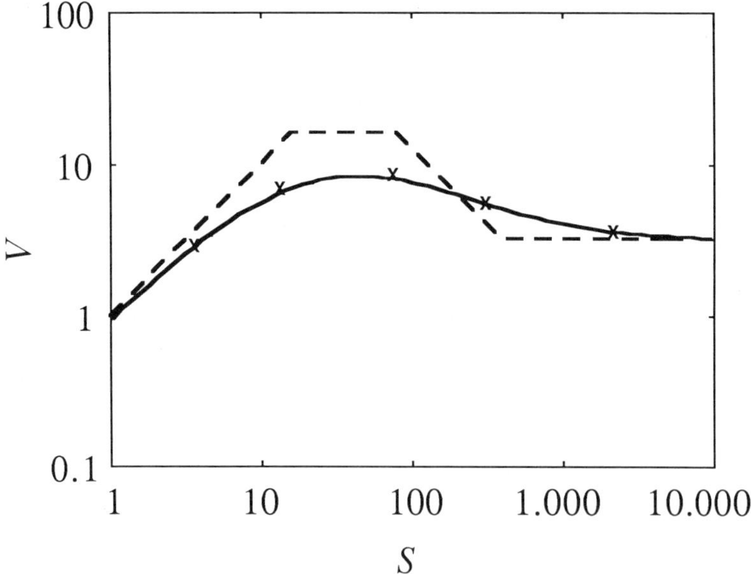

FIGURE 1.1. Modified Bode Plot. Log-log plot of rate versus concentration. Asymptotes determined by individual terms in the factored form of the rate law [2].

Bode showed that the transfer function was accurately approximated by straight-line segments over a rather wide range of frequencies in the log-log plot. It was possible to analyze a given network by constructing a mathematical representation, determining the asymptotes of the Bode plot, and quickly sketching the transfer function. Similarly, one could synthesize a network of physical components to match a desired transfer function by sketching an appropriate set of asymptotes in the Bode plot, dissecting out the mathematical terms that correspond to each asymptotic element, and then identifying each of these with an appropriate physical component.

The log-log plot was no longer a sort of mathematical artifact, but it could reveal physical meaning. This became a very important idea in my thinking. Each of the separate terms in the rational function can be given specific physical meaning. When you try and create some understanding of the function represented, you can break it into separate components with separate identities and then add them back together to make a composite that is constructed of straight-line asymptotic regions. There are some very simple rules for the behavior around the break points, where two straight lines meet, and by using these rules and the asymptotes you simply sketch in the resulting curve.

Recall that these developments occurred in the early days of computers, before their use became widespread. Thus, most of these calculations were done by hand in the fifties and early sixties. Bode analysis was a great advance in our ability to think about complex rational functions. But it was fairly tedious to implement without computers, as you can imagine. As a student, I was happy to be finished with Bode analysis, but the seeds were sown, and after several years germinating, Bode analysis would bloom again in a variant form.

The course that followed network theory introduced me to the nonlinear world, and it was a very messy world at that time. Nearly everyone was trying to avoid it because there really were no general tools to examine nonlinear phenomena systematically. There was still the "lure of the linear." Every time you encountered a nonlinear system, you tried to force it into some linear form with which you could work. Of course, the real reason for doing this was that the power of linear mathematics could then be brought to bear on the problem (Table 1.2). Linear mathematics is probably the paradigm of successful analysis. It has a very regular structure, and so it is very systematic; it is general and simple and very revealing. In principle, linear problems are completely soluble. Linear analysis is guaranteed to be an accurate representation, even of nonlinear systems, at least over some limited range of variation in the system variables.

The disadvantage of linear mathematics is that it does not capture the essential nonlinear properties that are inherent in natural phenomena. It is these nonlinearities that ultimately are responsible for the most interesting properties that we associate with living systems. Nevertheless, at the time, linear analysis was the paradigm to be emulated. The modest extension to piecewise linearity was

Table 1.2. The Linear Formalism

Two-variable example

$$dX_1/dt = a_{11}X_1 + a_{12}X_2 + b_1$$
$$dX_2/dt = a_{21}X_1 + a_{22}X_2 + b_2$$

or

$$dX/dt] = [A]X] + b]$$

Advantages

Systematic
General
Simple
Revealing (completely solvable)

Disadvantages

Fails to capture essential properties
Limited range of accuracy

about as far away as one ventured. In this approach one pieced together a sequence of straight-line segments to represent the nonlinear relationships. This was an obvious next step, but very little work was done to address the nonlinearities in a really serious way.

1962–1966

The principle of similarity enunciated by Galileo provided my first encounter with the use of power-law functions in science; it also was the first example that I had seen where biology and mathematics were united in a fundamental way. However, in the years that followed I was to learn that keeping these two together is always difficult, in large part because of the traditional differences between these disciplines. I learned that there are nearly as many forces tending to pull them apart as there are tending to bring them together.

The next event in my training that brought these disciplines together occurred in graduate school when I was studying physiology. This was to be my second memorable encounter with the power-law function in biology. If you take phylogenetically different organisms of different sizes and graph their oxygen consumption, or metabolic rate, as a function of body weight, you obtain a nice straight-line relationship in a log-log plot. This is a familiar observation that goes back to Rameaux & Sarrus and Bergmann (Table 1.1). At that time no one really understood what it meant, of course, but it was a very firm observation that has been repeated through the years. You also can take a given organism at different sizes during its development, and its metabolic rate can be shown to vary in a similar fashion. These relationships are now seen to be a manifestation of the similarity principle; you can't change the size of an organism and keep everything else about the system the same. These observations were to reawaken in me thoughts about mathematical regularities in biology and how they might be explained in terms of underlying mechanisms.

About midway through my Ph.D. training I had by chance encountered the work of the physiologist Ludwig von Bertalanffy. He introduced the notions of open systems and steady-state behavior. He was among the first to focus our attention on biological *systems* as such. As far back as the early 1920s, he emphasized that organisms are integrated systems; one cannot simply reduce them to their elements and thereby address the fundamental issues in biology. He showed that every property of an intact biological system is in principle dependent on all of the components that constitute the system. He emphasized the fundamental task in biology, which is to discover the laws—or what I like to call the design principles—of biological systems and not just describe biological phenomena in all their beauty and diversity.

> *[A system consists of] a set of elements standing in interaction, . . . a complex of components in mutual interaction, . . . a complex of interacting elements, . . . a dynamic order of parts and processes standing in mutual interaction. . . . The fundamental task of biology [is] the discovery of the laws of biological systems.*
>
> —Bertalanffy (1928)

He was a very seminal thinker with a broad-scale approach [60]. He was thinking of systems not only in biology, but also in sociology and philosophy (see [51]). Reading his papers was a very intense experience for me, because he made connections that other biologists normally don't make, and he made them for the first time in a systematic way. He made these points in relative isolation, and, in recent years, his work has been ignored for the most part by mainstream biologists. There were of course a few others who shared this systems perspective in biology. The geneticist Chetverikov, who is thought of as the father of physiological genetics, comes to mind. He independently expressed similar ideas, noting that one cannot define the function of a gene in isolation, but one really has to look at the complex of interactions in the intact system:

> *"Each inherited trait . . . is determined by not just some one gene but by their whole aggregate, their complex. The very same gene will manifest itself differently, depending on the complex of the other genes in which it finds itself. The phenotypic expression of each gene is determined by its "genotypic milieu."*
>
> —Chetverikov (1926)

I was studying the traditional physiology curriculum, but as I read the older literature and Bertalanffy especially, I found myself more and more stimulated by this systems perspective. At the same time there was the practical task of mastering the physiology that I was being taught. It soon became apparent to me that physiologists were doing systems analysis without the appropriate tools, and what's more, they preferred it that way! I can recall taking an exam and being penalized for trying to use mathematics. We learned that changes in the acid-base balance altered reactions in the blood, and the products of these reactions were sensed by chemoreceptors that led to an increased respiration rate, that altered dissolved CO_2 in blood, that counteracted the changes in pH, *etc*. You were expected to write an essay describing these sequences of cause-effect linkages; you were not supposed to use symbols or mathematical reasoning. This experience almost turned me away from biology. In retrospect, it was the darkest moment before the dawn—I was ready to give it up.

Then I discovered Jacob and Monod [64]. This for me was biology as it ought to be practiced; there were circuits with well-defined interactions, beautiful models, and logical experiments to test a general theory. Their concept of allosteric interactions, which for the first time allowed a distinction to be made between the "catalytic structure" and the "regulatory structure" of a biological system, was to have a profound effect on thought patterns in molecular biology. The logical precision that I came to associate with Monod's thought was immediately understandable to me when I read of his early mathematical training [69]. It is still a very stimulating experience to read these papers.

As I struggled with my experimental system (heart muscle cells in tissue culture) and began formulating my first molecular models, I was naturally attracted to the model of Monod, Wyman and Changeux [70] for representing regulatory interactions. However, I soon learned that this rational function model was much too complex if one hoped to do any sort of general symbolic analysis and deduce general properties of a system that contained a fair number of such reactions. As I groped for suitable alternatives I considered the Hill equation as a simplified rational function. This choice was influenced by Brian Goodwin [62] who had pioneered the use of this simple function in his models for the control of gene expression. In what has become known as the Goodwin model, one uses a simple linear representation for all of the interactions in the system except for the feedback inhibition, which is represented by a simple rational function of the Hill type. This approximation facilitated analysis, but it eventually proved restrictive in my case because I wished to consider situations in which nonlinearity was distributed throughout a model. If one considered many of the processes to be nonlinear, then rational functions of the Hill type were also too complex for symbolic analysis.

1966–1970

As I struggled with the representation of nonlinearities in my model I recalled my classes in communication theory. In particular, I remembered an elegant analysis of the signal-to-noise ratio in nonlinear systems represented by power-law equations. I tried to emulate this representation in my biochemical system. It was immediately apparent that a simple power-law function is the limiting case of the Hill rate law in those regions where synergism or saturation dominate. This simple idea led to my first power-law model of gene expression, which I included as an appendix to my thesis. The simplicity of the analysis was seductive, and I immediately fell in love with the idea. One could represent the nonlinear interactions in gene regulation by simple power-law functions; these could readily be solved and used to make rather general predictions that did not require numerical solutions, as is usually the case with nonlinear models. One could show immediately how the degree of cooperativity in the regulatory interaction influenced the steady-state behavior of the system.

As I said, I fell in love with the idea and wrote it up as a short note for submission to the *Journal of Theoretical Biology*. A few weeks later it came back rejected, and I was crestfallen. This turned out to be a very important event in my scientific career.

I was lucky enough to have received the benefit of sound advice from an insightful individual who was also conscientious about his responsibilities as a reviewer. The review was well thought out and brutally honest. Perhaps most important for my development at the time and for the subsequent development of the power-law formalism, it also offered constructive criticism. The reviewer said: ''The power-law approach is not sufficiently developed; nevertheless, one can see that there is the germ of a good idea in this paper. The author should be encouraged to develop it more thoroughly.'' In retrospect, it was wise advice. If that paper had been accepted, I might well have continued to be pleased and willing to consider it the end of the story.

Of course as a student I still considered it a rejection, and I didn't pay much attention to the advice at the time. Instead, I went back to the important paper by Monod, Wyman and Changeux. They had developed a complex rational function to describe allosteric interactions in proteins. As I was thinking about these rational functions, I was struck by the similarities to the transfer functions of network theory, and then all the ideas of Bode flooded my mind. It all began to come together. Here were these rational functions that are formally similar to those found in the transfer functions of network theory. I saw how one could combine these notions from biochemistry and network theory to provide a new method for analyzing enzyme-kinetic data (see Fig. 1.1). I wrote up a manuscript showing how one could modify Bode analysis and make this general approach available for enzyme-kinetic analysis. The major advantage of this approach is that it is not restricted to rate laws that produce double-reciprocal plots that are linear, as was true then of the conventional enzyme-kinetic methods (and in many ways is still true today). Also the error structure of experimental data is naturally accounted for in Bode analysis, so no special weighting of the data is required. In principle, this method allowed one to analyze more complex rational functions and extract parameter values systematically. Well, to make a long story short, it was rejected! This time, a review by a well-known enzyme kineticist was to provide another kind of object lesson. In this case, the reviewer was threatened by an idea that was unfamiliar to him and he reacted in a very defensive, and for me destructive, manner. Again, this may have been a blessing in disguise. Had this idea been easily accepted, I might well have been drawn irreversibly into the arcane world of enzyme kinetics.

Later, I published the modified Bode method in the *Journal of Theoretical Biology* [8] and, with somewhat more detail, in an appendix to my book on Biochemical Systems Analysis [2]. However, at that moment of rejection, I decided to go back and reconsider the power-law representation. I went back with this new view of Bode analysis in my mind. I recalled how nicely nonlinear functions are "straightened out" and "separated" in the log-log plots of Bode. The notion occurred to me that Bode's plots could be well represented over a considerable range of variation by a linearization in a logarithmic space.

From my early student training I was imbued with the philosophy of analysis that is based on Taylor's theory and ideas of successive approximation. But I had never seen this type of analysis carried out in a logarithmic space, and so I thought, "Let's see where it leads." In fact, the first-order approximation in a logarithmic space leads to a power function in Cartesian coordinates. Fig. 1.2

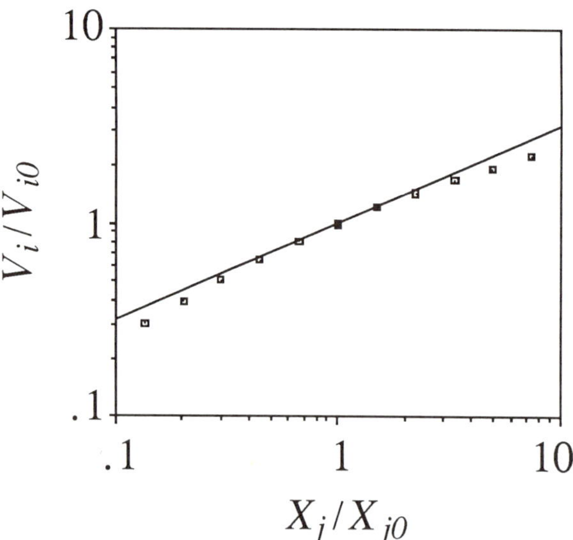

FIGURE 1.2. Estimation of parameter values in the power-law formalism. Log-log plot of rate versus concentration. Rate and concentration are normalized with respect to their nominal steady-state values.

provides an example of experimental data for a process that operates about some nominal steady state in the cell. Each point represents a different steady state about the nominal value and the data are plotted as the logarithm of the normalized velocity versus the logarithm of the normalized concentration. The straight-line tangent to the curve at the nominal operating point is the best linear representation, which consists of the first two terms for the Taylor series in this logarithmic coordinate system (Table 1.3). The parameter values in this representation can be readily extracted from the data. No complex methods of nonlinear parameter estimation are required; simple linear regression methods will do. One can estimate the slope and intercept in a straightforward manner, and these have definite meaning. The slope (g_{ij} in Table 1.3) corresponds to the derivative in terms of logarithmic variables or to the relative derivative of rate versus concentration, and it is well known in chemical and biochemical kinetics as the *kinetic order* of the reaction. When the first-order representation in logarithmic coordinates is transformed back into Cartesian coordinates, one obtains a power-law function in which the exponent corresponds to the kinetic order. The multiplicative parameter in this representation also has a well-known meaning from chemical and biochemical kinetics; it is known as the *rate constant*.

When there are several variables affecting the rate law one obtains a simple sum in which the additional terms consist of logarithms of the variables multiplied by the corresponding kinetic orders, or in Cartesian coordinates, the representation consists of a product of simple power-law functions, one for each variable that influences the rate (e.g., see Table 1.4). Parameter estimation in the general case is readily accomplished by multiple linear regression.

This product of power-law functions is a very simple mathematical representation. But the real questions were whether or not it would lend itself to further analysis and, if so, how well it would represent biochemical reality. To address these key issues one had to go from a representation of the isolated component to a representation of the intact system that would incorporate an arbitrary arrangement of such elements. The most natural and obvious way to do this was,

Table 1.3. Component Representation

Linear Regression in Logarithmic Space	*Representation in Cartesian Space*
$\log V_i = \log \alpha_i + g_{ij} \log X_j$	$V_i = \alpha_i X_j^{g_{ij}}$
Kinetic Order	*Rate Constant*
$g_{ij} = \left(\frac{\partial V_i}{\partial X_j}\right)_0 \left(\frac{X_j}{V_i}\right)_0$	$\alpha_i = V_{i0} X_{j0}^{-g_{ij}}$

Table 1.4. Generalized Mass Action Representation

Kirchhoff's Node Equations

$$dX_2/dt = v_{12} + v_{42} - v_{25} = V_{21} + V_{22} - V_{-2}$$

System Representation in the Power-Law Formalism

$$dX_2/dt = \alpha_{21} X_1^{g_{211}} X_2^{g_{221}} X_0^{g_{201}} + \alpha_{22} X_2^{g_{222}} X_3^{g_{232}} X_4^{g_{242}} - \beta_2 X_2^{h_{22}} X_5^{h_{25}} X_6^{h_{26}}$$

Steady-State Equations

$$0 = \alpha_{21} X_1^{g_{211}} X_2^{g_{221}} X_0^{g_{201}} + \alpha_{22} X_2^{g_{222}} X_3^{g_{232}} X_4^{g_{242}} - \beta_2 X_2^{h_{22}} X_5^{h_{25}} X_6^{h_{26}}$$

of course, to represent each and every reaction by the product of power-law functions. This corresponds to a generalization of traditional mass action kinetics, although we are not primarily interested in elementary chemical kinetic systems but rather large systems of enzyme-catalyzed reactions.

As an example, consider the network of reactions shown in Fig. 1.3, and within this network let us focus our attention on the intermediate X_2. Kirchhoff's node law says that the difference between the incoming flux, represented by the conversion of X_1 to X_2 (v_{12}) and X_4 to X_2 (v_{42}), and the outgoing flux, represented by the conversion of X_2 to X_5 (v_{25}), must accumulate at the node with the rate dX_2/dt (Table 1.4). Each of these three rate laws is then represented by the appropriate product of power-law functions. Similar equations apply for each of the other dependent variables in the system.

As I did this, it immediately became apparent that there would be difficulties

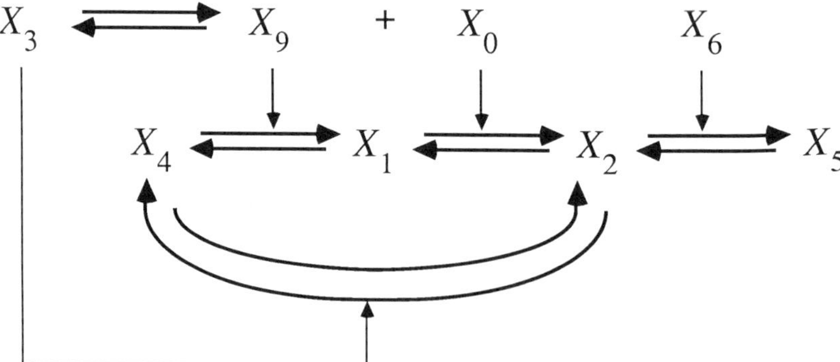

FIGURE 1.3. System that involves enzyme-enzyme interactions and channeling of metabolic flux. X_4 and X_5 are independent metabolite concentrations; X_1 and X_2 are dependent metabolite concentrations; X_6, X_9 and X_0 are concentrations of "free" enzymes; X_3 is the concentration of the multienzyme complex; $X_7 = X_3 + X_9$ is the total concentration of the first enzyme; and $X_8 = X_3 + X_0$ is the total concentration of the second enzyme.

using this generalized mass action representation for analysis of the integrated system. The simplest part of the analysis in most cases is to determine the steady states of the system, by setting the derivatives to zero and solving the resulting system of algebraic equations. However, the algebraic equations for this representation are still horribly nonlinear and there is no analytical solution in general. This sobering fact put a stop to the beautiful idea—at least for a while.

What got me out of this state was a paper that appeared in an ecology journal. This may seem like an insignificant event, but for me it illustrates an interesting point. This article was obscure in many ways. I don't remember who the authors were, nor do I remember the journal, and as far as I can tell, the rest of the world also didn't pay any attention to it. But to me it was critical, although I didn't realize it at the time.

As I recall, the authors were attempting to transform their data in such a way that one could make it accessible to analysis with conventional linear mathematics. It was a rather forced effort of the type that I had seen often in my early student days when it seemed that everyone was trying to force nonlinearities into a linear straightjacket. The authors produced a nonlinear representation that was really a distorted linear system in the sense that one could transform the variables and the underlying linear structure became apparent. The idea hardly registered at the time I read the paper. I was busy with a series of kinetic experiments on fatty-acid metabolism and I lost track of it.

When I returned to thinking about power laws, I was still stuck on the generalized mass action representation. I started trying all sorts of representations. Intermittently, I went back to the Michaelis-Menten formalism in which the individual reactions were represented by rational functions, which of course brought back in spades the difficult problem of having no analytical solutions. I was getting nowhere.

At this point I thought about the ecology paper and tried as best I could to express its key ideas in a biochemical context. It went something like this (Table 1.5): Lump all of the rate laws on the right-hand side into a single function. Transform the variables to the logarithms of the variables. The derivative of the logarithm of a given X variable would then be some function of the logarithms of all the X variables in the system. This complex function would then be written as a linear expression in the logarithms of the variables. The result is a (distorted) linear system. One could then analyze the system with linear mathematics and in the end exponentiate all the variables to produce the behavior in terms of the original X variables. For example, when the linear equations describe a sinusoidal oscillation, exponentiation of the variables produces the type of scalloped curves that are often seen in predator-prey data. The results suggested that this type of analysis might be more accurate than conventional linear analysis for ecological systems. It was a very simple idea. However, it was basically a linear system and so it would be unable to capture the essential nonlinear character of complex nonlinear systems. For example, it could not

Table 1.5. Distorted Linear Representation

Kirchhoff's Node Equations

$$dX_2/dt = (v_{12} + v_{42} - v_{25}) = V_2$$

Distorted Linear Equations

$$d \log X_2/dt = \log \alpha_2 + g_{20} \log X_0 + g_{21} \log X_1 + g_{22} \log X_2 + g_{23} \log X_3 + g_{24} \log X_4 + g_{25} \log X_5 + g_{26} \log X_6$$

Steady-State Equations

$$0 = \log \alpha_2 + g_{20} \log X_0 + g_{21} \log X_1 + g_{22} \log X_2 + g_{23} \log X_3 + g_{24} \log X_4 + g_{25} \log X_5 + g_{26} \log X_6$$

represent a stable limit cycle. It had the single redeeming feature that it allowed one to do some analysis.

The linear representation in logarithmic coordinates was of course familiar to me from my derivation of the power-law representation that used the Taylor series in logarithmic coordinates, but this systemic description differed from what I had been examining in two ways. It used the derivative of the logarithms and a single aggregate rate law in each equation. So I thought maybe I could use similar simplifications in my representation.

I first went back to Kirchhoff's equations and the notion of lumping the rate laws together to form an aggregate rate law. If the individual rate laws were rational functions, as I had shown was true in the conventional Michaelis-Menten formalism, then the sum of such functions also would be a rational function. At this point nothing had changed mathematically. But if we represented the aggregate rational function by a product of power-law functions, we obtained equations like those shown in Table 1.6.

If one took the logarithm of the right-hand side of this equation, one obtained a linear function in the logarithms, just like the distorted linear function in the

Table 1.6. Half-System Representation

Kirchhoff's Node Equations

$$dX_2/dt = (v_{12} + v_{42} - v_{25}) = V_2$$

System Representation in the Power-Law Formalism

$$dX_2/dt = \alpha_2 X_0^{g_{20}} X_1^{g_{21}} X_2^{g_{22}} X_3^{g_{23}} X_4^{g_{24}} X_5^{g_{25}} X_6^{g_{26}}$$

No Steady State

ecology paper. However, the left side didn't quite fit because the logarithm of the derivative was not the same as the derivative of the logarithm. In other words, these operations do not commute. I wrestled for a while trying to make my equations fit into a linear form, but it didn't work. Another problem became apparent. Since the concentration variables are all positive, by physical necessity, the power-law functions will all be positive real numbers, and therefore the derivatives must all have fixed signs, either positive or negative. This meant that there was no possibility for these equations to represent a steady state. Every concentration in the system must either be increasing or decreasing. This didn't seem very useful, so I went on to other things. [Later developments were to renew my interest in this representation, and I spent a good fraction of a sabbatical year in 1983 studying their properties. But that's another part of the story.]

Then, all of a sudden, the notion occurred to me that the simplest representation of a steady state is the balance between just *two* terms. So I arranged Kirchhoff's node equations into two terms on the right-hand side—one represented the aggregation of the incoming fluxes, the other represented the aggregation of the outgoing fluxes. Each of these aggregate rate laws for net synthesis and net degradation was then represented by a product of power-law functions.

$$dX_i/dt = \alpha_i \prod_{j=1}^{n+m} X_j^{g_{ij}} - \beta_i \prod_{j=1}^{n+m} X_j^{h_{ij}} \qquad i = 1, \ldots, n \qquad (1.1)$$

As soon as I looked at the steady-state equations I saw that something very nice had happened (Table 1.7). When you set the derivatives to zero and rearrange the algebraic equations, they again form a product of power-laws, and these can be transformed into a linear system by taking logarithms. More importantly,

Table 1.7. S-System Representation

Kirchhoff's Node Equations

$$dX_2/dt = (v_{12} + v_{42}) - v_{25} = V_2 - V_{-2}$$

System Representation in the Power-Law Formalism

$$dX_2/dt = \alpha_2 X_1^{g_{21}} X_2^{g_{22}} X_3^{g_{23}} X_4^{g_{24}} X_0^{g_{20}} - \beta_2 X_2^{h_{22}} X_5^{h_{25}} X_6^{h_{26}}$$

Steady-State Equations

$$g_{21} y_1 + a_{22} y_2 + g'_{23} y_3 = (b_2 - g_{24} y_4 + h_{25} y_5 + h_{26} y_6 - g_{28} y_8)$$

where

$b_i = \log(\beta_i/\alpha_i) \qquad a_{ij} = g_{ij} - h_{ij} \quad y_i = \log X_i$

this was not just a distorted linear system because the differential equations themselves remained fundamentally nonlinear.

This is the point at which the representation that we subsequently called a *synergistic system* (later an *S-system*) was born. From this point on, the power-law formalism really began to develop rapidly because the S-system representation had opened up the possibility of explicit solutions and symbolic analysis, at least for steady-state behavior. The formalism was systematically developed and expressed in conventional matrix notation (Table 1.8). Kirchhoff's node equations are written for each dependent concentration variable in the system; they are always the difference between net incoming flux and net outgoing flux, and these aggregate expressions are each represented by products of power-law functions. The α and g parameters are used to represent incoming fluxes whereas the β and h parameters are used for outgoing fluxes. This representation has many of the features familiar from linear analysis [9]; it is very systematic, general, and simple. It is revealing in the sense that an explicit steady-state solution can be readily obtained. However, it has a much wider range of accurate representation than the linear formalism, and it captures the rich repertoire of behavior characteristic of nonlinear systems. The differential equations are, as I said, still nonlinear, but their systematic form has allowed us to develop efficient algorithms for their numerical solution (as we shall see later).

The steady-state behavior is characterized by the simple matrix equation

$$[A]y] = b] \tag{1.2}$$

Table 1.8. Power-Law Formalism

Two-variable example

$$dX_1/dt = \alpha_1 X_1^{g_{11}} X_2^{g_{12}} - \beta_1 X_1^{h_{11}} X_2^{h_{12}}$$
$$dX_2/dt = \alpha_2 X_1^{g_{21}} X_2^{g_{22}} - \beta_2 X_1^{h_{21}} X_2^{h_{22}}$$

Advantages

Systematic
General
Simple
Revealing (explicit steady-state solution)
Captures most essential properties
Wide range of accuracy
Efficient methods for dynamic solution

Disadvantages

Lack of familiarity
Limits still being explored

where the elements of the $(n + m) \times n$ matrix of kinetic orders are given by $a_{ij} = g_{ij} - h_{ij}$, $y_j = \log X_j$, and $b_i = \log(\beta_i/\alpha_i)$. Separation of independent and dependent variables allows the steady-state equations to be written as

$$[A]_d y]_d = -[A]_i y]_i + b] \tag{1.3}$$

where the subscript d signifies that the matrix $[A]_d$ contains only kinetic orders with respect to dependent concentrations and the vector $y]_d$ contains only logarithms of dependent variables. The subscript i has the same interpretation but it applies to the independent variables.

The steady-state equations for the S-system of Eq. (1.1) correspond to a linear algebra in the logarithms of the dependent variables. In this linear algebra, the matrix of kinetic orders, $[A]_d$, is always a square matrix of dimension nxn. A fundamental axiom of linear algebra states that for all matrices $[A]_d$ with non-zero determinants there exists an inverse operator, $[A]_d^{-1}$, defined as

$$[A]_d^{-1}[A]_d = [I] \tag{1.4}$$

where $[I]$ is the nxn identity matrix. In order to simplify the notation, we let the inverse operator $[A]_d^{-1} = [M]$ [12]. As a consequence of the definition of this operator, it follows that $[A]_d$ also will be the inverse operator of $[M]$. That is

$$[M][A]_d = [I] = [A]_d[M] \tag{1.5}$$

Hence, the inverse operator $[M]$ is a matrix whose row and column vectors are *orthogonal* to the column and row vectors of $[A]_d$, respectively. This also can be written as

$$\sum_{j=1}^{n} M_{ij}a_{jk} = \delta_{jk} \qquad i, k = 1, \ldots, n \tag{1.6}$$

or, since the multiplication of these elements commutes,

$$\sum_{j=1}^{n} a_{ij}M_{jk} = \delta_{ik} \qquad i, k = 1, \ldots, n \tag{1.7}$$

where δ_{ik} is the Kronecker delta equal to 1 when $i = k$ and 0 when $i \neq k$.

Several years later this relationship, although it was not recognized as the orthogonality principal of linear mathematics at the time, was to become the basis of the work by Kacser and Burns [63] and Heinrich and Rapoport [66].

If the conditions for the existence of the inverse operator of the system matrix $[A]_d$ are fulfilled, the explicit steady-state solution can be obtained by premultiplying Eq. (1.3) by $[M]$. The explicit steady-state solution for the S-system in Eq. (1.1) then can be written [12]:

$$y]_d = -[M][A]_i y]_i + [M]b] = [L]y]_i + [M]b] \tag{1.8}$$

In this equation the solution for the logarithms of the dependent concentrations $y]_d(y_j, j = 1, \ldots, n)$ are divided into two parts. The first exhibits the linear dependence on the logarithms of the independent concentrations $y]_i(y_j, j = n + 1, \ldots, n + m)$; the second exhibits the linear dependence on the logarithms of the rate constants $b]$ ($b_j = \log(\beta_j/\alpha_j), j = 1, \ldots, n$).

The flux through any pool X_i in steady state is obtained by a simple secondary calculation involving the known concentrations in steady state:

$$(\log V_+)] = (\log \alpha)] + [G]y] \text{ or } (\log V_-)] = (\log \beta)] + [H]y] \tag{1.9}$$

Thus, the explicit solution in Eqs. (1.8) and (1.9) gives the complete relationship between the steady-state values of the dependent variables and the values of the independent variables and parameters of the system.

The question of parameter estimation can be addressed at different levels. At one level are the isolated components, for example, the reaction for the synthesis of X_1, which biochemists typically study. The relevant parameter values can be obtained directly by plotting rate versus concentration data in a log-log coordinate system and using linear regression. The slope yields the kinetic order and the intercept yields the rate constant (see Fig. 1.2 and Table 1.3). From network theory it is known that systemic factors, like gain and parameter sensitivity, are also related to the underlying parameters of the system and can be used to estimate these parameter values. For example, the explicit steady-state solution predicts a power-law relationship among the concentration variables of the system. The logarithmic gain factors defined by these relationships allow one to estimate parameter values from systemic measurements at various steady states. Thus, parameter estimation can be viewed from the systemic point of view as well as the elemental point of view. These principles were well known to anyone familiar with network theory; I was simply adapting them to a new formalism and the specific conditions that prevail in biochemical systems.

Although much of this development of the power-law formalism followed directly from established practice in network theory and sensitivity theory, most biochemists were unwilling to accept its relevance to the problems of biochemistry. This was made very clear when I had a rather unpleasant encounter with a well-known biochemist whose interests were in the area of metabolic regulation. I went to him expecting that we would have something in common and that he might have some interest in efforts to develop a more systematic approach to the analysis of biochemical systems. I was taken aback when he immediately attacked my views and suggested that what was needed was *more* biochemistry and *less* mathematics, control theory, *etc*. I was already convinced that more biochemistry would be required, but it was painfully obvious to me that the very tools he had renounced were desperately needed if we were to make sense of the detailed biochemical information that was threatening to engulf us even in the late sixties. Again, I was to learn that authorities with a vested interest in a certain set of ideas seldom take kindly to an alternative approach, even if that approach might be helpful in advancing their own agenda.

The characterization of steady-state properties had come to a certain level of completion, and I turned to the dynamic behavior of these nonlinear equations. My exploration of their dynamic repertoire was facilitated by the introduction of an interactive, time-shared computer system by the Stanford University Medical School. While writing my first interactive program I realized that the logarithmic transformation, which simplified the steady-state analysis, also improved the efficiency of the dynamic solutions. ''Interaction'' at this time consisted of manually plotting points as the data emerged from the computer and, when appropriate, interrupting the solution to restart it again from an alternative set of conditions selected to redirect the exploration. With this program, primitive as it was by today's standards, I quickly showed that this formalism was sufficiently general to capture the essential characteristics associated with nonlinear systems.

At this time I also changed the experimental focus of my work. The work with heart muscle in tissue culture was frustrating because there was so little material to work with and the results were extremely variable. My advisor, John Steward, had helped me through these difficulties in my Ph.D. program. His good sense and infectious enthusiasm were invaluable to me. His training was in immunology and microbiology and on several occasions he suggested that I should look at bacteria. They would really provide a much better system for the kind of quantitative work I was doing. By the time I returned to Stanford I had come to realize that he was right, and John generously gave again of his time to teach me about the wonders of microbiology. Having a more appropriate experimental system in hand, I turned my attention to the biological goals for which this formalism was developed, namely, understanding the integrated behavior of organizationally complex biological systems in terms of their underlying molecular determinants.

Let me digress here to consider the distinctive features of these complex systems (Table 1.9). The most distinctive feature of any living process is its self-replication. This tremendously complicated process occurs within a very small space, typically the size of a cell about ten microns in extent.

From the size of the proteins and the genome for the bacterium *Escherichia coli*, we know that there are on the order of 10,000 elements or dynamic variables with which one must deal. These elements differ in kind and cannot be simply averaged as electrons that flow through the wires of a linear electronic circuit. Each element of the system has an identity of its own. An enzyme and its substrate are very different in size, weight, and composition. For example, the molecular weight of a typical sugar is about 150 and that for a typical enzyme is about 40,000—a ratio of 1 : 267. One must develop a separate method to monitor each of the constituents in the cell, and this makes the experimental study of such complex systems very difficult.

The more significant interactions among the numerous components of biological systems result from association rather than addition. An enzyme and its substrate must associate to form a complex that yields activation of substrate and subsequent conversion to product. Antibody-antigen complexes must form before an invading pathogen can be sensed and cleared by the immune system of the host organism. In predator-prey systems, different species must be in the same location at the same time in order for the prey to be captured. In all of these systems, the components are different in kind and they must associate—they must form complexes, at least temporarily—to bring about the appropriate result. Associative interactions among the components of biological systems produce highly nonlinear behavior that is extremely difficult to predict intuitively. Properties of the intact system often can change in a dramatic qualitative fashion in response to a slight quantitative change in one of its component prop-

Table 1.9. Some Distinctive Features of the Cell

Self-Replicating	
Small Size	10 μ
Organizationally Complex	
Many variables	10,000
Elements differ in kind	Substrate/enzyme (1/250)
Associative interactions	Nonadditive Nonlinear
Rich networks	Strongly coupled Highly organized

erties. The range of nonlinear behavior includes synergism, saturation, thresholds, stable oscillations, and deterministic chaos.

There is a large number of components and processes in such systems and these are highly connected and strongly interacting. This situation is unlike a system of "loosely coupled oscillators" where one can characterize a representative oscillator and then perform some type of averaging over a large ensemble to predict their aggregate behavior. Organizationally complex systems must be treated in such a way that all the interactions constituting the organizational integrity of the system are taken into account.

When faced with this kind of complexity, people tend to respond in one of two ways. As an antidote for those who argue that complexity is only a reflection of our ignorance and that nature is basically simple, I quote Yates:

Complexity is not simplicity in masquerade.

I believe there are such things as inherently complex systems and that we have to deal with such systems in biology. On the other hand, for those who argue that all our representations are inadequate and that nature is inherently too complex to be understood, I quote Laplace:

Half the battle in understanding is having the right representation.

This fits my sense that while these systems are inherently complex and cannot be made simple, they are not beyond understanding. With an appropriate representation they can be made comprehensible if not reduced to simplicity. As is often the case, the real answer lies somewhere in between. To repeat Einstein:

We should make things as simple as possible, but not simpler.

From the early development of the power-law representation we knew that it provided a reasonable approximation for an individual enzyme-catalyzed reaction—though not as good an approximation as the conventional Michaelis-Menten formalism. However, the purpose in developing the power-law representation was not to deal with the kinetics of isolated enzymes, but to address organizationally complex systems consisting of many enzymes. Thus, one had to examine intact biochemical systems in order to determine the validity of this new representation.

The steady-state equations representing the behavior of these systems were identical to those of the linear resistive networks that had been completely characterized years ago by Bode and the other network theorists. So once again, to paraphrase Newton, we were able to "stand on the shoulders" of these giants.

They had shown that, although the explicit steady-state solution provides all the available information about the steady-state behavior of the system, it also is useful to characterize the system in terms of standard factors that relate its global properties to its underlying molecular determinants, namely, gain factors [53, 58, 67, 83] and parameter sensitivities [53, 59, 67, 83]. I had determined earlier that these factors, which characterize the systemic response to change in specific independent concentrations or parameters, can be determined readily from the explicit steady-state solution [11, 12, 76].

When environmental constituents (independent variables) are changed in concentration, these changes are detected and propagated throughout the cell. Various cellular processes (dependent variables) are influenced to one degree or another. The extent to which such signals are amplified or attenuated as they are propagated throughout the cell is conventionally represented by a set of gain factors. In the power-law formalism the variables are more naturally represented as the logarithms of concentrations, and thus these gain factors were defined as *logarithmic gains* [12]. The logarithmic gain between a given independent and dependent variable is not determined by any one component, or even by a small number of components in the system. Rather it is a systemic property that in general depends upon all of the components and their specific organization or design.

A change in any given cellular component (parameter) is ultimately manifested as a resultant change in behavior of the integrated system. The extent to which such perturbations in the components of the system influence its dependent variables is conventionally represented by a set of parameter sensitivities. In the power-law formalism the *rate-constant sensitivities* are defined as the percentage change of a dependent concentration going from one steady state to another as a result of a one-percent increase in a rate constant [11, 12].

In a similar fashion, one can differentiate the explicit steady-state solution with respect to one of the kinetic orders in the system and thereby determine the percentage change in a dependent concentration resulting from a one-percent change in a kinetic order. These factors are defined as *kinetic-order sensitivities*, and they are related in several different ways [5, 12, 34].

The logarithmic gains and parameter sensitivities are systemic factors that can be measured in the intact system. They also are functions of the rate constants and kinetic orders that characterize the molecular elements of the system, and so they can be used to relate behavior at the molecular level to behavior at the system level.

With these tools in hand I began an analysis of cascade mechanisms, which had interested me from the time of my thesis work and which were relevant to the experimental work that I was then doing on gene regulation of the threonine synthetase system in *E. coli*. This application of the power-law formalism showed three important and very encouraging results [8, 11, 12]. First, it showed that the analysis was very straightforward, drawing as it does on well-established methods from network theory and sensitivity theory. Second, it provided a method for quantitatively characterizing the effectiveness of feedback inhibition. Third, there were abundant physiological data for a number of cascade systems, so we could begin to estimate the range of variation over which the S-system representation accurately described these systems. The results showed that the range was very wide, unexpectedly wide in some cases, and much greater than that for the isolated components. The analysis also suggested a reason why the range of accurate representation would be greater than it is for the isolated components. This phenomenon, which might at first seem counterintuitive, is due to the functional simplifications that result from the presence of feedback control. This is analogous to the linearization of highly nonlinear characteristics by feedback in electronic amplifiers.

1970–1974

Throughout this period I was trying to come to grips with the difference in scientific worldview between biologists, on the one hand, and mathematicians and physical scientists on the other. It was clear that the latter group had the more rigorous quantitative methodology, but I felt uneasy with their simplifications. Biologists, on the other hand, clearly had the longest tradition of dealing with organizationally complex systems and they seemed to understand the concept of a control experiment better than any other group, but it was difficult to accept their Luddite approach to all things mathematical. Was there a way to combine the best aspects of each tradition? There undoubtedly were many individuals that we might call mathematical biologists who wrestled with this question. In my case, I dwelt on a constellation of ideas that would constitute the mathematical equivalent of a well-controlled comparison—an approach that would include the mathematical equivalent of the experimentalist's control experiment. This led to the notions of *internal equivalence* and *external equivalence* and what we now call *mathematically controlled comparisons*.

I applied the concept of mathematically controlled comparisons to one of the simplest of biochemical systems—biosynthetic pathways subject to feedback inhibition. The analysis of biosynthetic pathways and cascade mechanisms showed a number of similarities and differences. It was shown that one could use the same general concepts of logarithmic gain and parameter sensitivity to quantitate the effectiveness of feedback control in these systems. Moreover, the analysis showed that some of the characteristics routinely attributed to feedback inhibition were not in fact inherently associated with this mechanism, while others were inseparable from it. Although this was a fairly simple application, it seemed to be the best vehicle for communicating these concepts to the biochemists.

As another specific application, I examined a newly discovered mechanism in amino acid biosynthetic pathways—inhibition of the cognate aminoacyl tRNA synthetase by an intermediate more proximal in the pathway. No one knew what this feedforward inhibition mechanism was doing in the cell. Its only known manifestation was a sort of "choking" response to the addition of pathway intermediates. Such an addition was expected to provide nutrients to the cell that, if anything, would allow the cell to grow somewhat faster. However, for some strains the addition inhibited growth. The inhibition was attributed to the presence of an intermediate in the pathway that inhibited the synthetase and thereby choked off the supply of amino acid required for growth. The analysis showed clearly that feedforward inhibition alone could not account for the self-starvation. Nor could this mechanism in conjunction with the better-known mechanism of feedback inhibition account for the dysfunctional behavior. However, if we included an alternative fate for the amino acid, thus acknowledging that the pathway was in fact a branched pathway, then we were able to account for the self-starvation. Indeed, the analysis yielded a mathematical condition among the parameters of the system: if the condition was satisfied then normal growth could be expected; if not, then self-starvation would result. The results of analysis also suggested that such a feedforward inhibition mechanism could be used by the cell to deflect flux preferentially into protein synthesis or into the alternative fate, depending on whether the nutrient source entered the pathway at a position before or after that of the feedforward modifier. Again, the results provided a satisfying rationalization for the behavior of the integrated system that was difficult and, in some aspects, impossible to obtain by other means.

These two applications were written up and submitted in 1970 to the *Journal of Biological Chemistry*. In the first review of these manuscripts the referees found many "provocative statements" but they wanted more clarification. After the revision they reported no "unexpected deductions." A year-long correspondence produced a thick file that I was later to name The Great JBC Debate. Ultimately it came down to a "philosophical difference," and a "policy decision" was made to reject the manuscripts. The editor felt compelled to conclude the correspondence with the following statement: "In returning your manuscripts, I do wish to emphasize that we have at no time made serious criticisms of their substance." From this point on I became much more conscious of different scientific worldviews and how they influence scientific progress.

The examination of unbranched biosynthetic pathways showed how precursor-product relationships implied constraints among the parameters of the S-system representation. However, it became clear that these constraints had to be modified when we included an alternative fate for the amino acid in our analysis of feedforward inhibition. This also was clear from our analysis of other branched pathways. In particular, when you aggregate flux through the pool at the branch point, the net flux representing the degradation of the branch-point metabolite is represented differently from the net fluxes representing the syntheses of the intermediates following the branch. This is not to say that these representations are unrelated. In fact, the parameters in one are not independent of those in the other. There must be constraints among the parameters of each because the alternative representations characterize the same physical fluxes.

An analysis of control by the nested pattern of feedback inhibition in branched pathways also demonstrated a mathematical constraint among the parameter values of the system. If the constraint in this case is satisfied, then there is a normal physiological interaction between the branches of the pathway. The addition of one end product leads, if anything, to a slight increase in the concentration of the others. But if the constraint is not satisfied, then the addition of one end product leads to a reduction in the concentration of the others. This response becomes pathological if the reduction is so severe that growth of the organism is curtailed. These results led to a precise rationale for the existence of macromolecular complexes among the regulatory enzymes in branched pathways controlled by the nested pattern of feedback inhibition.

By this time I had written up the older work on cascade mechanisms. The results of this analysis could be understood directly in terms of the experimental data. They did not require the biochemist to understand more powerful and more abstract concepts, so I thought this paper might be more suitable for submission to a biochemical journal. However, I was not convinced by my previous experience with *JBC* that it would be appropriate. I decided instead to try the *Archives of Biochemistry and Biophysics*, thinking that the ''biophysics'' in the title might imply a more rigorous mathematical and physical predisposition. This time I was successful [12], although it probably had nothing to do with any such predisposition. I consider this one of my best early papers.

With some of these early successes I began to see more clearly how the power-law formalism (Table 1.8) differed from the other two formalisms that had been used more extensively in biochemistry, namely, the linear formalism (Table 1.2) and the Michaelis-Menten formalism (Table 1.10). I compared these at some length, showing the complementary advantages and disadvantages of each. These comparisons are still valid and illuminating, although their implications have not been widely appreciated. In retrospect, it is clear that I failed to take into account the deeper issues that often are at work when there is a clash of paradigms. In a recent article [80] I have made an effort to rectify this omission and to extend the comparisons with more current evidence.

Most of the early work to this point was reviewed in [1]. In terms of biological applications of the power-law formalism, I turned more and more to questions of function, design and evolution—first of biosynthetic pathways and then of regulatory circuits for gene expression.

Following the discovery of end-product inhibition in the late 1950s, numerous studies documented the existence of this regulatory mechanism. In fact, it

Table 1.10. The Michaelis-Menten Formalism

Two-variable example

$$dX_1/dt = V_1/(X_1 + K_1) - V_2X_1^2/[(X_1^2 + K_2^2)(X_2 + K_3)]$$
$$dX_2/dt = V_2X_1^2/[(X_1^2 + K_2^2)(X_2 + K_3)] - V_4X_2/(X_2^2 + K_4^2)$$

Advantages

Captures many essential properties
Wide range of accuracy

Disadvantages

Ad hoc
Specific
Complex
Opaque (no explicit solution in general)

seemed that this was a nearly universal mechanism of regulation at the molecular level in unbranched biosynthetic pathways. It was surprising that no one seemed to ask why this might be so. To anyone familiar with control theory it was obvious that other patterns of feedback control could accomplish the functions that were being attributed to end-product inhibition, and some of these alternatives were known to have some advantages not possessed by the simple pattern. I already had begun to analyze specific alternatives back in 1970. I found that the simple pattern of end-product inhibition was better than the alternatives on the basis of several steady-state criteria for functional effectiveness. These analyses were broadened to consider arbitrary patterns of control by feedback inhibition, and to my surprise, the steady-state results for the specific comparisons could be generalized to an arbitrary pattern of feedback inhibitions. These results provided the first detailed rationale for the evolution of end-product inhibition in unbranched biosynthetic pathways [1].

I then broadened the analysis of feedback inhibition in biosynthetic pathways to include dynamic criteria as well. Most of my previous work had emphasized steady-state behavior, in part because this behavior was governed by nice linear equations in the power-law formalism. One also could linearize the systems and apply concepts from linear stability analysis. After doing this repeatedly, I saw that one could do it once and for all, thereby expressing the conditions for local stability directly in terms of the parameters of the power-law formalism. This is possible because the formalism is very systematically structured. I used these results in practically all our subsequent analyses and found that the simple condition necessary for local stability, which could be expressed as an inequality involving the determinant of the difference in kinetic orders, must often be used for a proper interpretation of steady-state behavior as well.

These results started me thinking about the factors involved in the design of biosynthetic pathways regulated by feedback inhibition. By enumerating the possible factors and carefully comparing the alternatives, one could come up with a convincing scenario [14, 16]. First, the number of steps would be roughly determined by the differences in structure between the initial substrate and the required product. Long pathways are more difficult to stabilize and so one could predict that mechanisms that shorten the effective length of unbranched segments in the pathway would evolve and that alternatives to long pathways would be used wherever possible. Bernard Davis [61] reviewed the organization of biosynthetic pathways in bacteria and mammals; the results showed that, indeed, long pathways are avoided wherever possible. In all cases, the amino acids are synthesized by a highly branched set of pathways in which the lengths of most unbranched segments are relatively short. Mammals derive half the amino acids from their diet and have no need for the corresponding biosynthetic pathways. All of these same amino acids that must be synthesized in bacteria involve long pathways.

The question remained, How do the bacteria deal with these long pathways that they cannot avoid in the way that mammals can? The analysis showed that the effective length can be shortened by kinetic means, even when the pathway cannot be shortened physically. This is accomplished by having values for the kinetic parameters distributed over a wide range. The kinetic behavior of the pathway is then dominated by a smaller number of temporally dominant reactions in the pathway. Stability also can be obtained by more complex patterns of feedback inhibition; in fact, it was shown that any pattern is better than end-product inhibition in this regard. However, the steady-state behavior is compromised by the more complex patterns of regulation. Stability also can be improved by diminishing the strength of end-product inhibition, but this leads to reduced effectiveness according to the steady-state criteria. Hence, the only strategy that preserves the steady-state advantages of end-product inhibition and at the same time promotes stability, is ''shortening kinetically'' the length of the pathway. The histidine biosynthetic pathway in *Salmonella typhimurium* was the most thoroughly investigated long pathway, and the experimental data were entirely in agreement with the expectations from the mathematical analysis.

It was some time before I realized that the concept of external equivalence could be made more rigorous. As I began to apply a higher standard in our mathematically controlled comparisons I would come to see that these early analyses of feedback patterns in unbranched biosynthetic pathways would have to be redone.

At this point I was beginning to see a certain pattern of expectation as to what a mathematical model should be in biology. For many, it should be focused on a single system. After all the basic phenomena are known and characterized, one formulates a model that encapsulates all this knowledge and produces an efficient description of the system that will faithfully mimic what is known. If the model and the experimental data agree, it is considered a good model.

For me this misses the point. Mathematics for me is the language of science; its use is none other than the scientific method. Models can be simple and located in one's head, or they can be complex and embodied in computer software. The form of the mathematics can be logic expressed in words, or differential equations expressed in symbols. The real point of applying mathematics in biology is to discover underlying mechanisms and to develop understanding, not to mimic biological phenomena.

I had come to see the method of mathematically controlled comparisons as a powerful tool for revealing the essential differences between alternative designs and to recognize the importance of always considering simultaneously at least two different designs or models for any given phenomenon. In some cases, where there appeared to be a nearly universal design that needed understanding, the alternative designs had to be hypothetical ones that may or may not have existed at some time in the past [14].

More often, it was clear from the biology that different designs had evolved in nature and that such differences appeared even in systems that seemed to be performing the same physiological function. One of these differences was the use of positive versus negative mechanisms to control gene expression. In the mid-1970s most molecular biologists believed that the difference was simply historical accident. The lactose (*lac*) operon happened upon a negative mechanism while the maltose (*mal*) operon happened upon a positive mechanism. They perform the same function—induction of enzymes that split disaccharides—and either mechanism would work equally well. I reasoned that this would be a good test for the method of mathematically controlled comparisons. One could construct models of each design and make them both internally equivalent and externally equivalent to the extent that this would be possible. A detailed analysis should then reveal any inherent differences between the two designs.

After considerable analysis I came to the conclusion that these two designs behaved nearly identically, as long as there were no changes in the parameters that characterize the systems. If there were to be any differences they would have to arise as a result of changes in the regulatory parameters. This allowed one to focus on differences in response to mutations in the regulatory mechanism. My first thought was that the design best buffered against harmful mutations would be selected, much the way one might select for fault-tolerant designs in technological products. Indeed, I had found this a useful criterion for comparing other biochemical mechanisms. This intuition prevented me from seeing that the result with these regulatory mechanisms was just the opposite. Since this is almost invariably the stumbling block that I have encountered in presenting this material, I have come to conclude that my initial intuition is shared by most investigators. However, when one thinks it through carefully, it becomes clear that defects in the regulatory mechanism must be expressed and selected against in order to maintain the functional regulatory mechanism.

A simple rule followed from this analysis: A positive mode of regulation is correlated with high demand for expression of the regulated genes, while a negative mode is correlated with low demand.

I immediately began to test this prediction. Here I was fortunate in having Rolf Freter in the lab next to mine. He is an authority on the ecology of intestinal flora, so over the years I had effortlessly absorbed a fair amount of information about the natural environment of enteric bacteria. As a result it had become clear to me that the conditions employed in the laboratory for the study of *E. coli* had little resemblance to what the organism experienced in its natural environment. I began searching the literature in intestinal absorption and gastroenterology. Data relevant to the catabolic systems that had been most thoroughly characterized at the molecular level were entirely consistent with the predicted correlation. This was the beginning of the *demand theory* of gene regulation [77], which over the years has been extended to many different areas of biological investigation. It now provides a simple unifying thread that has implications at nearly every level of biological organization [79].

In retrospect, this work has taken on a life of its own and no longer is closely related to the power-law formalism. Nevertheless, if it had not been for the rigorous analysis that initially excluded many of the alternative hypotheses, I doubt that I would have come to focus so intently on the remaining hypothesis that eventually led to the emergence of demand theory.

Back in 1972 or 1973 Gerald Smith gave a seminar to our department in which he described his studies of the histidine utilization (*hut*) operon in *Salmonella typhimurium*. He had clearly demonstrated that the regulator of the *hut* system was a repressor protein whose structural gene was located within one of the transcriptional units controlled by the *hut* repressor. Accordingly, the expression of the regulator and the regulated functions varied coordinately during induction of the *hut* system. This was the first well-documented example at the molecular level of what was subsequently called an autogenous circuit. This design is quite different from that of the *lac* operon in which the regulator is located in a separate transcriptional unit and varies little during induction while the level of the regulated functions can vary by 2,000-fold. In this design the regulator and effector circuits are essentially uncoupled in a manner that has been called the classical circuit.

Mathematically controlled comparisons of the classical and autogenous circuits of gene regulation demonstrated quite clear differences in function. Based on these differences one could predict that inducible systems governed by a positive regulator should utilize the classical circuit whereas those governed by a negative regulator should utilize the autogenous circuit [12]. Most of the data available at the time were consistent with these expectations. There was, however, as Whitehead would say, one happy exception—the *lac* operon. [It was to take another ten years before we developed a more general circuit theory of gene regulation in which we began to understand why *lac* is different.]

This type of analysis was quickly extended to repressible biosynthetic pathways where the predictions were just the opposite. Positively regulated systems could be expected to utilize the autogenous circuit whereas negatively regulated systems could be expected to utilize the classical circuit [17]. Although the analysis was somewhat less conclusive, and there were fewer data available for comparison, the results were in general agreement.

1974–1978

During my sabbatical in Göttingen I spent most of my time thinking about the implications of the power-law formalism for large systems—much larger systems than those I had been examining to this point. All through the year I kept thinking of the analogy to statistical mechanics. One learns as a physics student that the molecules of a gas can be modeled at the microscopic level by a simple and admittedly inaccurate description—point masses that undergo perfectly elastic collisions, *etc.*—that, nevertheless, upon averaging over millions of molecules and countless collisions, yields the gas laws as a macroscopic description.

Brian Goodwin had explored a similar idea in his book *Temporal Organization in Cells* [62]. He modeled elementary gene circuits as nonlinear oscillators and then averaged their behavior to deduce macroscopic quantities related to what he called "talandic temperature." While the idea was suggestive, there was difficulty in relating this macroscopic property to anything specific that was known about cells. In the end, it didn't seem to work, and Brian himself seemed to have dropped this direction of research. I kept asking myself, Why is the success story of statistical mechanics, as applied to a gas, not applicable to cellular systems?

I had come to believe that cells exhibit a fundamentally different kind of complexity from that of a gas. I termed this organizational complexity. One had to respect the organizational integrity of the cell in the process of relating the molecular properties to the macroscopic properties of the cell. It seemed to me that simple averaging procedures ignored this organizational integrity. It occurred to me that the power-law formalism might allow one to draw general conclusions about organizationally complex systems because it was systematic enough to account for major organizational features while still being mathematically tractable.

Let us assume that this nonlinear formalism provides a reasonable representation for the molecular elements of such systems. If we consider an arbitrary assembly of such components with rich interactions, what would emerge at the system level? What might be the appropriate macroscopic laws, analogous to the gas laws, that we might hope to deduce from the microscopic representation? I concluded that the most promising candidates were the growth laws and the allometric relationships. These were among the most regular quantitative features of organizationally complex systems. There also were abundant data that had been collected over centuries.

But how does one get from the differential equations of the S-system representation to the observed growth laws and allometric relationships? First, one had to represent aggregate variables, such as the weight of the organisms, and other general constraint relationships in the power-law formalism. This was a straightforward application of the Taylor series approach in logarithmic coordinates that I had first used to represent the rate laws. Second, I already had seen that complex systems ranging through a series of steady states had variables that were related to each other by a straight-line relationship in a log-log plot. These were in fact allometric relationships! Thus, if most of the variables in organizationally complex systems were in quasi-steady state, then the allometric relationships would naturally follow. Third, it is well known that the processes in an organism occur on widely separated time scales and that most growth curves have monotonously simple shapes that suggest no more than two or three time-constants. So it was natural to explore the cases where just one or two of the cellular processes are temporally dominant.

When the fast processes are considered to be in quasi-steady state, one can solve for the fast variables in terms of the temporally dominant ones. The algebraic relationships, as noted above, are products of power-law functions. When these algebraic relationships are substituted for the fast variables in the differential equations, one obtains a reduced S-system with exactly the same form as the original S-system but with a smaller number of variables—the number that are temporally dominant in the original system. This was called the telescopic property of the equations.

Most nonlinear formalisms do not exhibit this nice telescopic property, which permits a reduced model of a complex system to be described by equations of exactly the same form as the more complex model or, conversely, permits a simplified model to be expanded with more detail and yet be described by equations of exactly the same form. The linear formalism has this telescopic property, but it cannot account for the known allometric relationships, which are nonlinear in general. On the other hand, the Michaelis-Menten formalism does not exhibit this telescopic property. There may be other nonlinear formalisms that also possess this important property, but at the moment the S-system representation in the power-law formalism is the only one that I know for which this has been rigorously demonstrated.

By this line of thought I discovered that low dimensional S-systems are a generalized growth law. Most of the well-known growth laws in the literature were easily shown to be special cases of this general growth law. In many cases, when the growth law had a differential form, all that one had to do was set certain parameters in the general growth law to zero or other specific values. In other cases, when the growth law had the form of an algebraic equation, one could differentiate the expression to produce the differential form, which then could be seen as a special case.

I also realized that any growth law that represents a monotonic increase to a maximum mature form can be viewed as a cumulative probability distribution. [In fact, I soon saw that this was an association that often had been made in the past.] So, I spent most of the summer and early fall looking through books on probability distributions; it turned out that nearly every one is a special case of the general growth law.

During these months it dawned on me that I was not representing these nonlinear mathematical functions in an approximate way, but rather the S-system representation was *exactly* equivalent to the original function. This was an entirely different way to view the S-system representation. It was no longer just an approximation. One could start with an empirical growth law, or a probability distribution, or indeed many other nonlinear functions, and systemically step-by-step convert them into the S-system form. The basic operation was the chain rule of differentiation that decomposed composite functions into products of power-law functions. One also had to negate or translate variables so as to make them positive definite, and to add and subtract variables in order to reduce sums of more than two terms to the standard difference between products of power laws. [In time we were to develop standard methods for dealing with these features.]

Later, in the fall of 1977, I attended a conference in Poland where I met Werner Ebeling. During a discussion, I mentioned this discovery of a general growth law, and he told me of a colleague in Rostock who also had developed a general growth law. I later learned that his was a polynomial equation that represented a broad family of growth laws. I could transform this equation into the S-system form; it was a special case of the general growth law that is an S-system. This motivated me to pay special attention to other classes of functions that had previously been shown to represent very general families of growth laws and cumulative probability distribution functions. In each case, I was able to show that these were special cases of the general growth law in S-system form.

1978–1982

During the early part of this period I worked with an undergraduate, Gerald Jacknow, refining and extending the analysis of feedforward inhibition in amino acid biosynthetic pathways. I first thought that feedforward control might be a mechanism recognized in biological systems before it had been discovered by the engineers. But then I came across a history of feedback control that pointed out that feedforward control had been discovered by the early millwrights [68]. I then read a paper by Jurgen [65] who pointed out that Black [52] rediscovered feedforward control at about the same time he was becoming famous for his use of feedback control. Because of technical problems, feedforward control remained a curiosity. Years later, when modern technology could produce the components of control systems with sufficient robustness, it was realized that feedforward control provided anticipatory control yielding a faster temporal response and greater stability for the system.

With these new insights, we began to examine the effect of a change in position within the pathway for the feedforward modifier. The results showed that stability and temporal responsiveness were enhanced as the position shifted further back in the pathway to achieve a greater degree of anticipatory signaling. This led us to predict that the first intermediate in the pathway would make the best feedforward modifier, that the first and last enzyme in the pathway had a special relationship to one another, and that these two enzymes would form a multienzyme complex for the purpose of improving the function of the regulatory mechanism [21]. Some five years later all of these predictions were confirmed experimentally for the isoleucine pathway of *E. coli* by Singer, Levinthal and Williams [81] at Purdue University.

The laboratory was relatively quiet for the first few years after I returned from my sabbatical. It was a useful period for consolidating ideas and catching up on my writing. However, within a few years the situation had changed dramatically. My association with some very talented colleagues was to make this period personally enjoyable and scientifically fruitful.

Eberhard Voit stands out in particular. He came for one year as a postdoctoral fellow and he was the first to take up the power-law formalism with real enthusiasm. He quickly mastered the fundamentals and went on to make substantial contributions of his own. For example, many investigators including myself had looked for analytical solutions to the simplest S-system equations. This had gone on for more than ten years with very little success. Within a short time after his arrival in the lab, Eberhard had found analytical solutions for the simplest S-system and for many other cases of the general growth law [24]. He also developed methods for estimating the parameters of an S-system from macroscopic observations [23]. In this one short year Eberhard not only made his mark on the development of S-systems, but he also mastered the nuances of American TV commercials while playing an active role in The University of Michigan glee club. His humor is still legendary in the lab. By the beginning of 1982 he had left, but not for good as we shall see. Eberhard had training in developmental biology and upon returning to his position at the University at Köln, he continued applying the S-system methodology to problems in cell growth and tissue regeneration.

1982–1986

Masahiro Okamoto arrived as Eberhard was leaving, and he stayed on for three years as a postdoctoral fellow. His work represented the first large-scale application of the S-system methodology although it wasn't really designed that way. His study of the isoleucyl tRNA synthetase system from *E. coli* built upon the traditional Michaelis-Menten approach to this problem, which had been followed by many others in our lab and elsewhere. However, the resulting differential equations turned out to be very stiff and therefore solutions were slow and expensive to compute. Masahiro showed that these equations could be approximated within the accuracy of experimental error by an S-system. In this form the fast equations were identified and used to develop the reduced S-system that then could be solved efficiently with the interactive computer program that we had developed to solve S-systems [72].

Masahiro also did a detailed sensitivity analysis in which every parameter was examined for its influence on the behavior of the system [71]. It was a beautiful example of how sensitivity analysis can be used to explore the internal consistency of a model and make predictions regarding the relative importance of each interaction in the model. Masahiro found that the existing experimental data for isoleucyl tRNA synthetase proofreading valyl-tRNA were inconsistent. There was no physically realizable solution when these data were incorporated into the model; there was clearly something wrong. The sensitivity analysis revealed the degree to which a change in the value of any given parameter would influence system behavior and presumably restore the possibility of a realistic solution. The parameters were ranked according to their degree of influence on the solution, and the experimental data for the top three were carefully reexamined. We discovered that the value reported in the literature for one of these parameters was in error by tenfold. When this error was corrected, the entire set of data became internally consistent and we were able to deduce a number of other important properties of this proofreading system.

At about this same time Doug Irvine came into the laboratory. Doug was finishing the first two years of medical school as part of a combined M.D.-Ph.D. program. He had taken my course in biochemical systems analysis and had begun to explore the behavior of immune networks using the same methodology. This proved to be very stimulating because none of us in the lab at that time had pursued problems in immunology despite the fact that we were surrounded by immunologists in the department. Doug showed us all that the immune system was a rich area for application because of its strong network structure. Doug had made a good start on his Ph.D. research when it was time for my second sabbatical.

I could recall the benefits of my first sabbatical and was therefore strongly committed to go once again. However, Doug was at an early stage of his research and it didn't seem appropriate to leave him on his own for the year. In the end it was decided that Doug and his new wife, Theresa, would join us in Canberra, Australia for the year.

Doug was applying the method of mathematically controlled comparisons to alternative interactions in the immune network. He was controlling a number of important variables and yet the results were not making sense. Slowly we began to discover that there were other properties of the system that we had not been controlling and that were subtle. This led us to refine the notion of mathematically controlled comparisons and we then produced very clear results. Doug went on to examine in detail the dynamic behavior of these alternative networks. This required a very large number of numerical solutions to draw general conclusions over the realizable parameter space. Doug soon discovered that our interactive computer program, which we knew to be relatively efficient, was consuming an inordinate amount of computer time. Up to this time we had used conventional subroutines in implementing our program; no one had gone deeply inside these to see if their performance could be improved. Doug began to look at alternatives. He found that the subroutine we were using to solve the nonlinear differential equations performed as well as any of the alternatives that were available. For our problems, these seemed the best we could do, and yet it was not good enough for Doug's purposes.

Instead of abandoning his problem and turning to something more tractable, Doug decided to get inside this subroutine and see if he could improve it in some way. Doug had seen how our existing algorithm took advantage of the special structure of S-systems to improve efficiency. He pushed this line of reasoning to the limit and came up with a novel strategy for solving S-systems that was to prove much more efficient than any of the conventional approaches. This allowed Doug to finish his comparative analysis of the dynamics of alternative immune networks [25, 26], which was one of the primary objectives of his thesis.

There were four other projects that occupied my time during the sabbatical year in Canberra. First, I worked with Peter Sands to generalize the existing theory of gene circuits. This proved more difficult than we had initially supposed. What we thought was a simplification in our approach in fact was making the analysis of stability nearly intractable. The fruits of this work appeared several years later, when by chance I removed the ''simplification'' and realistically represented the reactions for the synthesis and degradation of the inducer. Second, I had observed over many years of testing local stability that the last two Routh criteria always proved to be critical, and so I set out to demonstrate this mathematically. I managed to prove this geometrically, and then sometime later I discovered a beautiful algebraic proof that had been buried in the older literature. [I would still like to write this up and make the appropriate connections between these proofs.] Third, I continued the search for more general analytical solutions, particularly for a subclass of S-systems that I began to call Half-systems because each equation consisted of a single term rather than the two terms found in the S-system representation. I showed that these Half-systems always have a single trajectory in the concentration space that can be obtained analytically and that this corresponds to the unique steady-state solution in the S-system form. I had hoped that this approach might lead to other solvable trajectories, but these proved as elusive as the analytical solutions in the S-system form. Finally, the majority of my time was spent working on a review of demand theory.

Since the beginning of the 1980s I had been giving seminars, lectures in short courses, and presentations at conferences extolling the virtues of S-systems as a canonical nonlinear form that is particularly appropriate for the analysis of organizationally complex biological systems. At first, the very idea that there might be such a thing as a canonical form for nonlinearities was greeted with skepticism. Certainly there was no proof in the mathematician's sense; there was only this growing catalog of specific cases that had been "recast" into the form of an S-system. I cannot say that I remember receiving any real support for these ideas until a conference in 1984, which otherwise was quite unremarkable. After I had given my talk and received the usual questions, I was surprised to find someone in the audience stand up and say that this was not only very interesting, but similar to results obtained independently by one of his colleagues. This someone was Werner Ebeling, who seven years earlier had provided a similar link between me and another of his colleagues. Each of us was unfamiliar with the other's work.

The first year following our return from sabbatical was exceptionally busy. It was Masahiro's last year and we were trying hard to finish the work on two-stage proofreading by isoleucyl tRNA synthetase, and Doug was in the process of converting our interactive program into a version for the IBM-PC based on his more efficient algorithm for solving differential equations in S-system form.

Eberhard and I were finding that infrequent correspondence was an unsatisfactory substitute for the intense collaboration we previously had enjoyed. So it was a delightful prospect for both of us when we learned that he would return to Ann Arbor in 1985. Within a few months we were hard at work expanding the repertoire of methods for recasting nonlinear equations into S-system form. There is a natural tendency to express the power-law formalisms in the generalized mass action representation; this tendency also is manifested in recasting. Therefore, recasting into S-system form required that one reduce a sum of three or more terms to a simple difference between two terms. Although we had a number of methods for doing this in specific cases, we still had no systematic approach to this reduction. Eberhard discovered that the product rule of differentiation could be used to accomplish this reduction systematically.

We set out to develop a catalog of recasting methods that experience had shown were sufficient to recast essentially any nonlinear function into S-system form. It soon became apparent that neither of us had our heart in this project. As I recall, Eberhard went off and began recasting equations into as many different forms as he could imagine. For example, if one could reduce three terms to two, why not reduce two to one? He had discovered the Half-systems that I had studied during my sabbatical. He also showed that one could reduce the exponents to ones and zeros by recasting, thereby producing a new form that we called Binary-systems. At one time or another, everyone in the lab had tried unsuccessfully to recast these equations as a linear system. Eberhard had gone as far as he could, but he could go no further. Some product nonlinearity always remained. In looking at his equations I saw that they could be put in the form of a generalized Volterra-system. Thus, within a few months of his return to Ann Arbor, we had shown that GMA-systems, S-systems, Half-systems, Binary-systems, and Volterra-systems were all mathematically equivalent canonical forms.

Perhaps the most surprising equivalence was that between S-systems, which involved all products and one sum per equation, and Volterra-systems, which involved all sums and one product per equation [45]. This suggested to me that all nonlinearities could be constructed out of products and sums, and that to a large extent one could exchange one for the other as long as neither was entirely eliminated.

While Eberhard had been exploring all these variations of recasting, I also had strayed from the catalog project and I was trying to automate the recasting procedure. Up to this point we had always been able to do it, but it was more art than science. It wasn't long after Eberhard discovered the use of the product rule that an algorithm emerged. There were three basic operations that one performed repeatedly until the equations terminated in the desired canonical form: *translocation* of variables into the positive cone, *decomposition* of composite functions by the chain rule of differentiation, and *reduction* of terms by the product rule of differentiation [46].

At about this time we received a copy of the book by Peschel and Mende [73], and we learned in detail that they had independently been traveling a very similar path. They were the first investigators we had found who had a deep appreciation for the possibilities of these canonical nonlinear forms. This initial contact led to an exchange of visits and we have subsequently continued to share information.

By the time Eberhard had returned, Doug had implemented his algorithm as a sophisticated user-friendly program for the PC. I am not sure when it dawned on us, but I suspect it was soon after Doug had demonstrated the efficiency of his numerical algorithm for solving S-systems, that Doug's algorithm became a general purpose algorithm when it was coupled with recasting. Simultaneously, we saw that recasting might have practical implications, above and beyond any theoretical or esthetic value it might have, when it was coupled with Doug's algorithm. One might be able to solve nonlinear differential equations more efficiently by first recasting them into S-system canonical form and then solving them with ESSYNS, the acronym we had adopted for the program. Doug had collected the standard benchmark problems used to test the efficiency of differential equation solvers, but he was busy writing his thesis and improving the implementation of ESSYNS.

Eberhard and I were writing a review of S-systems and recasting. Although we felt great benefit might result from the exploration of these canonical nonlinear forms, the realization of these benefits might have to await the further development of methods for their analysis. The manuscript needed a little something, and I was anxious to test the notion that recasting coupled with ESSYNS might prove to have an immediate practical benefit. Doug was not in a position to help, so I suggested that we divide up the benchmark problems and start recasting. I started with a bunch of the simpler ones and Eberhard finished off with the more difficult cases. Eberhard ran ESSYNS in the "parallel mode" (i.e., he loaded it on a dozen PC's in the medical center's computing facility), which allowed us to solve the problems simultaneously, and we soon had the results for comparison. The combination of recasting and ESSYNS was much faster than the conventional solutions for most problems and error tolerances, as much as 100-fold for some requiring high accuracy. In the few cases where it was slower, the difference was less than 2-fold. Although we knew that ESSYNS was fast for problems in S-system form, recasting generally introduces additional variables that might make solving the recast equations slower. The results showed clearly that this cost of recasting was more than recouped by the efficiency of ESSYNS [46].

Doug's algorithm is so efficient that we now solve dynamic problems on an IBM-PC that previously required the university's mainframe computer. His contributions to the field of scientific computing could well have given him a second thesis; as it was, the unusual breadth and depth of his work contributed to his receiving a very competitive Distinguished Thesis award.

In the midst of these exciting developments, we had to take some time out to go back and restate theoretical results that had been obtained ten and fifteen years earlier [27, 28]. This had become necessary because restricted versions of this theory were being described as a completely general theory with new names and origins; the logical relationships to the earlier work were not recognized and the advantages of the S-system representation were being obscured. In particular, there was a great deal of confusion regarding the relative accuracy of the different representations within the power-law formalism. Most investigators recognized that the S-system representation was an approximation, many thought the generalized mass action representation was a more accurate approximation, and a few considered the latter to be "exact."

I have to accept some of the responsibility for this misunderstanding. In my early work I had shown that the power-law formalism systematically overestimates a saturating function while it systematically underestimates a sum [2]. The S-system representation is more accurate because one makes both types of errors and there is some cancellation; in contrast, no such cancellation occurs with the GMA representation because one makes only a single type of error.

This result was not sufficiently emphasized because it was only one of several reasons why we had chosen to focus on the S-system form. When some people began to claim that methods based on the GMA representation were "exact" while those based on the S-system representation were "only" an approximation, it was clear that something had to be done to set the record straight.

Eberhard did a detailed study of accuracy for the prototype situations involving converging and diverging reactions [29]. He developed a number of clever ways to characterize objectively the accuracy of various representations, and from all these different perspectives he demonstrated that the power-law representation of the aggregate flux through metabolic pools (the S-system) is nearly always more accurate than the corresponding representation without aggregation (the GMA system). With the completion of this work, and the end in sight for the recasting review, Eberhard left for Charleston where he had been appointed an associate professor at the Medical University of South Carolina. He has developed new applications of the S-systems methodology in the areas of plant growth, population dynamics, and purine metabolism. He also has continued developments on the theoretical front with studies of the group structure of S-system equations. These are some of his recent activities that will be found among others elsewhere in this volume.

1986–1990

Eberhard had gone to Charleston, and Doug was doing his clinical rotations, so once again things were relatively quiet in the laboratory for about a year. This period also introduced me to several other investigators who were doing some very nice work in the field. In 1986 I learned of Tom Johnson's work at North Carolina State University. He had been applying the power-law formalism to problems of animal growth in agriculture and developing new methods for parameter estimation. A year later we were on the program of a symposium together, and I had the pleasure of meeting him personally. [You will hear about his recent work elsewhere in this volume.]

Albert Sorribas arrived in 1987 for a year-long stay in Ann Arbor as a postdoctoral fellow. It was only one year, but it was an extremely active and fruitful year. After some initial exploration of several topics, Albert quickly mastered the fundamentals of the power-law formalism and its use in analyzing biochemical systems. He then began a comparative study of a model system exhibiting enzyme-enzyme interactions [35, 36]. His work showed quite clearly the logical relationships among the various representations that have been based on the underlying power-law formalism. His exhaustive analysis demonstrated the superior accuracy of the S-system representation in the context of this intact biochemical system. It also showed that the S-system methodology can be applied in a straightforward manner, whereas other approaches—based on so-called summation and connectivity methodology—are invalid for such systems.

Albert made the first systematic analysis of alternative strategies for representing amphibolic pathways [37]. He showed that the strategy used by nearly everyone in the field is qualitatively inappropriate for pathways near equilibrium and for pathways that actually undergo a reversal of flux under physiological conditions. In my view, the series of three papers that were published in 1989 set a new standard for clarity and thoroughness in this field. Albert also developed the beginnings of a new method for estimating parameter values from macroscopic measurements on the system.

Albert has a light-hearted, unassuming manner that masks a deep commitment to fundamental problems in biology. Who else could have produced, directed, and starred in the internationally acclaimed ''An S-System's Analysis of Paella''! He is a genuinely warm human being with a delightful sense of humor who is sure to make a lasting mark on the field. He returned to the University of Barcelona, where he now holds an appointment as an assistant professor in the department of biostatistics.

Eberhard's arrival in Charleston generated an immediate interest in S-systems among several of his new colleagues. One of these individuals, Phil Rust, is now using S-systems and ESSYNS to advance the area of computational statistics, which is one of the fastest growing areas in modern statistics. Ever since my early work recasting probability distributions and showing that these may be considered special cases of a suprasystem defined by S-systems, and Eberhard's early work on regeneration and cell-cycle statistics, I have considered this to be one of the most promising areas for development. Phil already has treated several of the more important distributions that are difficult to study by conventional means. He has demonstrated that one can obtain practically any property of these distributions, and to a high degree of accuracy, simply by recasting the appropriate equations and solving them with ESSYNS. Again, there will be more on this topic elsewhere in this volume.

Most recently, the field has benefited from the entry of an outspoken and critical spirit in Marta Cascante. She already had come to appreciate the limitations in other approaches to the study of metabolic control, and had used sensitivity theory to generalize one of these approaches [55, 56], before she became aware that it is a variant of the implicit GMA approach within the power-law formalism [35, 36]. Marta and Albert are collaborating on some extensions of the power-law approach to the dynamics of biochemical systems, so one can safely predict that the field will continue to benefit from Marta's incisive inquiry and enthusiasm.

Finally, at the end of this period Doug returned from clinical rotations to complete his manuscripts on ESSYNS. These are beautiful papers in which one can see the influence of the methods we developed to make mathematically controlled comparisons. His careful comparisons in this area—numerical methods for solving nonlinear differential equations—have drawn on ideas that stem from the biologist's notion of a control experiment. The thoroughness with which these papers are written is characteristic of Doug's work. He completely redid the recasting of the benchmark problems that Eberhard and I had done earlier. In the process he improved the recasting of some problems and completely redid the solutions using more rigorous conditions for comparing the alternative methods. This was done in such a way that he could carefully dissect out the relative contribution to the overall efficiency that is made by each of the techniques that he had implemented in ESSYNS. I believe these papers set a new standard for the comparison of numerical algorithms in this area; they also have introduced concepts and methods that I expect will become widely used.

CONCLUSION

I had first intended to include a section on the next twenty years, but there are so many exciting developments that are currently being pursued that this section would have doubled an already lengthy account. Instead, let me conclude with just a few observations.

Over the years I have come to appreciate that Bertalanffy had deep insight and was well ahead of his time:

> *"Classical" system theory applies classical mathematics, i.e., calculus. Its aim is to state principles which apply to systems in general or defined subclasses (e.g., closed and open* [, *chemical and ecological, etc.*] *systems), to provide techniques for their investigation and description, and to apply these to concrete cases.*
>
> —Bertalanffy (1967)

In this definition of classical systems theory he begins with the kinds of tools that are used. He states very clearly that the aim is to get at basic principles that apply to systems in general and to specific subclasses. But these are not to be vague generalizations and slogans; the aim is to develop techniques for investigation and characterization of these principles and to apply these to concrete cases. This is clearly seen as a complete program and not some kind of specialized esoteric activity. In looking back over the past twenty years of S-system development, it has become clear to me how well this program fits Bertalanffy's definition of a classical systems theory.

Secondly, for me, one of the most satisfying aspects of this work has been the rich cross-fertilization of biology and mathematics. Biological insights have resulted from the application of mathematics to generic as well as specific systems, and new mathematical developments have been suggested by the biology itself. Of course, this is common knowledge when you go back to physics, especially that of the seventeenth and eighteenth centuries, where problems in physics inspired much of what is now known as classical mathematics and the physics that emerged would have been impossible without the concomitant mathematical development.

However, with the increasing specialization and division of labor in modern science, investigators in the various disciplines have become so intellectually isolated from one other that it has become nearly impossible for them to communicate. They have lost the sense of a common enterprise, that different areas can enrich each other. Indeed, one can see evidence of just the opposite; effort is expended in protecting ''turf'' and preventing the intrusion of ideas from other areas. This is echoed in the comments of biologists who insist that these are problems for biology and that mathematics has nothing to contribute. It also is seen among mathematicians who think that biology has nothing to offer but problems for which mathematics already has the tools to solve.

Modern science has become complex and fragmented with the explosion of information. In this context the erection of barriers to the exchange of information can be seen as a defensive strategy for coping. (I am reminded of the story about a graduate student who became ecstatic at being able to define a thesis topic that dated from a ''single paper'' of five years before, which meant that he would only have a five-year history of papers to read.) But in the long term this is not a solution. In many cases this leads to frequent but inadvertent ''reinvention of the wheel,'' and in some cases to a conscious repackaging of results obtained in other disciplines without any trace of their origins. These practices are justified on the grounds that the areas, ''being so different,'' require specialized versions for each particular audience anyway. (I recently had the frustrating experience of going to a modern textbook and finding an interesting lead on some topic, only to realize that I could follow it no further because there were no references in the entire book!) This compartmentalization and duplication denies the next generation access to the deep roots and rich traditions that underlie the concepts of their discipline.

The advance of science demands that the best from our disciplines be integrated in an effective synthesis that will counter the centripetal influences that are propelled by the reductionist paradigm. John Casti in the conclusion of his book *Paradigms Lost* [57] advocates ''a return to the Great Problems to reverse the trend toward fragmentation and incoherence in the sciences. . . . [These problems] force one to learn about the mutual interrelationships of many things.'' Indeed, his book is a case study of this approach, and, I might add, a rich source of references for whose who wish to pursue it further.

Finally, let me bring this rambling account to a close with the famous quote from Pascal:

> *I have made this letter longer than usual because I lacked the time to make it short.*

Part I

General Methods of Canonical Modeling

2. Introduction to S-Systems

Eberhard O. Voit, Michael A. Savageau, and Douglas H. Irvine

In the analysis of natural processes or complex technological systems, methods based on linear mathematics often cannot capture essential nonlinear characteristics. For example, living systems typically display a broad repertoire of nonlinear responses such as synergism, saturation, and stable oscillations. For natural processes, these types of responses are the rule rather than the exception. They are intrinsically nonlinear and difficult even to mimic when restricted to linear mathematics. Nevertheless, the first attempt to understand a complex system often is an analysis of a linearized model of the system because of the depth and robustness of analyses possible with linear mathematics. A variety of theorems and practical methods help to determine the existence of solutions for linear systems and provide a straightforward means to evaluate steady states, local asymptotic stability, and parameter sensitivity. In contrast, nonlinear systems analysis lacks such powerful systematic methodology. For instance, deriving analytical solutions for nonlinear ordinary differential equations is a sophisticated art limited to so few special cases that it is insufficient for general nonlinear systems analysis.

One fundamental difference between linear and nonlinear functions and differential equations is that linear representations all have the same underlying structure and are completely determined by a set of parameter values, whereas there is no unifying structure for nonlinear representations. Consequently, it is difficult to compare and to classify nonlinear functions or to develop generally applicable methods for nonlinear analysis. What is needed is a nonlinear theory based on a language powerful enough to treat nonlinearities in an objective fashion, yet structured enough to make them amenable to systematic analysis.

Ideally, such a theory would rival linear theory in power and methodology but apply to the incomparably richer domain of nonlinear phenomena. While still far from this goal, the *power-law formalism* with its associated *S-system* differential equations shows remarkable potential. S-systems provide systematic methods for analysis of nonlinear systems. In addition, they have the striking property that virtually any differentiable nonlinear function or ordinary differential equation can be expressed in S-system form.

POWER-LAW FORMALISM; S-SYSTEMS

Biological and other organizationally complex systems are comprised of numerous, richly interacting components. The details of the processes that govern expressions, interconversions, and interactions of these components usually are not known, but these processes almost always are nonlinear. Consequently, their description requires a representation general enough to capture the essence of the observed response and, at the same time, it should be mathematically manageable to permit detailed analysis. One approach that satisfies these requirements is the power-law formalism. The seeds for this formalism came from Bode's observation that many nonlinearities are well approximated piece by piece with straight lines in logarithmic space, and also from the observation that many functional relationships in nature are linear in logarithmic space (cf. Chapter 1 and [2, 8, 19]).

There are only two basic requirements for the application of the power-law formalism. They are that each component of the system can be quantified by positive values and that the logarithm of the function describing each process within the system is differentiable. Twenty years of experience with the formalism have demonstrated that neither of these requirements is unduly restrictive. The dynamics of a system can be represented by formulating the change in each system component X_i as the difference between all influences that increase X_i (contribute to net accumulation) and all influences that decrease X_i (contribute to net clearance). Symbolically, the change in X_i is represented as

$$\dot{X}_i = F_i^+(X_1, \cdots, X_n) - F_i^-(X_1, \cdots, X_n) \qquad i \in \{1, \cdots, n\}$$

X_1 through X_n are the system constituents and $\dot{X}_i = dX_i/dt$ is the derivative of X_i with respect to the independent variable t, which usually is interpreted as *time*. F_i^+ and F_i^- are positive-valued differentiable functions for each system component and describe the net accumulation and net clearance of X_i. F_i^+ is nonlinearly approximated by expressing it in terms of the logarithms of the system components $[F_i^+(Y_1, \cdots Y_n)$ with $Y_i = \log X_i]$, approximating its logarithm by the constant and linear terms of the Taylor series, and finally translating back into the Cartesian space [2, 8]. The same type of approximation is made for F_i^-. The resulting functions are nonlinear local representation for F_i^+ and F_i^-, each in the form of a product of power-law functions:

$$F_i^+ \approx \alpha_i \prod_{j=1}^{n} X_j^{g_{ij}}$$

$$F_i^- \approx \beta_i \prod_{j=1}^{n} X_j^{h_{ij}}$$

The equations that describe the dynamics of the system are thus

$$\dot{X}_i = \alpha_i \prod_{j=1}^{n} X_j^{g_{ij}} - \beta_i \prod_{j=1}^{n} X_j^{h_{ij}} \qquad i \in \{1, \cdots, n\}$$

This regularly structured system of nonlinear ordinary differential equations is called an *S-system* to indicate its ability to capture *S*aturable and *S*ynergistic characteristics of nonlinear systems. In a biochemical context, the nonnegative parameters α_i and β_i are called *rate constants*, and the real-valued exponents g_{ij} and h_{ij} are referred to as *kinetic orders*.

From this mathematical derivation it is apparent that the S-system representation is only one possible realization among several others within the power-law formalism. For example, when the change in X_i is represented as

$$\dot{X}_i = \sum_{k=1}^{p} F_{ik}^+(X_1, \cdots, X_n) - \sum_{k=1}^{q} F_{ik}^-(X_1, \cdots, X_n)$$

and each function in each sum is approximated with a product of power-law functions, the resulting set of equations has the form

$$\dot{X}_i = \sum_{k=1}^{p} \alpha_{ik} \prod_{j=1}^{n} X_j^{g_{ijk}} - \sum_{k=1}^{q} \beta_{ik} \prod_{j=1}^{n} X_j^{h_{ijk}}$$

which has been called a *Generalized Mass Action system.* This type of power-law approximation is employed implicitly in a method called *metabolic control theory* which is used to study the distribution of steady-state parameter sensitivities, or control coefficients, in biochemical pathways (cf. Chapters 1, 3, and 4).

As another alternative to S-systems and Generalized Mass Action systems, one can define

$$G_i = F_i^+ - F_i^-$$

and approximate G_i with a single product of power-law functions to generate the system description

$$\dot{X}_i = \gamma_i \prod_{j=1}^{n} X_j^{g_{ij}}$$

in which each X_i is positive and all parameters can be positive, zero, or negative. This type of system is called a *Half-system* (cf. Chapters 10, 12, and [46]).

These different representations within the power-law formalism provide alternative mathematical structures with slightly different properties (cf. Chapters 1, 3, 10, and 12). They also can be interpreted as alternative modeling strategies, each addressing a different level of organization (cf. [29]).

While S-systems, Generalized Mass Action systems and Half-systems are all viable modeling alternatives, S-systems are of particular interest. They provide the only type of power-law model with nonzero steady states that can be calculated by symbolic, analytical means [27]. Furthermore, analysis of biochemical pathways has shown that S-systems yield more accurate representations than competing Mass Action systems [28, 35–37]. Half-systems have an appealingly simple mathematical structure and are important for the classification of nonlinearities (cf. Chapter 12). In cases of high redundancies among the S-system parameters, Half-systems are also a valuable tool for parameter estimation (cf. Chapter 10). However, the use of Half-systems as power-law approximations of biological systems often is limited, because they can only represent steady states in which at least one variable vanishes.

MODELING WITH S-SYSTEMS

Due to the structural homogeneity of S-system differential equations, all steps required for description and analysis of complex biological networks are straightforward. Systematic procedures for system representation, parameter estimation, steady-state solution, stability analysis, and sensitivity analysis have been developed. These methods, in combination with efficient methods for evaluating the dynamic responses of the system, provide a powerful set of tools for understanding the integrated function of complex nonlinear systems such as molecular and cellular regulatory networks.

System Representation

The initial procedure for any type of systems analysis is to represent the system of interest. For S-systems, this process can be divided into three steps.

1. Identify which variables are important and should be included explicitly in the model, and which variables are less important and do not need to be included. Then enumerate the interrelationships among the variables, that is, decide whether or not variable X_i directly influences accumulation or clearance of variable X_j. In contrast to most other nonlinear approaches, an appropriate S-system representation can be formulated from this information, according to a few simple rules: For each variable X_i, $\dot{X}_i$ is equal to the difference between two terms, one for net accumulation (α-term) and one for net clearance (β-term). Each term is a product of power-law functions containing every relevant system variable that affects the process.
2. Once all relevant variables and interrelationships have been identified, draw an arrow diagram to illustrate graphically the variables and their interrelationships. For instance, the diagram in the figure below could represent a biochemical system with six metabolites in which X_1 is used to synthesize X_2, X_2 is used to synthesize X_3, and X_3 is used to synthesize both X_4 and X_6. X_1 and X_4 combine to form X_5, which is removed from the system. X_3 inhibits production of X_2, and X_4 activates its own synthesis. The darker arrows show the flow of material in the system. The arrow originating from both X_1 and X_4 and leading to X_5 represents a single bimolecular reaction. (Two dark arrows, from X_1 to X_5 and from X_4 to X_5, would represent two separate reactions.) The lighter arrows heading from a variable to the center of a darker arrow ($X_j \rightarrow X_k$) represent modulation of the flow between X_j and X_k; a *plus* sign next to these arrows indicates activation, and a *minus* sign indicates inhibition.

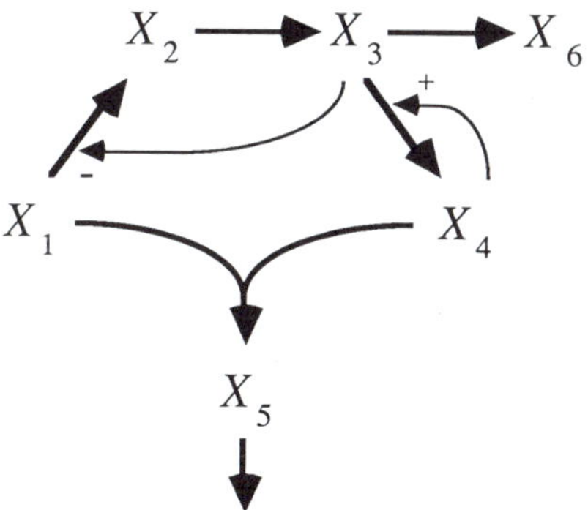

FIGURE 2.1. Arrow Diagram of Interactions

3. Write the corresponding S-system equations for the diagram. There is one equation for every system variable X_i. In each equation, the change in X_i with time (dX_i/dt written as $\dot{X}_i$) is equal to the difference between two

terms, an α-term and a β-term, which are both products of power-law functions. The heavier arrows leading to a variable, say X_i, correspond to the net accumulation term (α_i-term); and those leading away correspond to the net clearance term (β_i-term). The double-tailed arrow from X_1 and X_4 to X_5 corresponds to inclusion of both X_1 and X_4 in the α-term for X_5 and in the β-terms for both X_1 and X_4. Inclusion of X_4 in the β-term of X_1 implies that X_5 cannot be formed from X_1 alone but it needs two components, X_1 and X_4, that come together in a bimolecular or sequential reaction. The lighter arrows from X_3 to $X_1 \rightarrow X_2$ and from X_4 to $X_3 \rightarrow X_4$ indicate modulation of the processes $X_1 \rightarrow X_2$ and $X_3 \rightarrow X_4$ and imply that X_3 and X_4 should be included in the corresponding α_i or β_i-terms. In our example, the α- and β-terms of the S-system differential equations therefore contain the following variables:

	α-term	β-term
$\dot{X}_1$	---	X_1, X_3, X_4
$\dot{X}_2$	X_1, X_3	X_2
$\dot{X}_4$	X_3, X_4	X_1, X_4
$\dot{X}_5$	X_1, X_4	X_5
$\dot{X}_6$	X_3	---

For each $\dot{X}_i$ there are exponents and multiplicative constants associated with the α_i- and β_i-terms. The exponents are represented by the double-subscripted parameters g_{ij} in the α_i-term and h_{ij} in the β_i-term. In each case, the first subscript is associated with the corresponding $\dot{X}_i$ and the second subscript is associated with the variable X_j that modulates the process. Hence, for Fig. 2.1 the S-system representation is

$$\begin{aligned}
\dot{X}_1 &= && -\beta_1 X_1^{h_{11}} X_3^{h_{13}} X_4^{h_{14}} \\
\dot{X}_2 &= \alpha_2 X_1^{g_{21}} X_3^{g_{23}} && -\beta_2 X_2^{h_{22}} \\
\dot{X}_3 &= \alpha_3 X_2^{g_{32}} && -\beta_3 X_3^{h_{33}} X_4^{h_{34}} \\
\dot{X}_4 &= \alpha_4 X_3^{g_{43}} X_4^{g_{44}} && -\beta_4 X_1^{h_{41}} X_4^{h_{44}} \\
\dot{X}_5 &= \alpha_5 X_1^{g_{51}} X_4^{g_{54}} && -\beta_5 X_5^{h_{55}} \\
\dot{X}_6 &= \alpha_6 X_3^{g_{63}} &&
\end{aligned}$$

Conservation of flow between some components means that the corresponding parameters in two terms actually have the same value. For example, the above figure shows that degradation of X_2 is the same as synthesis of X_3. Hence, the β-term of X_2 and the α-term of X_3 are equal, and the parameters α_3 and g_{32} are the same as β_2 and h_{22}. Similarly, α_5, g_{51} and g_{54} are the same as β_4, h_{41} and

h_{44}, respectively. (See [35] for more complicated constraint relationships among parameters.)

Note that for the changes in X_1, X_2, X_3, and X_4 it is unimportant whether the fates of X_3 and X_5 are explicitly specified with other variables like X_6. For instance, one could drop X_6, or on the other hand, one could represent the fate of X_5 using an additional variable, X_7. When X_7 is not included, the system is called *open*, since X_5 leaves the system. If the equation

$$\dot{X}_7 = \alpha_7 X_5^{g_{75}} = \beta_5 X_5^{h_{55}}$$

is added to the above six equations, the system is called *closed* since all material is accounted for within the system. In each case, however, neither X_6 nor X_7 appears in any of the first five equations, which means that the dynamic responses of X_1 through X_5 are not affected by including or excluding X_6 or X_7.

The literature contains various examples of how to set up the S-system differential equations from an arrow diagram (see also Chapters 3, 5, 6, 9, and 14). In many cases this process is uniquely determined by the power-law formalism. However, exceptions exist in which different modeling strategies are possible. A specific example is a reversible biochemical pathway in which fluxes through pools can be aggregated in different ways before the power-law approximation is performed. This example has been analyzed in detail by Sorribas and Savageau [37].

Parameter Values

Three different approaches are available for estimation of parameter values. They involve rather different experimental strategies that complement each other.

1. ***Estimates from kinetic data* in vitro.** In biochemistry, most data come from kinetic experiments *in vitro*. The methods for extracting values of the S-system parameters from this type of data are well established (cf. [2]): One can plot the logarithm of the reaction rate as a function of the logarithm of the relevant reactant or modifier concentration, while holding all other variables that affect this process constant at their normal operating values *in situ*. Conventional linear regression in these logarithmic coordinates yields values for the kinetic order (slope) and the rate constant (intercept).

Alternatively, when kinetic data already are represented in mathematical form as a conventional steady-state rate law, values for the parameters in the power-law representation are determined by applying straightforward calculations based on the definitions of S-system parameters [2, 12, 27, 29, 46]. To approximate the function $V(X)$ with a power-law representation

$$V(X) \approx \alpha X^g$$

one computes the constant and linear terms of the Taylor series of log (V) as a function of log (X):

$$\log V = \log V_0 + \left.\frac{d(\log V)}{d(\log X)}\right|_{X=X_0} (\log X - \log X_0)$$

where the index 0 indicates the operating point. Upon collecting terms, the equation can be written as

$$\log V \approx \log \alpha + g \log X$$

where

$$\log \alpha = \log V_0 - \left.\frac{d(\log V)}{d(\log X)}\right|_{X=X_0} \log X_0$$

and

$$g = \left.\frac{d(\log V)}{d(\log X)}\right|_{X=X_0}$$

According to the differentiation rule for the logarithmic function, the parameter g is thus

$$g = \left.\frac{dV}{dX}\frac{X}{V}\right|_{X=X_0}$$

The appropriate rate constant, α, is most easily obtained by equating the function V and its power-law approximation at the operating point and then dividing both sides of the equation by X_0^g:

$$\alpha = V_0 X_0^{-g}$$

It should be noted that, independent of the particular form of the function V, the power-law parameters g and α always are defined in this manner. Therefore, one can directly use the two previous equations, without actually evoking the Taylor series.

As an illustration, say we want to determine S-system parameters g and α that represent a Hill rate law of type $V = V_m X^n/(X^n + K_m^n)$, where V_m is the maximal velocity, K_m the substrate concentration that produces half-maximal binding, and n the so-called Hill coefficient. Let's assume that experiments have shown that n ranges between 1.5 and 3.0. The kinetic order for this process in S-system representation is given by the relative derivative of the Hill rate law; thus,

$$\begin{aligned} g &= \left.\frac{dV}{dX}\frac{X}{V}\right|_{X=X_0} \\ &= nK_m^n/(K_m^n + X_0^n) \end{aligned}$$

where X_0 is the operating point which is the substrate concentration *in situ* for which the S-system representation gives the same result as the Hill equation. When the operating point X_0 is chosen as K_m, the expression for g reduces to

$$g = nK_m^n/(K_m^n + K_m^n) = n/2$$

and α becomes

$$\alpha = \tfrac{1}{2} V_m K_m^{-n/2}$$

Thus, for a system operating near the K_m with a Hill coefficient between 1.5 and 3, the value for g is between 0.75 and 1.5 and the rate constant α falls within a corresponding range that is given by the last equation.

For a process appropriately described by a Michaelis-Menten rate law operating near the K_m, the corresponding S-system exponent is 0.5. This is easy to confirm since the Michaelis-Menten rate law is the special case of the Hill rate law where $n = 1$.

When V is a function of two or more variables $X_1, \cdots, X_n$, its power-law approximation is

$$V = \alpha X_1^{g_1} X_2^{g_2} \cdots X_n^{g_n}$$

Again, the Taylor series in logarithmic coordinates is used to determine the exponent associated with X_i, g_i, but since V in this case depends on several variables, the appropriate derivative is the *partial* derivative

$$g_i = \left.\frac{\partial(\log V)}{\partial(\log X_i)}\right|_{X_i = X_0} = \left.\frac{X_i}{V}\frac{\partial V}{\partial X_i}\right|_{X_i = X_{i0}}$$

α is defined in the same fashion as before, that is, by equating the rate law and its power-law representation at the operating point and solving for α:

$$\alpha = V_0 X_{10}^{-g_1} X_{20}^{-g_2} \cdot \cdot \cdot X_{n0}^{-g_n}$$

Again, these definitions are independent of the particular function V and can be employed without actually developing the Taylor series. An example is presented in Chapter 9.

2. ***Estimates from steady-state experiments* in situ.** Most of what is known about the integrated responses of biological networks comes from steady-state experiments. The experimental system is perturbed by a persistent change in an input variable and the response of the systems constituents is measured. Of particular interest are changes in the steady-state fluxes and concentrations. In a second type of steady-state experiment, a change in the activity of an enzyme is introduced (e.g., by mutation of the corresponding structural gene), and, again, resultant changes in fluxes and concentrations are measured. Both types of experimental changes are easy to examine using S-systems (cf. [2, 25–26, 35–37]).

The response of a system to a change in input (i.e., an independent variable) can be described by the so-called *logarithmic gain*, which is easy to calculate (cf. the next Section). The logarithmic gains can be determined by plotting experimental data in logarithmic coordinates and then performing linear regression, or by calculating the corresponding logarithmic gain in the power-law formalism using the steady-state solution of the underlying kinetic equations. The latter calculation can be done numerically with a given set of parameter values or symbolically without having to assume particular values for the parameters. These types of general symbolic relationships have yielded insight into the functional significance of a variety of molecular and cellular control mechanisms (e.g., see [2, 25–26, 35–37]).

Determining one logarithmic gain usually does not suffice to estimate the value for a particular kinetic order, but rather it yields a constraint equation involving several kinetic orders. This constraint equation allows for one kinetic order to be written in terms of the others; substituting the result throughout the system thus eliminates one kinetic order and reduces by one the number of remaining parameters to be estimated (cf. Chapter 4).

The response of the system to a change in enzyme activity can be characterized by the established concept of *parameter sensitivity* (cf. [11]). Sensitivity analysis has proven very useful for the comparative analysis of alternative molecular and cellular regulatory mechanisms (e.g., see Chapters 4, 5, and [2, 25–26, 35–37]). Like logarithmic gains, parameter sensitivities can be determined directly from measurements made with the intact system, or indirectly in terms of the underlying parameters. Again, determination of one parameter sensitivity yields one constraint equation and reduces by one the number of independent parameters that must be estimated for a full characterization of the intact network (cf. Chapter 4).

3. ***Estimation from dynamic experiments.*** Dynamic experiments are technically difficult and can be tedious to perform for intact biological systems (whereas they are an efficient and fun part of computer simulations). In the laboratory, one needs the technology to sample important components in a time series without disrupting the natural response of the system. Generally, this noninvasive technology takes considerable time and effort to implement. The mathematical estimation of parameter values in dynamic equations usually is a complicated process that requires extensive computational effort. In many cases it involves a nonlinear regression procedure, combined with a numerical integrator. The literature on nonlinear regression is quite extensive, and algorithms like the Gauss-Newton and Levenberg-Marquardt methods are implemented in standard software libraries, such as IMSL and BMDP. Nevertheless, even for moderately large systems, these algorithms can be problematic. Often they are very sensitive with respect to initial parameter guesses and they fail when the system parameters are redundant. In some cases the estimation problem can be reduced from one involving nonlinear differential equations to one involving just nonlinear algebraic equations, which simplifies the iterative computations significantly (cf. Chapter 10 and [22–23]).

Steady-State Analysis

Calculation of steady states and their local asymptotic stability is straightforward for mathematical models expressed in S-system form. Complete steady-state analyses can be performed with paper and pencil, symbolically with computer algebra programs (Chapter 6), or numerically (Chapter 7). The symbolic solutions are perhaps the most important feature of S-systems because such methods for computation of steady states are very seldom available for other general classes of nonlinear ordinary differential equations.

1. ***Steady-States.*** The simplicity of steady-state computations is due to the regular structure of S-systems. By setting derivatives equal to zero it is clear that, at steady state, each α-term must equal the corresponding β-term:

$$\alpha_i \prod_{j=1}^{n} X_j^{g_{ij}} = \beta_i \prod_{j=1}^{n} X_j^{h_{ij}} \qquad i \in \{1, \cdots, n\}$$

If none of the α or β terms is equal to zero, simple logarithmic transformation of the system variables reduces the equations to a set of linear algebraic equations:

$$\sum_{j=1}^{n} (g_{ij} - h_{ij}) \log X_j = \log \left(\frac{\beta_i}{\alpha_i}\right) \qquad i \in \{1, \cdots, n\}$$

Upon identifying $Y_i = \log X_i$, $a_{ij} = g_{ij} - h_{ij}$, $b_i = \log(\beta_i/\alpha_i)$, this system has the standard form of a system of linear algebraic equations, and if the system matrix is regular, calculation of nonzero steady-state values for the X_i's reduce to standard matrix inversion [2, 9]. In cases where one or more of the α or β terms equals zero, the system has no finite steady state or at least one of the steady-state values is equal to zero (see [2]). In the latter case, other steady-state values can be found by inspection or, in some cases, by exploiting the symmetry structure of the system (cf. Chapter 15). In the degenerate case of $\dot{X}_i = 0$, X_i can assume any steady-state value.

As an example, let's compute the steady state of an S-system with two dependent variables. Setting the two time derivatives equal to zero ($\dot{X}_1 = \dot{X}_2 = 0$), we obtain the equations

$$\alpha_1 X_1^{g_{11}} X_2^{g_{12}} = \beta_1 X_1^{h_{11}} X_2^{h_{12}}$$

$$\alpha_2 X_1^{g_{21}} X_2^{g_{22}} = \beta_2 X_1^{h_{21}} X_2^{h_{22}}$$

Assuming that none of the terms vanishes, we take logarithms and obtain a pair of linear algebraic equations

$$a_{11} Y_1 + a_{12} Y_2 = b_1$$

$$a_{21} Y_1 + a_{22} Y_2 = b_2$$

The solution to these equations in terms of the original variables X_1 and X_2 is

$$X_{1S} = \left[(\beta_1/\alpha_1)^{a_{22}} (\alpha_2/\beta_2)^{a_{12}}\right]^{1/\Delta}$$

$$X_{2S} = \left[(\beta_2/\alpha_2)^{a_{11}} (\alpha_1/\beta_1)^{a_{21}}\right]^{1/\Delta}$$

where the subscript S signifies steady-state values and $\Delta = a_{11}a_{22} - a_{12}a_{21}$ is the determinant of the linear system (cf. [2: p. 126]).

2. ***Stability.*** To test asymptotic local stability, one computes so-called *F-factors*, F_i. These factors come from linearizing the S-system at the steady-state point $(X_{1S}, X_{2S}, \cdots, X_{nS})$ and are defined as

$$F_i = \alpha_i X_{1S}^{g_{i1}} X_{2S}^{g_{i2}} \cdots X_{iS}^{g_{ii}-1} \cdots X_{nS}^{g_{in}}$$
$$= X_{iS}^{-1}\left(\alpha_i \prod_{j=1}^{n} X_{jS}^{g_{ij}}\right)$$
$$= X_{iS}^{-1}\left(\beta_i \prod_{j=1}^{n} X_{jS}^{h_{ij}}\right)$$

(cf. [2: pp. 126–129]). Each F_i characterizes the relative flux through the system at the steady-state point or, in other words, the turnover number in unit time. In a biochemical context, F_i is the ratio of the pathway's rate to the concentration X_i in steady state. *F*-factors can also be understood within the framework of Michaelis-Menten kinetics: If the K_m for the *i*-th intermediate reaction is doubled, the concentration of the substrate for that reaction will increase until it also is doubled and the steady-state rate for the pathway is reestablished. The corresponding value for F_i is therefore halved while the remaining system parameters are unchanged; i.e., F_i is proportional to $1/K_m$ (cf. [2: p. 186]).

For the above two-variable system, the *F*-factors are

$$F_1 = \alpha_1 X_{1S}^{g_{11}-1} X_{2S}^{g_{12}}$$
$$F_2 = \alpha_2 X_{1S}^{g_{21}} X_{2S}^{g_{22}-1}$$

and one can show that the system is locally stable if

$$F_1 a_{11} + F_2 a_{22} < 0$$
$$a_{11} a_{22} - a_{12} a_{21} > 0$$

(cf. [2: p. 128]).

The methods of steady-state computation and stability analysis in S-systems extend directly to systems with more than two variables and, in principle, always can be performed with paper and pencil. Of course, the computational effort increases and, in particular, the rather simple stability conditions become conditions on characteristic equations in the form of higher-order polynomials. General steady-state and stability analyses have been implemented in numerical or symbolic computer programs (Chapters 6 and 7).

For an S-system with *n* variables, local stability at a regular steady state is determined by the roots or eigenvalues λ_i of the characteristic equation, which is given by equating the system determinant to zero:

$$\begin{vmatrix} (F_1a_{11}-\lambda) & F_1a_{12} & F_1a_{13} & \cdots & F_1a_{1n} \\ F_2a_{21} & (F_2a_{22}-\lambda) & F_2a_{23} & \cdots & F_2a_{2n} \\ \vdots & \cdots & \cdots\cdots & & \vdots \\ F_na_{n1} & F_na_{n2} & F_na_{n3} & \cdots & (F_na_{nn}-\lambda) \end{vmatrix} = 0$$

Solution of this determinant equation leads to a polynomial of order n. A variety of methods can be employed to find the roots of this polynomial. For local asymptotic stability considerations, it is sufficient to determine if any eigenvalues have positive real parts. If that is the case, the system is unstable. Perhaps the easiest method for this purpose is the Routh-Hurwitz procedure [86, 89] which determines with elementary operations whether a polynomial of the form

$$\lambda^n + \phi_1\lambda^{n-1} + \phi_2\lambda^{n-2} + \cdots + \phi_{n-1}\lambda + \phi_n = 0$$

has roots with positive real parts. The procedure begins with setting up the following array of the coefficients ϕ_i and of numbers Δ_{ij} that are computed from the ϕ_i:

$$\begin{array}{llllll} 1 & \phi_2 & \phi_4 & \phi_6 & \phi_8 & \cdots \\ \phi_1 & \phi_3 & \phi_5 & \phi_7 & \phi_9 & \cdots \\ \Delta_{11} & \Delta_{13} & \Delta_{15} & \Delta_{17} & \Delta_{19} & \cdots \\ \Delta_{21} & \Delta_{23} & \Delta_{25} & \Delta_{27} & \Delta_{29} & \cdots \\ \cdots \\ \\ \Delta_{i1} & \Delta_{i3} & \Delta_{i5} & \cdots \\ \cdots \\ \Delta_{n-3,1} & \Delta_{n-3,3} \\ \Delta_{n-2,1} \\ \Delta_{n-1,1} \end{array}$$

where the Δ_{ij} are formed from the two elements just above it in the column just to the right and from the two corresponding elements in the first column:

$$\begin{aligned} \Delta_{11} &= (\phi_1\phi_2 - \phi_3)\phi_1^{-1} \\ \Delta_{13} &= (\phi_1\phi_4 - \phi_5)\phi_1^{-1} \\ &\cdots \\ \Delta_{21} &= (\Delta_{11}\phi_3 - \phi_1\Delta_{13})\Delta_{11}^{-1} \\ \Delta_{23} &= (\Delta_{11}\phi_5 - \phi_1\Delta_{15})\Delta_{11}^{-1} \\ &\cdots \end{aligned}$$

Once the array is completed, one counts the number of sign changes among the elements in the first column. This number is equal to the number of roots with positive real parts. Thus, for a system to be stable, all elements in the first column must be positive since the first element of the array is equal to 1.

As an example, let's evaluate the polynomial

$$\lambda^4 + 4\lambda^3 - 3\lambda^2 - 16\lambda - 12 = 0$$

with the roots -3, -2, -1, and $+2$. The ϕ coefficients are simply given as $\phi_1 = 4$, $\phi_2 = -3$, $\phi_3 = -16$, $\phi_4 = -12$, and we compute the Δ coefficients according to the definitions above. We obtain $\Delta_{11} = 1$, $\Delta_{13} = -12$, $\Delta_{21} = 32$, and $\Delta_{31} = (\Delta_{21}\Delta_{13} - \Delta_{11}\Delta_{23})\Delta_{21}^{-1} = -12$, since $\Delta_{23} = 0$. Thus, the complete array is

$$\begin{array}{ccc} 1 & -3 & -12 \\ 4 & -16 & \\ 1 & -12 & \\ 32 & & \\ -12 & & \end{array}$$

which has one sign change indicating that there is one root with a positive real part. Thus, a system with the above polynomial as the characteristic equation is unstable.

The Routh-Hurwitz procedure fails if some of the powers of λ are missing. In this case, the system is either marginally stable or unstable (cf. [86]).

3. ***Sensitivities.*** Another set of calculations that can be executed analytically for S-systems is computation of system sensitivity to changes in parameter values or input variables [2, 11–12]. These concepts are of interest in biology, since living systems are exposed continually to internal changes and fluctuations in the environment, and they must adapt to changes while buffering out the harmful effects of fluctuations. The ease of steady-state characterization with S-systems has allowed implementation of sensitivity analysis in numeric and symbolic computer programs (Chapters 6 and 7). Sensitivities are discussed extensively in Chapter 4, and only for completeness of this section do we briefly introduce their definitions.

Three types of sensitivity are informative. The first quantifies the relative change in system variable X_i in response to a change in input variable X_0. This quantity is called *logarithmic gain*, and is defined as

$$\frac{\partial Y_i}{\partial Y_0} = \frac{\partial(\ln X_i)}{\partial(\ln X_0)} = \frac{\partial X_i}{X_i}\frac{X_0}{\partial X_0}$$

The second type of sensitivity quantifies the relative change in variable X_i in response to a change in one of the exponents g_{jk} or h_{jk}, and is defined as

$$S(Y_i, g_{jk}) = \frac{\partial Y_i}{\partial g_{jk}} \frac{g_{jk}}{Y_i} \quad \text{or} \quad S(Y_i, h_{jk}) = \frac{\partial Y_i}{\partial h_{jk}} \frac{h_{jk}}{Y_i}$$

The third type of sensitivity quantifies the relative change in variable X_i in response to a change in a rate constant α_j or β_j, and is defined as

$$S(X_i, \alpha_j) = \frac{\partial X_i}{\partial \alpha_j} \frac{\alpha_j}{X_i} \quad \text{or} \quad S(X_i, \beta_j) = \frac{\partial X_i}{\partial \beta_j} \frac{\beta_j}{X_i}$$

Many important insights have been gained from the study of logarithmic gains and parameter sensitivities (cf. [2]).

Straightforward formulation of the system equations and symbolic analytical solutions make the S-system approach a versatile tool for qualitative and quantitative nonlinear analysis. Instead of using the large numbers of computer simulations necessary to explore the steady-state characteristics of other nonlinear descriptions numerically, the procedures outlined above for S-systems allow nonlinear systems to be characterized more extensively with more general symbolic solutions. The analytical methods also allow objective comparisons of alternative mechanisms governing a particular process, thus revealing principles that underlie biological regulatory mechanisms. Examples of these types of analyses are sketched in the next section and discussed in detail in Chapter 5 and in [4, 46].

Examples of Symbolic Steady-State Analysis

As mentioned above, perhaps the most important feature of S-systems is that they allow symbolic analysis, making it possible to formulate specific conclusions and predictions independent of parameter values [2, 14, 16, 21, 25–26]. Often these predictions can be tested by independent experimental work (cf. Chapter 8), and the following two examples are just intended to give the flavor of some of the questions that can be addressed.

Savageau [1–2] has analyzed nested patterns of regulation in branched pathways. In these pathways, each end product inhibits the first enzyme involved in its synthesis as well as one of the isozymes common to the synthesis of several end products. Analysis of the integrated response of such pathways reveals the critical importance of a balance between the two enzymes inhibited by a given end product: Breakdown of this balance either leads to self-starvation for one of the other end products or to wasteful excretion of intermediates and overproduction of the other end products. Examples of these pathologies have been demonstrated experimentally in bacteria with mutations affecting the regulatory enzymes of branched pathways for biosynthesis of the aspartate family of amino acids [90]. Furthermore, the existence of multienzyme complexes or bifunctional enzymes to ensure maintenance of balanced inhibition in branched pathways was predicted from the results of a mathematical analysis using S-systems [2]. Well-documented examples of multienzyme complexes that are not involved in substrate channelling in branched pathways support these predictions [85, 87].

Another example is the comparative analysis of alternative regulatory interactions in the immune response to foreign antigens (cf. Chapter 5 and [25–26]). A foreign organism that invades the host can grow autocatalytically and stimulate production of antigen-specific effector lymphocytes. Effector lymphocytes produce effector molecules that can inhibit further invasion and growth, or they can facilitate clearance of the foreign organism from the system. Alternative regulatory topologies involving suppression have been analyzed and compared by modeling the corresponding immune networks as S-systems. The results showed, for example, that an immune network with suppression of the production of effector lymphocytes is superior on the basis of multiple criteria for functional effectiveness. Experimental results published by others support many of the predictions, and this analysis also has highlighted the importance of some new experiments.

Other general conclusions and predictions have been obtained through the use of the S-system methodology applied to alternative mechanisms for gene expression (cf. Chapter 8); in each case they are supported by independent experimental evidence.

Dynamic Responses

Exploring the dynamics of an S-system not at steady state usually requires solutions of the differential equations (for an exception, see Chapter 16). While some particular classes of S-system differential equations can be solved in the form of known explicit functions [24], most S-systems have no closed-form solution and must be solved numerically. Again, the regular structure of the equations is a considerable asset: The numerical differential equation solver in ESSYNS exploits the mathematical properties of S-systems to solve the nonlinear differential equations one to two orders of magnitude faster than conventional methods (cf. Chapter 7 and [30, 38, 46]).

Accuracy of S-System Representations

Throughout the development of the power-law formalism, there has been an effort to evaluate the accuracy of alternative representations. These evaluations have been performed at different hierarchical levels of organization. At one extreme, S-system representations of individual rate laws have been examined [9, 88], and at the other extreme S-system representations of macroscopic growth have been shown to be accurate [19, 33].

Different types of error analyses have been performed for power-law representations of biochemical pathways. Predicted induction characteristics for gene expression were tested against experimental data and shown to agree over several orders of magnitude in concentration, an unusually wide range (e.g., see [2]). Linear representations and alternative types of power-law representations were compared as models for branched pathways. In virtually all cases involving converging or diverging pathways, the S-system model is most accurate [29]. These findings were confirmed and extended in independent studies with a completely different approach ([35–37]; see also Chapter 5). In yet another type of analysis, the dynamics of cell populations and biochemical mediators were studied to compare alternative model designs for accuracy of fit to experimental data. The best S-system representation produced accurate fits for all components of the system [23].

The factors that contribute to the unusually wide range of accurate representation achieved with S-systems are not yet fully understood, but a few contributing factors can be enumerated.

1. To some extent, this wide range is due to the mathematical nature of the power-law formalism: The Taylor series approximation is performed in logarithmic space and yields a nonlinear representation that is accurate over a much wider range than conventional linear approximation. However, this argument does not suffice to explain sizes as large as 100- or 1,000-fold for the range of validity.
2. Another explanation is the multitude of homeostatic regulatory mechanisms, like feedback loops, that are found in every biological system. These mechanisms effectively maintain many concentration variables within narrow limits, which tends to promote the accuracy of any representation based upon a Taylor series approximation [2].
3. Such regulatory mechanisms also tend to simplify the analytical form of processes in the intact system, as opposed to the form that would be obtained in isolation [2: pp. 81–83]. In technological systems, this simplification often is linear [83], but in biological systems, the resulting simplification is that of a power law.
4. The aggregation of interactions and modulations into just one difference of products of power-law functions, that is, the formulation in S-system form, improves the accuracy of representation over alternative formulations in which the various influences are individually approximated with a product of power-law functions, which results in the structure of a Generalized Mass Action system [29].

5. Finally, the exact recasting of nonlinearities in S-systems (cf. the next Section and Chapter 12) shows that virtually all relevant functional relationships in biology can be represented in the canonical form of an S-system which illustrates the appropriateness of the power-law formalism from a different perspective.

Of course, if a system deviates too far from the chosen operating point, the power-law approximation may no longer provide sufficient accuracy. But even in these cases, S-system representations can be very useful. On one hand, one can employ power-law approximations at a series of operating points that are chosen densely enough to guarantee sufficient accuracy over the entire range. On the other hand, one can use methods of exact recasting (cf. the next Section and Chapters 12 and 14).

EXACT S-SYSTEM REPRESENTATION; RECASTING

While the power-law formalism and its associated S-system differential equations originally were developed for the analysis of organizationally complex systems, a second aspect of S-systems is taking on an importance of its own: All available evidence indicates that any finite system of ordinary differential equations comprised of elementary functions can be *recast* exactly as a finite S-system (Chapter 12 and [46]). It is also known that higher-order differential equations and special functions, for instance, Bessel functions and elliptic functions, can be recast in S-system form. Any finite Taylor series or Chebyshev polynomial can be recast as a finite set of equations in S-system form, so any function that is represented well in these forms also is represented well in S-system form. The full range of functions that can be recast remains to be explored more exhaustively. In any event, recasting provides a new approach for classification, generalization, and extension of models representing nonlinear systems.

The recasting process for a set of ordinary differential equations is straightforward and basically utilizes the chain rule and the product rule of differentiation (Chapter 12 and [46]). Nonlinearities not expressed in the form of ordinary differential equations but as integral equations or as power series often can also be recast as S-systems (cf. [41–42, 48–49]).

Recasting in combination with the computer program ESSYNS (Chapter 7) has resulted in efficient numerical solutions of high accuracy [38, 46]. Recasting is still an emerging aspect of S-system analysis but already it has found application in diverse areas such as numerical analysis [30, 38], computational statistics (Chapter 13 and [48–49]), and mathematical modeling (Chapter 14 and [39, 47]).

SUMMARY

Organizationally complex systems are difficult to understand intuitively because they contain a large number of components, and the dynamic interrelationships among components usually are nonlinear. A particularly useful mathematical representation for the analysis of complex systems is a set of nonlinear ordinary differential equations that has been called an *S-system*, indicating its ability to capture the responses of *S*ynergistic, *S*aturable systems. S-systems have a regular mathematical structure that offers various advantages. Of prominent importance are straightforward recipes for setting up S-system equations and techniques for the symbolic evaluation of steady states, local asymptotic stability, and parameter sensitivities. S-systems also are well suited for extensive numerical analysis of nonlinear dynamics. S-systems have the intriguing property that virtually any ordinary differential equation can be recast exactly in S-system form. Recasting, coupled with efficient computer methods for numerical solution, provides a new means of analyzing and classifying complex nonlinearities.

3. Optimal Strategies for Modeling Biochemical Systems

ALBERT SORRIBAS

Mathematical models are becoming an essential tool in studying biochemical systems. This is due to the fact that understanding complexity escapes intuition and demands an approach able to deal with a high number of interacting elements.

In selecting a mathematical model, it is fundamental to develop methods that produce the most valuable results in an easily understandable fashion. Defining suitable formalisms to deal with biochemical systems has been an open problem since the turn of the century.

A common approach is based on mathematical representations of individual rate laws, obtained for each component in isolation. Embedded in the tradition of enzymology, this approach leads to mechanistic system representations. A significant drawback is that, even for relatively small biochemical networks, this approach results in complex mathematical models that lack a clear systematic structure that would facilitate analysis and comparison (cf. [91, 93]). Moreover, the experimental approach necessary to characterize the model is strictly based on isolated parts, but there is no guarantee that the integrated behavior of the system is preserved when information about isolated component parts is used to reconstruct the whole system.

To alleviate these problems, we need a mathematical representation of biochemical systems that is based on a systematic formalism. In search of such a formalism, we must evaluate which aspects of biochemical systems ought to be analyzed.

First, there is a definite need for an *integrated* representation that enables us to study the function of the system as a whole. Ideas of this kind are well established in network theory and, in fact, it is not far fetched to consider a biochemical pathway as a network of reactions.

Second, in an intact biochemical system each component is known to operate within a certain range. A mathematical formalism should be an *appropriate* representation of intact biochemical pathways over ranges that are relevant *in vivo*. These ranges typically are rather restricted (cf. the discussion on p. 189 of [35]) and much narrower than ranges that can be studied *in vitro*. It is therefore feasible to use good *approximations* for individual processes, as long as they cover the relevant ranges *in vivo*, instead of searching for unknown "exact" representations that exhaust wider ranges that are possible *in vitro* but irrelevant *in vivo*. All our physical and chemical "laws" are approximations that are based on simplifying assumptions, and thus, the concept of employing a good approximation in biochemical systems analysis is a natural way of dealing with complexity and should not be considered with suspicion.

Third, the desired formalism should provide a *systematic* representation that facilitates the comparison between different systems. It should provide the framework for a theory of integrated metabolic systems that allows us to discuss issues of optimal design and optimal performance. Such a theory is a prerequisite for a true understanding of biochemical systems.

The power-law formalism provides the basis for a systematic approach to analyzing intact systems. It opens new perspectives for analyzing the design principles in biochemical pathways and for addressing questions of their regulation and dynamic properties. Among its many applications, the qualitative analyses of feedback regulation (Chapters 8, 9, and [2, 14, 16]), genetic regulation (Chapter 8 and [92]), and immune networks (Chapter 5 and [25–26]) may best illustrate the power and potential of this formalism.

In this chapter, I shall contrast the power-law formalism to the classical kinetic concept of mechanistic rate laws. Also, I shall discuss alternative options for constructing power-law representations and provide criteria for optimal modeling strategies.

SCOPE OF THE POWER-LAW FORMALISM

Most biochemists, especially those engaged in enzymology, accept the Michaelis-Menten rate law

$$v = V_m S/(K_m + S) \tag{3.1}$$

as an accurate mathematical representation for the initial rate, v, of a monosubstrate enzyme-catalyzed reaction, in which the substrate, S, is in excess with respect to the enzyme. The parameters V_m and K_m have a definitive meaning, and Eq. 3.1 is often considered to be "the actual rate law." For reactions that involve more complex mechanisms and many effectors, the resulting rate laws again are rational functions, but these are much more complicated; procedures for deriving these types of rate laws are well known and need not be discussed here.

No matter which types of pathways are studied within the general framework of Michaelis-Menten kinetics, the resulting rate laws focus on individual reactions, and the description of an intact biochemical system is based on combining all individual rate laws in an appropriate way. The rate laws are derived under a definite set of simplifying assumptions and conditions and, consequently, they are approximations to the actual behavior that is observed experimentally.

The power-law formalism (cf. Chapters 1 and 2) provides alternative representations of rate laws, derived from a different point of view and with the prime objective of yielding an *integrated* model for the *intact* system. In this formalism, an individual rate law is represented by the power-law function

$$v = \alpha S^g \tag{3.2}$$

in which v is the velocity of the reaction, α is the *rate constant*, and g is the *kinetic order*. This type of rate law reminds us of elemental chemical kinetics, but it is generalized in the power-law formalism since the kinetic order g can assume any real value (cf. Chapter 2).

In comparison with Michaelis-Menten rate laws or other rate laws based on rational functions, this representation offers the significant advantage that it always has the same mathematical structure, independent of the mechanisms involved in the reaction. It also extends directly to more complex pathways: For more than one metabolite contributing to the rate v, the general rate law expressed in the power-law formalism simply is a product of power-law functions of all metabolites involved:

$$v_i = \alpha_i \prod_{j=1}^{n+m} X_j^{g_{ij}} \tag{3.3}$$

(cf. [9]). In this equation, the variables X_j ($j = 1, \ldots, n + m$) account for all metabolites affecting the rate v_i, α_i is the rate constant, and g_{ij} is the kinetic order. Even for arbitrarily complex biochemical pathways, the mathematical structure is preserved, and this offers enormous advantages for the analysis of biochemical systems that consist of a relevant number of constituents. In contrast, rate laws of a generalized Michaelis-Menten type rapidly grow in complexity, up to a point where the rate law no longer yields any insight. For instance, the rate law for an enzyme like glutamine synthetase might involve 500 terms, even under simplifying assumptions (cf. [2: p. 85]).

The parameters that appear in Eq. (3.3) characterize the mechanism and the operating point at which the power-law representation is derived (cf. Chapter 2). At this point, the numerical value of the rate will be the same for traditional rate laws and the power-law function (see [35] for a numerical example). In the vicinity of the operating point, both will provide similar values, but for large perturbations, the two representations differ. Thus, when it is necessary to represent isolated components over wide ranges of variation in substrate, for example, under *in vitro* conditions of an enzymological study, the power-law formalism may not represent the behavior of the reaction over the entire range with sufficient accuracy. In such cases, however, the power-law formalism still can be employed: one simply uses a series of representations at different operating points, or one can resort to techniques of exact recasting (cf. Chapters 13, 15).

The accuracy provided by power-law representations usually is sufficient to cover the entire relevant range of variation in metabolite concentrations *in vivo* ([2: Chapters 13–14]; [35–36]). In carefully analyzed examples, power-law representations turned out to be valid over about 20-*fold* variations [35, 37], which is quite amazing when one considers that these representations are approximations. Several possible explanations for the exceptional accuracy have been discussed in the literature ([2: pp. 80–83]; [3, 29]).

POWER-LAW REPRESENTATION OF METABOLIC PATHWAYS

To compare alternative representations for biochemical pathways, let's consider the specific example in Fig. 3.1, which could model the synthesis of two amino acids from a common precursor [2, 14]. The pathway consists of four dependent metabolites, X_1 through X_4, and three independent metabolites (a source metabolite, X_7, and two sink metabolites, X_5 and X_6). Furthermore, there are six

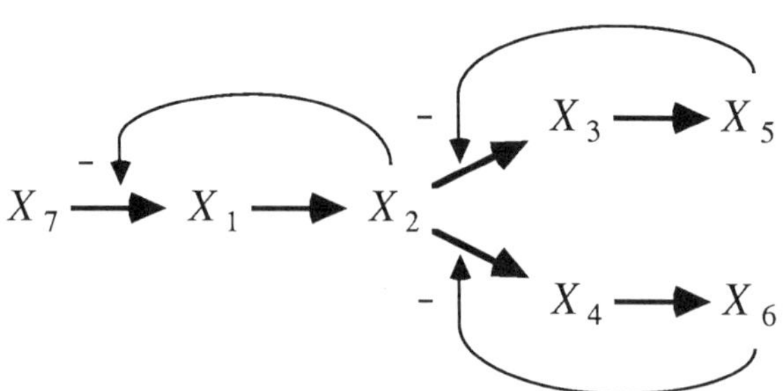

FIGURE 3.1. Simple Branched Pathway with Feedback

enzyme-catalyzed reactions and three feedback signals that are represented with light arrows and labeled with minus signs.

To construct a mathematical model of the dynamics of the system, one commonly uses a set of differential equations of the general form

$$\dot{X}_i = f_i(X_1, \ldots, X_{n+m}) \qquad i = 1, \ldots, n \tag{3.4}$$

in which $\dot{X}_i = dX_i/dt$ and f_i is a real function that reflects changes in X_i. In the example of Fig. 3.1 we identify $n = 4$, representing the dependent metabolites, and $m = 3$, representing the independent metabolites.

In metabolic pathways, the function f_i is a combination of the rates of the different reactions acting on X_i, both producing or degrading this metabolite. So, a more specific, biochemically motivated model is

$$\dot{X}_i = \sum_j \sum_k v_{jik} - \sum_j \sum_k v_{ijk} \quad i = 1, \ldots, n \tag{3.5}$$

in which v_{jik} is the rate of production of X_i from X_j via the k-th parallel reaction. For instance, the change in X_2 of the pathway in Fig. 3.1 is represented as

$$\dot{X}_2 = v_{12} - v_{23} - v_{24} \tag{3.6}$$

Each individual rate v_{ij} depends on several metabolites whose concentrations determine the value of v_{ij} in the differential equation at any given time. Employing power-law rate laws, according to Eq. (3.3), the change in X_2 becomes

$$\dot{X}_2 = \alpha X_1^{g_{21}} - \beta_{21} X_2^{h_{221}} X_5^{h_{251}} - \beta_{22} X_2^{h_{222}} X_6^{h_{262}} \tag{3.7}$$

This type of equation is called a *Generalized Mass Action* (GMA) representation (cf. Chapters 2, 12, and [36]). Its main characteristic is that individual reactions are used as units for the power-law representation, that is, there is a separate power-law function for each rate. In terms of GMA systems, the pathway in Fig. 3.1 is formulated as

$$\begin{aligned}
\dot{X}_1 &= \alpha_1 X_2^{g_{12}} X_7^{g_{17}} - \beta_1 X_1^{h_{11}} \\
\dot{X}_2 &= \alpha_2 X_1^{g_{21}} - \beta_{21} X_2^{h_{221}} X_5^{h_{251}} - \beta_{22} X_2^{h_{222}} X_6^{h_{262}} \\
\dot{X}_3 &= \beta_{21} X_2^{h_{221}} X_5^{h_{251}} - \beta_3 X_3^{h_{33}} \\
\dot{X}_4 &= \beta_{22} X_2^{h_{222}} X_6^{h_{262}} - \beta_4 X_4^{h_{44}}
\end{aligned} \tag{3.8}$$

This representation allows numerical analysis of the dynamics of the system and of system responses to changes in any of the parameters (see [36] for details).

S-SYSTEMS: A LANDMARK CONCEPT IN BIOCHEMICAL SYSTEMS THEORY

The GMA representation is based on choosing each individual reaction as the unit for representation. However, this is not the only possible formulation of a power-law model (cf. Chapter 2). If one focuses on each dependent *metabolite* as the unit for representation, the value of $\dot{X}_2$ at each time point is determined by indivdual processes producing or degrading this metabolite. Hence, an alternative system description is

$$\dot{X}_i = V_i - V_{-i} \tag{3.9}$$

in which the different processes affecting X_i have been *aggregated* to define two net rates, one of synthesis (V_i), and one of degradation (V_{-i}) [8–9]. Following this strategy, a power-law term is formulated for each of the resulting aggregated functions. For the pathway in Fig. 3.1 one obtains

$$\begin{aligned}
\dot{X}_1 &= v_{71} - v_{12} = V_1 - V_{-1} \\
\dot{X}_2 &= v_{12} - (v_{23} + v_{24}) = V_2 - V_{-2} \\
\dot{X}_3 &= v_{23} - v_{35} = V_3 - V_{-3} \\
\dot{X}_4 &= v_{24} - v_{46} = V_4 - V_{-4}
\end{aligned} \tag{3.10}$$

and, using power-law representations for aggregated rates,

$$\begin{aligned}
\dot{X}_1 &= \alpha_1 X_2^{g_{12}} X_7^{g_{17}} - \beta_1 X_1^{h_{11}} \\
\dot{X}_2 &= \alpha_2 X_1^{g_{21}} - \beta_2 X_2^{h_{22}} X_5^{h_{25}} X_6^{h_{26}} \\
\dot{X}_3 &= \alpha_3 X_2^{g_{32}} X_5^{g_{35}} - \beta_3 X_3^{h_{33}} \\
\dot{X}_4 &= \alpha_4 X_2^{g_{42}} X_6^{g_{46}} - \beta_4 X_4^{h_{44}}
\end{aligned} \tag{3.11}$$

This strategy can easily be derived for any metabolic scheme, following the steps described above, and it produces a set of equations that is known as an *S-system*.

For the description of biochemical pathways, the main conceptual difference between the S-system form and the GMA form lies in the fact that S-systems use pools rather than reactions as the units for representation. This difference is not trivial, since the resulting models have different mathematical properties and describe the dynamic behavior of the system with different accuracy [27–29, 35–36].

The structural differences between S-system equations and GMA system equations have significant implications for the types of analyses that can be performed. Most importantly, the S-system form allows us to obtain steady-state solutions in explicit form. This is a major advantage for characterizing the behavior of the system in response to changes in parameters [1–2, 11, 14, 25–26, 35–37] and it is essential in the comparative analysis of design and regulation of biochemical systems. Many aspects of the method of "controlled mathematical comparison" (cf. Chapter 5 and [2, 14, 16, 25–26]) are based on steady-state analyses, and this method has produced very general results about feedback regulation [2, 14], genetic control [2, 92], and immune networks [25–26].

OPTIMAL MODELING STRATEGIES

The way in which the different processes affecting a dependent metabolite can be aggregated to obtain an S-system representation is not unique. In the pathway of Fig. 3.1, for instance, the rates affecting X_2 could be aggregated differently from those appearing in Eq. (3.10). Specifically, one could aggregate the rates v_{12} and v_{23} and thus formulate the change in X_2 as

$$\dot{X}_2 = (v_{12} - v_{23}) - v_{24} = V_2 - V_{-2} \tag{3.12}$$

The resulting S-system then would be

$$\dot{X}_2 = \alpha_2 X_1^{g_{21}} X_2^{g_{22}} X_5^{g_{25}} - \beta_2 X_2^{h_{22}} X_6^{h_{26}} \tag{3.13}$$

which obviously is different from the second equation of (3.11). Thus, a question arises, Which of the two strategies would provide the better representation for the integrated system? The answer depends on criteria of optimality of representation.

Criteria for judging alternative strategies include [37]:

1. Accuracy in predicting values of the concentration variables at steady states around the nominal operating point;
2. Accuracy in predicting values of steady-state fluxes around the nominal operating point;
3. Accuracy in predicting transient responses between two such steady states; and
4. Robustness of representation.

Amphibolic pathways provide a class of systems in which the alternative S-systems can be tested in a particularly critical situation. These biochemical pathways cannot be considered essentially irreversible, and under some physiological conditions, the flux through some reactions can even change its direction.

Two major alternative strategies for representing amphibolic pathways are obvious: The "irreversible" strategy involves aggregation of forward and reverse fluxes through each reaction; the "reversible" strategy, in contrast, involves aggregation of incoming fluxes separately from outgoing fluxes for each metabolite.

Both strategies produce the same steady-state behavior for the metabolite concentrations, and one could be tempted to disregard all further considerations if only the steady-state responses of the concentration variables were of interest. However, the results for the behavior of the flux in the system are significantly different: Only the reversible strategy is capable of accurately representing the flux at or near thermodynamic equilibrium (cf. [37]). Futhermore, representations based on the irreversible strategy are extremely inaccurate and sensitive to changes in parameter values when the operating point is near equilibrium. Under these conditions one can obtain distorted results that deviate far from the correct results; the equilibrium point itself is a singularity and the system cannot be represented at this point. The implications of these results for the analysis of experimental data are clear: The response of a biochemical pathway to a realistic perturbation can be very different from the response to a theoretical, infinitesimal perturbation and the comparison between theory and experiment may be meaningless. Even when the operating point is far from equilibrium, the irreversible strategy leads to inferior representations, and the reversible strategy is always more accurate. These considerations unambiguously demonstrate that the accuracy of the reversible strategy is superior to that of the irreversible strategy in representing amphibolic pathways; other advantages are discussed at length in [37].

CONCLUSIONS

The power-law formalism offers alternative strategies for representing an intact biochemical system. Careful comparisons have shown that the S-system form has clear advantages over other canonical forms, such as Half-systems and GMA systems [8–9, 27–29, 35–37]. This optimal form is based on aggregation into processes of net synthesis and net degradation.

For the canonical nonlinear modeling of metabolic pathways, maybe the most outstanding characteristic of the S-system form is the fact that the steady-state solutions can be determined and evaluated with analytical means. The characterization of a steady state is an integral part of controlled comparisons between alternative designs (cf. Chapter 5). These comparisons, in turn, elucidate essential properties of biochemical pathways and help predict their responses to external stimuli. They yield insight into the design of metabolic networks and advance our understanding of complexity in living systems.

The comparison of alternative strategies for representing biochemical pathways suggests the following rules for constructing power-law models for intact biochemical sytems:

1. The S-system representation usually is superior to alternative representations, such as GMA representations. Consequently, it is recommended to aggregate rates by focussing on metabolites rather than individual reactions.
2. In cases where alternative S-system representations are possible (e.g., for amphibolic pathways), the best strategy is to aggregate individual processes of synthesis into V_{+i}, and individual processes of degradation into V_{-i}.
3. Each term of the S-system model must include all variables that directly affect production or degradation of a metabolite, respectively. When an enzyme is affected by other system variables, as is the case in cascade mechanisms, it must be considered a dependent variable and represented with a differential equation. Constraint equations, as they represent mass conservation of free enzymes and enzyme complexes should be included (cf. [35]).
4. The S-system, derived at a nominal operating point, produces an accurate approximation of the behavior of the system around that point. Since accuracy is best in the vicinity of the operating point, this point should be chosen with particular care.

SUMMARY

The power-law formalism provides a general theoretical framework for addressing questions about regulation and dynamic properties of biochemical pathways. Among different approaches that can be derived from this general framework the S-system representation has clear advantages. This chapter discusses the role of the power-law formalism in biochemistry and its use for deriving mathematical representations of biochemical pathways. Some general rules for selecting optimal power-law representations are suggested.

4. Sensitivity Analysis

A Common Foundation of Theories for the Quantitative Study of Metabolic Control

MARTA CASCANTE, RAFAEL FRANCO, AND
ENRIC I. CANELA

The previous chapters introduced power-law representations of complex biochemical systems. They discussed alternative methods of formulating appropriate system equations and summarized basic methods of analysis. In this chapter, we focus on one particular subject of biochemical systems theory: the analysis of the system's *sensitivities* to changes in parameters or external variables. While this scope may seem rather restricted, we will see that sensitivity analysis actually is a key concept for achieving a premier goal of biochemical systems analysis, namely, the establishment of relationships between the properties of individual system components and the responses of the integrated system (cf. [1, 2, 11, 63, 66]). When one is interested in predicting the dynamics of a biochemical system from the knowledge of its component parts, such relationships can be very difficult to determine. However, when the system remains close to a steady state, sensitivity analysis provides general theorems that relate steady-state concentrations and fluxes. These theorems yield insight into the structure of a biochemical system and can be employed in the evaluation of typical biochemical experiments.

The idea of sensitivity analysis derives from the fact that key parameters of mathematical models are typically estimated from experimental measurements that automatically limit their accuracy. Even when parameters are estimated from different samples of the same population, the results normally differ. This intrinsic uncertainty of numerical specifications of a system representation raises the question about how the differences in estimated parameter values affect the dynamics of the system. Of course, the experimentalist hopes that "similar" parameter values will correspond to "similar" dynamic responses of the system. However, it is well known that in some systems even minute changes in parameter values can change the system response drastically. The system may approach its steady state faster than before, or oscillations may become more pronounced. If the parameter value changes across a bifurcation point, even the *qualitative* behavior may change: a stable system may become unstable, or out of a steady-state point, a limit cycle may come into existence (cf. Chapter 16).

Sensitivity analysis helps to predict how the system performance is affected by changes in parameter values. It distinguishes situations where inaccuracies in parameter values are almost irrelevant from those where noise or error in the numerical estimates can lead to almost unpredictable results. Sensitivities can be used as quantitative measures that indicate how fast the responses of any system components change with changes in one of the parameter values, and they can be used in symbolic form to establish theoretical relationships between different system components and properties.

The mathematical approach to investigating sensitivities is based on the analysis of the differences between true and disturbed parameter values and their effects on the system (cf. [59, 94–96]). Specifically, let's consider a differential equation of the general form

$$\dot{\boldsymbol{X}} = \boldsymbol{F}(\boldsymbol{X}, \boldsymbol{p}, t) \qquad \boldsymbol{X}(t_0) = \boldsymbol{X}_0$$

where the dynamics is given by a function $\boldsymbol{F}$ of the system variables $X_1, \ldots, X_n$ that are collectively written as the vector $\boldsymbol{X}$, and the parameters $p_1, \ldots, p_m$ that constitute the vector $\boldsymbol{p}$. Next, we define the differences between the true and the disturbed initial conditions and parameter values, respectively, as

$$\Delta \boldsymbol{X}_0 = \boldsymbol{X}_0 - \tilde{\boldsymbol{X}}_0$$

and

$$\Delta \boldsymbol{p} = \boldsymbol{p} - \tilde{\boldsymbol{p}}$$

where $\tilde{}$ denotes the disturbed quantity. The variation in the initial conditions or parameter values causes a change in the solution which can be formulated as

$$\Delta X(t) = X(t) - \tilde{X}(t).$$

The Taylor series expansion of $\Delta X(t)$ is

$$\Delta X(t) = \left[\frac{\partial X}{\partial X_0}\right] \Delta X_0 + \left[\frac{\partial X}{\partial p}\right] \Delta p + \cdots$$

where the dots indicate higher-order terms, which subsequently are ignored. The expressions in brackets,

$$S(X, X_0) = \left[\frac{\partial X}{\partial X_0}\right]$$

and

$$S(X, p) = \left[\frac{\partial X}{\partial p}\right]$$

are evaluated for $\tilde{X}_0$ and $\tilde{p}$, respectively. They are called *sensitivity coefficients* and quantify how $X(t)$ changes as a result of perturbations in initial conditions or in parameters.

In biochemical systems theory it is often useful to distinguish whether a change in $X(t)$ is caused by a change in a true system parameter, such as a rate constant, or by a change in an experimentally controlled concentration of an external metabolite that, mathematically, can be considered a parameter too. In the first case, the sensitivity is specifically called *parameter sensitivity*; in the second case it is called *gain* (cf. [1, 11]). Sensitivity coefficients contain all information necessary to characterize the biochemical system at a steady state.

In addition to absolute sensitivities it is convenient to consider the *relative* change in a system component that is caused by a *relative* change in a parameter value. Relative sensitivity coefficients are defined as

$$\bar{S}(X, X_0) = \left[\frac{\partial X/X}{\partial X_0/X_0}\right] = \left[\frac{\partial \log X}{\partial \log X_0}\right]$$

and

$$\bar{S}(X, p) = \left[\frac{\partial X/X}{\partial p/p}\right] = \left[\frac{\partial \log X}{\partial \log p}\right]$$

Sensitivity analysis is the prominent and common method of two competing approaches to understanding biochemical systems. The first approach is based on S-systems and has been called *Biochemical Systems Theory* (BST); the second approach originates from the analysis of conservation laws of biochemical systems and has been called *Metabolic Control Theory* (MCT). Since their inception, BST and MCT have gained considerable interest among experimentalists and have complemented, or sometimes even replaced, traditional concepts such as rate limiting steps or key enzymes. Because of differences in emphasis and terminology, the two approaches have been studied separately, but recent comparisons have begun to elucidate their relatedness [e.g., see 27–29, 34–37].

In this chapter, we discuss the importance of analyzing sensitivities in biochemical systems and show how essential results of BST and MCT can be derived and compared within the framework of general sensitivity analysis. The work presented here complements our recent analyses of complex metabolic control systems [55, 56].

REPRESENTATION OF METABOLIC SYSTEMS

The development of a mathematical representation for biochemical pathways requires that we distinguish between independent and dependent variables. Typical independent variables are inputs, outputs, and external pool concentrations, whereas dependent (state) variables represent internal metabolite concentrations. In addition to concentrations, a biochemical system representation needs to address global properties and responses to stimuli. Particularly useful concepts in this class are individual and aggregated rate laws, fluxes, and sensitivities.

The following terms will be central to our discussion:

X_d = vector of dependent variables (concentrations of internal metabolites)
X_e = vector of external (independent) variables, such as inputs, outputs, and enzyme concentrations
v_{sj} = individual rate law of synthesis
v_{dj} = individual rate law of degradation
F_j = input or output flux
J_j = net flux through an individual step

It is possible and often convenient to employ alternative representations of the dynamics of a metabolic system with these characteristics. The selection of a particular representation depends on the primary focus of investigation. One can distinguish three basic categories of representation.

In the first category, one focuses on *fluxes of internal metabolites*. The change in an internal metabolite concentration is given as the sum of all individual rate laws that contribute to the synthesis of the metabolite, minus the sum of all rate laws that contribute to its degradation. Using the above nomenclature one obtains:

$$\dot{X}_j = f_j(\boldsymbol{X}_d, \boldsymbol{X}_e, t) = \Sigma\, v_{sj} - \Sigma\, v_{dj} = V_{+j} - V_{-j}$$

($j = 1, \ldots, n$), where n is the number of dependent variables of the metabolic system.

In the second category, one focuses on *fluxes of input and output metabolites*. This type of representation can be formalized as

$$\dot{X}_j = F_j = g_j(\boldsymbol{X}_d, \boldsymbol{X}_e, t) = V_{F_j^+}$$

($j = 1, \ldots, m$), where m denotes the number of inputs or outputs of the metabolic system.

In the third category, one focuses on *net fluxes through individual steps*. In mathematical terminology this representation reads

$$J_j = h_j(\boldsymbol{X}_d, \boldsymbol{X}_e, t) = V_{sj}$$

($j = 1, \ldots, r$), where r represents the number of steps of the metabolic system.

The functions f_j, g_j and h_j in general are unknown and may be very complicated; however, in most cases they are differentiable rational functions.

Each of the above representations describes the dynamics of the metabolic system. Without further information about the functions f_j, g_j and h_j, the dynamics could include almost any temporal behavior, including unrealizable responses. These can be excluded when one accounts for steady-state conditions that the metabolic system has to satisfy and that constrain the range of potential behaviors. Typical constraints of this sort mean that the concentrations of internal metabolites are in a steady state or that the fluxes are constant. Such constraints can be formalized as

$$\boldsymbol{X}_d = \boldsymbol{0}$$

$$\boldsymbol{F} = \text{constant}$$

$$\boldsymbol{J} = \text{constant}$$

Constraints of this type restrict the possible responses of mathematical representations of biochemical systems. They are useful for theoretical considerations and can be employed to reduce the number of parameters that need to be estimated in quantitative data analyses because one can solve each constraint in terms of an unknown parameter and subsequently eliminate this parameter.

While the employment of constraints significantly reduces the estimation effort, it is usually not sufficient for a complete determination of parameter values and it has to be complemented with numerical methods to estimate parameter values from experimental measurements. Several types of experiments are particularly useful and complement each other. They measure the following properties of the system:

- Concentrations of inputs, outputs, internal metabolites, and enzymes of the system.
- Fluxes through each reaction of the pathway or aggregated fluxes of synthesis and degradation of a system metabolite.
- Local properties (individual processes); specifically, infinitesimal relative changes in a local rate law of the system caused by infinitesimal relative changes in one of the variables.
- Global properties (whole system); specifically, infinitesimal relative changes in global responses of the system caused by infinitesimal relative changes in an independent variable of the system.

Relationships Between Global and Local Properties of a Biochemical System

With this terminology we are now in a position to relate properties of the integrated biochemical system to properties of its components. Let's consider a system described by two sets of equations:

$$\dot{X}_j = f_j(\mathbf{X}_d, \mathbf{X}_e, t)$$

$$J_j = V_{sj} = h_j(\mathbf{X}_d, \mathbf{X}_e, t)$$

By differentiating the first equation with respect to an independent variable X_i, one obtains so-called *absolute sensitivity functions* of the form

$$\frac{d}{dt}\frac{\partial X_j}{\partial X_i} = \frac{\partial f_j}{\partial \mathbf{X}_d}\frac{\partial \mathbf{X}_d}{\partial X_i} + \frac{\partial f_j}{\partial X_i}$$

Assuming that the system is in a steady state, the left-hand side is equal to zero, and the equation reduces to

$$\frac{\partial f_j}{\partial X_i} = -\frac{\partial f_j}{\partial \mathbf{X}_d}\frac{\partial \mathbf{X}_d}{\partial X_i}$$

When these operations are applied to all dependent and external variables, one obtains the matrix equation

$$\frac{\partial \boldsymbol{f}}{\partial \boldsymbol{X}_e} = -\frac{\partial \boldsymbol{f}}{\partial \boldsymbol{X}_d}\frac{\partial \boldsymbol{X}_d}{\partial \boldsymbol{X}_e}$$

or, equivalently,

$$\frac{\partial \boldsymbol{X}_d}{\partial \boldsymbol{X}_e} = -\left[\frac{\partial \boldsymbol{f}}{\partial \boldsymbol{X}_d}\right]^{-1}\frac{\partial \boldsymbol{f}}{\partial \boldsymbol{X}_e}$$

In a similar manner, one can differentiate the fluxes through the steps of the pathway with respect to the vector of external variables, $\boldsymbol{X}_e$, and can obtain the corresponding absolute sensitivity function with respect to the fluxes:

$$\frac{\partial \boldsymbol{J}}{\partial \boldsymbol{X}_e} = \frac{\partial \boldsymbol{h}}{\partial \boldsymbol{X}_d}\frac{\partial \boldsymbol{X}_d}{\partial \boldsymbol{X}_e} + \frac{\partial \boldsymbol{h}}{\partial \boldsymbol{X}_e}$$

These two sensitivity functions permit the derivation of all relationships between local and global properties in biochemical systems, as we will see next.

Biochemical Systems Theory (BST)

In BST all rate laws for individual steps of the system that tend to increase a given metabolite are aggregated into a *net* or *aggregate rate law of synthesis*. Similarly, all rate laws that tend to diminish the metabolite are aggregated into a *net* or *aggregate rate law of degradation* (cf. Chapters 1–3). The change in $\boldsymbol{X}_d$, is given by the change in each internal metabolite X_i and thus represented as

$$\dot{X}_i = V_{+i} - V_{-i}$$

Then, power-law functions are employed as explicit representations of the net rate laws of synthesis, V_{+i}, and degradation, V_{-i}, and the change in X_i is explicitly given as

$$\dot{X}_i = \alpha_i \prod_{j=1}^{k} X_j^{g_{ij}} - \beta_i \prod_{j=1}^{k} X_j^{h_{ij}} \qquad i \in \{1, \ldots, n\}$$

(cf. Chapters 1–3).

The first information obtained from the study of sensitivities is that the kinetic orders g_{ij} and h_{ij} in fact are relative sensitivity coefficients of the aggregated rate laws V_{+i} and V_{-i} with respect to variable X_j:

$$g_{ij} = \frac{\partial V_{+i}}{\partial X_j} \frac{X_j}{V_{+i}} = \bar{S}(V_{+i}, X_j)$$

$$h_{ij} = \frac{\partial V_{-i}}{\partial X_j} \frac{X_j}{V_{-i}} = \bar{S}(V_{-i}, X_j).$$

Next, we relate the system matrix A with elements $a_{ij} = g_{ij} - h_{ij}$ to the logarithmic gains, that is, to relative system sensitivities with respect to changes in external variables. Two types of these gains are of interest here: one type addresses relative changes in internal metabolite concentrations and is denoted as $L(X_j, X_i)$ or, in matrix form, as $\boldsymbol{L}(\boldsymbol{X}, \boldsymbol{X})$; the other one addresses relative changes in fluxes and is denoted as $L(V_{+j}, X_i)$ or, in matrix form, as $\boldsymbol{L}(\boldsymbol{V}_+, \boldsymbol{X})$:

$$L(X_j, X_i) = \frac{\partial \log X_j}{\partial \log X_i}$$

$$L(V_{+j}, X_i) = \frac{\partial \log V_{+j}}{\partial \log X_i}.$$

Without giving a proof, we state the desired relationships as

$$\boldsymbol{L}(\boldsymbol{X}, \boldsymbol{X}) = -\boldsymbol{A}_d^{-1} * \boldsymbol{A}_e$$

$$\boldsymbol{L}(\boldsymbol{V}_+, \boldsymbol{X}) = \boldsymbol{G}_d * \boldsymbol{L}(\boldsymbol{X}, \boldsymbol{X}) + \boldsymbol{G}_e$$

where the $\boldsymbol{G}$ matrices contain the kinetic orders (relative sensitivity coefficients) and the subscripts d and e refer to the dependent and external variables (cf. [34]).

The steady-state values of the system also can be expressed in terms of logarithmic gains and parameter sensitivities. Using the notation $Y_i = logX_i$, $b_i = log(\beta_i/\alpha_i)$ and $\boldsymbol{M} = (M_{ij}) = S(X_i, \beta_j)$, one can prove the equation

$$\boldsymbol{Y}_d = \boldsymbol{L}(\boldsymbol{X}, \boldsymbol{X}) * \boldsymbol{Y}_e + \boldsymbol{M} * \boldsymbol{b}$$

(cf. [12, 27–28, 34]). With these relationships the connection is made between properties of individual system components and the integrated system behavior. In particular, the steady-state concentrations of internal metabolites $(\boldsymbol{Y}_d)$ are related to external concentrations and sensitivities.

Metabolic Control Theory (MCT)

As an alternative to BST, Metabolic Control Theory was designed on the basis of rate laws in implicit form [63, 66]. In contrast to BST, where metabolite concentrations are the focus, MCT uses the net flux for individual enzyme-catalyzed reactions as the unit of aggregation. MCT is based on the following conventions: (1) Only infinitesimal variations around the steady-state are considered. (2) Each rate law is a homogeneous function of degree one with regard to the enzyme that catalyzes the step. (3) The internal metabolite concentrations are homogeneous functions of degree zero with respect to enzymes. Under these assumptions, a number of relationships can be developed that relate global system properties to properties of the system components. These relationships are expressed in terms of three types of relative sensitivity coefficients. The first is called *elasticity* and is defined as

$$\epsilon_{X_d}^{V_s} = \frac{\partial \log V_s}{\partial \log X_d} = \bar{S}(V_s, X_d)$$

Elasticities are analogous to the kinetic orders in BST. The other two sensitivity coefficients of interest are the *flux control coefficient* and the *concentration control coefficient*, which are defined as

$$C_E^F = \frac{\partial \log F}{\partial \log E} = \bar{S}(F, E)$$

$$C_E^{X_d} = \frac{\partial \log X_d}{\partial \log E} = \bar{S}(X_d, E)$$

where E is the vector of enzyme concentrations. With this notation, the global and local properties can be related in the form of the generic matrix equation

$$\begin{bmatrix} C_E^F \\ C_E^{X_d} \end{bmatrix} = [\theta \ \ -\epsilon_{X_d}^{V_s}]^{-1}$$

The matrix on the left-hand side contains several rows of flux control coefficients in the upper half and several rows of concentration control coefficients in the lower half. In the matrix on the right-hand side, all rows are of the same type and consist of two parts: some coefficients (θ) in the first columns and elasticities in the right columns. In this generic representation the entries are given in qualitative form, and the numbers of rows or columns are not further quantified. Specific examples of matrix equations of this type can be found, for instance, in [55] and [56].

EQUIVALENCE OF SENSITIVITY EQUATIONS IN BST AND MCT

It is not obvious that the results deriving from BST, on the one hand, and MCT, on the other, are equivalent. Several comparisons between BST and MCT have appeared in the recent literature [27–29, 34–37], but still it appears to be difficult to see similarities and differences between the two approaches. Thus, the remainder of this chapter employs sensitivity equations to demonstrate that the relationships between global and local properties are in fact the same in BST and MCT.

Derivation of BST Equations from Sensitivity Analysis

Differentiating the internal metabolite concentrations in $\boldsymbol{X}_d$ with regard to the independent variables in $\boldsymbol{X}_e$, we obtain the following matrix equation:

$$\left[\frac{\partial}{\partial \boldsymbol{X}_e}\frac{d\boldsymbol{X}_d}{dt}\right] = \left[\frac{d}{dt}\frac{\partial \boldsymbol{X}_d}{\partial \boldsymbol{X}_e}\right]$$

$$= \left[\frac{\partial f}{\partial \boldsymbol{X}_d}\right]\left[\frac{\partial \boldsymbol{X}_d}{\partial \boldsymbol{X}_e}\right] + \left[\frac{\partial f}{\partial \boldsymbol{X}_e}\right]$$

At steady state, the left-hand side can be set to zero. When we substitute f with $\boldsymbol{V}_+ - \boldsymbol{V}_-$ and rearrange our terms, we obtain

$$\left[\frac{\partial \boldsymbol{X}_d}{\partial \boldsymbol{X}_e}\right] = -\left[\frac{\partial \boldsymbol{V}_+}{\partial \boldsymbol{X}_d} - \frac{\partial \boldsymbol{V}_-}{\partial \boldsymbol{X}_d}\right]^{-1} * \left[\frac{\partial \boldsymbol{V}_+}{\partial \boldsymbol{X}_e} - \frac{\partial \boldsymbol{V}_-}{\partial \boldsymbol{X}_e}\right]$$

This equation can be expressed in terms of absolute sensitivity coefficients as:

$$[S(\boldsymbol{X}_d, \boldsymbol{X}_e)] = [S(\boldsymbol{V}_+, \boldsymbol{X}_d) - S(\boldsymbol{V}_-, \boldsymbol{X}_d)]^{-1} * [S(\boldsymbol{V}_+, \boldsymbol{X}_e) - S(\boldsymbol{V}_-, \boldsymbol{X}_e)]$$

which means

$$[\text{Matrix of } \boldsymbol{X}_d \text{ Sensitivities}] = [\text{Jacobian of } \boldsymbol{X}_d]^{-1} * [\text{Jacobian of } \boldsymbol{X}_e]$$

The equation can also be expressed in terms of relative sensitivity coefficients:

$$[\bar{S}(\boldsymbol{X}_d, \boldsymbol{X}_e)] = -[\bar{S}(\boldsymbol{V}_+, \boldsymbol{X}_d) - \bar{S}(\boldsymbol{V}_-, \boldsymbol{X}_d)]^{-1}$$
$$* [\bar{S}(\boldsymbol{V}_+, \boldsymbol{X}_e) - \bar{S}(\boldsymbol{V}_-, \boldsymbol{X}_e)]$$

This matrix equation expresses properties of the steady state of the system as a function of local properties. It is the analogue to the equation that results from aggregation and power-law approximation in BST:

$$L(X, X) = -A_d^{-1} * A_e$$

In a similar manner, one can differentiate the fluxes of inputs or outputs. Differentiation of the vector F with respect to the external variables in X_e yields the matrix equation

$$\left[\frac{\partial F}{\partial X_e}\right] = \left[\frac{\partial g}{\partial X_d}\right]\left[\frac{\partial X_d}{\partial X_e}\right] + \left[\frac{\partial g}{\partial X_e}\right]$$

Substituting g with V_{F+}, we obtain

$$\left[\frac{\partial F}{\partial X_e}\right] = \left[\frac{\partial V_{F+}}{\partial X_d}\right]\left[\frac{\partial X_d}{\partial X_e}\right] + \left[\frac{\partial V_{F+}}{\partial X_e}\right]$$

which can be expressed in terms of absolute sensitivity coefficients as:

$$[S(F, X_e)] = [S(V_{F+}, X_d)]\,[S(X_d, X_e)] + [S(V_{F+}, X_e)]$$

or in terms of relative sensitivity coefficients as:

$$[\bar{S}(F, X_e)] = [\bar{S}(V_{F+}, X_d)]\,[\bar{S}(X_d, X_e)] + [\bar{S}(V_{F+}, X_e)]$$

Substitution of $[\bar{S}(X_d, X_e)]$ according to

$$[\bar{S}(X_d, X_e)] = -[\bar{S}(V_+, X_d) - \bar{S}(V_-, X_d)]^{-1} * [\bar{S}(V_+, X_e) - \bar{S}(V_-, X_e)]$$

yields

$$\begin{aligned}[\bar{S}(F, X_e)] = {} & -[\bar{S}(V_{F+}, X_d)] * [\bar{S}(V_+, X_d) - \bar{S}(V_-, X_d)]^{-1} \\ & * [\bar{S}(V_+, X_e) - \bar{S}(V_-, X_e)] + [\bar{S}(V_{F+}, X_e)]\end{aligned}$$

This matrix equation allows us in general to express some steady-state properties of the system as a function of local properties. In fact, the equation is equivalent to the corresponding equation that results from aggregation and power-law approximation in BST:

$$L(V_+, X) = G_d * L(X, X) + G_e$$

Derivation of MCT Equations from Sensitivity Analysis

Differentiating the vector of net fluxes through individual steps, $\boldsymbol{J}$, with regard to the external variables in $\boldsymbol{X}_e$, we obtain the following matrix equation:

$$\left[\frac{\partial \boldsymbol{J}}{\partial \boldsymbol{X}_e}\right] = \left[\frac{\partial \boldsymbol{h}}{\partial \boldsymbol{X}_d}\right]\left[\frac{\partial \boldsymbol{X}_d}{\partial \boldsymbol{X}_e}\right] + \left[\frac{\partial \boldsymbol{h}}{\partial \boldsymbol{X}_e}\right]$$

Substitution of $\boldsymbol{h}$ with $\boldsymbol{V}_s$ yields

$$\left[\frac{\partial \boldsymbol{J}}{\partial \boldsymbol{X}_e}\right] = \left[\frac{\partial \boldsymbol{V}_s}{\partial \boldsymbol{X}_d}\right]\left[\frac{\partial \boldsymbol{X}_d}{\partial \boldsymbol{X}_e}\right] + \left[\frac{\partial \boldsymbol{V}_s}{\partial \boldsymbol{X}_e}\right]$$

Expressed in terms of absolute sensitivity coefficients, this equation reads:

$$[S(\boldsymbol{J}, \boldsymbol{X}_e)] = [S(\boldsymbol{V}_s, \boldsymbol{X}_d)]\,[S(\boldsymbol{X}_d, \boldsymbol{X}_e)] + [S(\boldsymbol{V}_s, \boldsymbol{X}_e)]$$

or in terms of relative sensitivity coefficients, it reads:

$$[\bar{S}(\boldsymbol{J}, \boldsymbol{X}_e)] = [\bar{S}(\boldsymbol{V}_s, \boldsymbol{X}_d)]\,[\bar{S}(\boldsymbol{X}_d, \boldsymbol{X}_e)] + [\bar{S}(\boldsymbol{V}_s, \boldsymbol{X}_e)]$$

Upon expressing the coefficients of matrix $[\bar{S}(\boldsymbol{J}, \boldsymbol{X}_e)]$ as a linear combination of coefficients of $[\bar{S}(\boldsymbol{F}, \boldsymbol{X}_e)]$, i.e.,

$$[\bar{S}(\boldsymbol{J}, \boldsymbol{X}_e)] = [\theta(\boldsymbol{J}, \boldsymbol{F})]\,[\bar{S}(\boldsymbol{F}, \boldsymbol{X}_e)]$$

and substituting and rearranging them, we obtain:

$$\begin{bmatrix} \bar{S}(\boldsymbol{F}, \boldsymbol{X}_e) \\ \bar{S}(\boldsymbol{X}_d, \boldsymbol{X}_e) \end{bmatrix} = -\,[\theta(\boldsymbol{J}, \boldsymbol{F}) - \bar{S}(\boldsymbol{V}_s, \boldsymbol{X}_d)]^{-1} * [\bar{S}(\boldsymbol{V}_s, \boldsymbol{X}_e)]$$

This equation relates system properties and local relative sensitivities of non-aggregated rate laws of the individual steps and includes theorems of MCT as special cases. When $[\bar{S}(\boldsymbol{V}_s, \boldsymbol{X}_e)]$ is the identity matrix (i.e., when the fluxes are homogeneous functions), this matrix equation is equivalent to the theorems in MCT expressed in matrix form.

Equations with aggregated rate laws (BST) can be converted into the above equation (MCT) by expressing relative sensitivity coefficients of aggregated rate laws in terms of fractional fluxes and relative sensitivity coefficients in terms of individual rate laws.

DERIVATION OF EXPLICIT STEADY-STATE EQUATIONS FROM SENSITIVITY EQUATIONS

Integration of the relative sensitivity equation

$$\left[\frac{\partial \log \boldsymbol{X}_d}{\partial \log \boldsymbol{X}_e}\right] = [\bar{S}(\boldsymbol{X}_d, \boldsymbol{X}_e)]$$

yields

$$\frac{\log \boldsymbol{X}_d}{\log \boldsymbol{X}_{d0}} = [\bar{S}(\boldsymbol{X}_d, \boldsymbol{X}_e)] \log \frac{\boldsymbol{X}_e}{\boldsymbol{X}_{e0}}$$

$[\bar{S}(\boldsymbol{X}_d, \boldsymbol{X}_e)]$ can be substituted with the following expression:

$$[\bar{S}(\boldsymbol{X}_d, \boldsymbol{X}_e)] = -[\bar{S}(\boldsymbol{V}_+, \boldsymbol{X}_d) - \bar{S}(\boldsymbol{V}_-, \boldsymbol{X}_d)]^{-1} * [\bar{S}(\boldsymbol{V}_+, \boldsymbol{X}_e) - \bar{S}(\boldsymbol{V}_-, \boldsymbol{X}_e)]$$

Furthermore, at steady-state, $\log \boldsymbol{X}_d$ can be expressed as

$$\begin{aligned}\log \boldsymbol{X}_d = {} & [\bar{S}(\boldsymbol{X}_d, \boldsymbol{X}_e)] \log\boldsymbol{X}_e + [\bar{S}(\boldsymbol{V}_+, \boldsymbol{X}_d) - (\bar{S}(\boldsymbol{V}_-, \boldsymbol{X}_d)]^{-1} * \\ & * \{[\bar{S}(\boldsymbol{V}_+, \boldsymbol{X}_d) - (\bar{S}(\boldsymbol{V}_-, \boldsymbol{X}_d)] \log \boldsymbol{X}_{d0} - \\ & - [\bar{S}(\boldsymbol{V}_+, \boldsymbol{X}_e) - \bar{S}(\boldsymbol{V}_-, \boldsymbol{X}_e)] * \log \boldsymbol{X}_{e0}\}\end{aligned}$$

That is, the steady-state concentrations of the internal metabolites can be related to the logarithmic gains and parameter sensitivities of the system. In fact, this matrix equation is the same as in BST:

$$\boldsymbol{Y}_d = \boldsymbol{L}(\boldsymbol{X}, \boldsymbol{X}) * \boldsymbol{Y}_e + \boldsymbol{M} * \boldsymbol{b}$$

DISCUSSION

BST and MCT provide matrix equations that relate global and local properties of a biochemical pathway. In both theories, the global properties are described as the relative sensitivity coefficients of fluxes or internal metabolites with respect to the independent variables of the system, even though they are named differently in the two theories: In MCT, they are called *control coefficients*; in BST they are referred to as *logarithmic gains*. In addition to global properties, each theory defines slightly different local properties. MCT uses *elasticity coefficients* defined as the relative sensitivity coefficients of the rate law of each individual step with regard to dependent variables or independent variables. In BST, each rate law derives from the convenient aggregation of individual rate laws. Local properties are called *kinetic orders* and are defined as the relative sensitivity coefficients of aggregated rate laws of synthesis or degradation of internal metabolites with respect to dependent or independent variables.

The steady-state equations in BST and MCT have a different appearance but, as we have shown, they follow directly from the same concepts of sensitivity analysis. For instance, the explicit expression for an internal metabolite concetration obtained in BST corresponds exactly to the equation obtained from integrating a general sensitivity matrix equation. Based on sensitivity analysis, the equivalence of BST and MCT at steady state is easy to see. For example, to represent a converging pathway with two individual reactions, v_k and v_i, BST uses the aggregated rate law of synthesis of the i-th metabolite, $V_{i+} = v_k + v_i$, whereas MCT employs the individual rate laws v_k and v_i. Even though aggregation in BST leads to a different representation that has major implications for the dynamics of the system, the relative sensitivity coefficients of the aggregated rate law (V_{i+}) are readily related to the relative sensitivity coefficients of the individual rate laws (v_k and v_i) according to the equation

$$\bar{S}(V_{i+}, X_d) = \bar{S}(v_k, X_d)\, v_k / V_{i+} + \bar{S}(v_j, X_d)\, v_k / V_{i+}$$

This case of a converging pathway is the paradigm difference in representation between BST and MCT. It is clear that pathways of more than two converging reactions or of diverging reactions can be treated in the same fashion. Thus, even though the results from BST and MCT have a different appearance, they can be traced back to the same roots of general sensitivity analysis. Moreover, when the analysis of a biochemical pathway is limited to responses close to steady state, results from one theory can be translated directly into the other.

SUMMARY

Sensitivity theory addresses the impact of changes in parameter values on responses of the integrated system. It thus relates (local) properties of individual system components to (global) properties of the intact system. We have applied general sensitivity theory to biochemical systems and we have derived matrix equations that fully characterize the systems at steady state. These equations are expressed in terms of gains, parameter sensitivities, fluxes, and concentrations of metabolites. We have also demonstrated how essential parts of two competing theories, Biochemical System Theory (BST) and Metabolic Control Theory (MCT), can be traced back to the same general sensitivity concepts. These theories have been developed almost independently of each other during the past two decades; they use different nomenclatures, and for the experimental biochemist, they are sometimes difficult to reconcile. Our results show that, despite seemingly significant differences, the fundamental quantitative coefficients and matrix equations in the two theories can easily be related to each other within the framework of sensitivity analysis.

5. The Method of Controlled Mathematical Comparison

DOUGLAS H. IRVINE

Given the extraordinary complexity of living systems, substantial progress has been made in genetics and molecular biology by developing the method of controlled comparison. To deal effectively with a complicated background consisting of numerous dynamic processes, the most revealing biological experiments use internal standards as parts of control experiments to nullify the effects of unpredictable variations in this background. Without these control experiments it would be virtually impossible to identify the inherent function of individual biological components in living cells or organisms, since even minor fluctuations in background events can introduce experimental artifacts that distort or obscure the true results.

The control experiment overcomes this problem by employing direct, side-by-side evaluation of systems with and without a single factor or interaction of interest. Whenever two systems are compared that are identical in every respect, except of course for the one factor of interest, background artifacts tend to be nullified, and there is less distortion in the results. Designing the most revealing control experiments can be a challenge, but only by doing so can one expect to identify the function of individual components in living systems.

The concept of a control experiment has been coupled with rigorous mathematical techniques from systems analysis to yield a new methodology called *controlled mathematical comparison*. The basic idea was described by Savageau [1], who used it for evaluating alternative regulatory patterns in biochemical pathways and genetic circuits (e.g., see [2]). This concept was extended and was applied to alternative regulatory interactions in immune networks by Irvine and Savageau [25–26]. Reviews contrasting this approach with other methods of comparison in science can be found in Savageau [5, 102].

The main purpose of this chapter is to show in detail how the method of controlled mathematical comparison is used with S-systems to obtain unique insight into the functional significance of regulatory interactions in living systems. An analysis of suppressor regulation of the immune response [25] is used to illustrate the concept of controlled mathematical comparison. First, the necessary background material is covered. Brief surveys of the elements and interactions in the immune network and the characteristics of suppressor regulation are given, followed by descriptions of the models and the criteria for functional effectiveness. Then, results for the uncontrolled and controlled mathematical comparisons are presented, and the utility of each approach is evaluated. Every calculation required for this type of analysis is shown, so the methods and results can be understood fully. However, actually working through the details of each calculation is optional; the important points to focus on are the types of general results that can be obtained with and without specially designed controls that eliminate functionally irrelevant background effects in alternative regulatory mechanisms.

This analysis amounts to a controlled comparison of uncontrolled and controlled mathematical comparisons. The uncontrolled mathematical comparison involves mathematical evaluation of alternative systems with and without a particular interaction of interest. The controlled mathematical comparison is similar, but it also involves application of special conditions that minimize irrelevant differences between the alternative systems. In particular, some characteristics that are not inherent to suppressor regulation of the immune response are left uncontrolled in the first comparison, but functionally irrelevant differences are eliminated in the second comparison by application of the appropriate experimental controls. Contrasting the results obtained with uncontrolled and controlled mathematical comparisons shows why experimental designs must be thought out carefully if the functional significance of regulatory interactions is to be determined without distortion from background artifacts.

ELEMENTS AND INTERACTIONS IN THE IMMUNE NETWORK

The immune system provides protection against unrestricted dissemination and growth of microorganisms or tumors. A large variety of potentially invasive microorganisms or tumors, each with characteristic molecular determinants or *antigens* on their surface, are recognized as foreign by specific cells and molecules in the immune system. Once recognized, a response is elicited that either inhibits invasion and growth, or enhances degradation and clearance of the foreign organism. The specificity of the immune response against potentially invasive microorganisms and tumors requires two essential features: (1) Cells within the immune system must have receptor-like molecules each capable of binding an antigen, and (2) the immune system must respond to foreign antigens while maintaining a state of nonresponsiveness or tolerance to antigens characteristic of healthy tissues of the host. Recognition of nonself *versus* self-antigens is a complex phenomenon that is not well understood even though several of the elements involved have been identified.

Effector cells in the immune system, including mature *B* and *T lymphocytes*, typically have receptors on their surface that strongly bind a particular antigen. T lymphocytes have molecules called *T-cell receptors* that bind antigens in association with proteins encoded by the major histocompatibility complex. B lymphocytes express *antibody molecules* that act not only as receptors on their surface but also as secreted effector molecules that, for example, can facilitate destruction of antigen-bearing cells or microorganisms. Both T-cell receptors and antibody molecules have *variable* and *constant domains* in their protein structures. The variable region is the antigen-binding portion of the molecule, and the constant region is associated with the function of the molecule.

Each antibody or T-cell receptor has an *idiotype* which simply is the molecular configuration, or antigenic identity, of the variable region, or binding portion, of the receptor molecule itself. Via these idiotypes a restricted number of B and T lymphocytes, antibodies, and possibly other elements of the immune system can interact specifically in response to a particular antigen. For example, antigen-specific effector lymphocytes or antibodies may be regulated by idiotype-specific *helper lymphocytes* that amplify the response, by idiotype-specific *suppressor lymphocytes* that inhibit the response, or by *anti-idiotypic antibodies* that either inhibit or amplify the response. Such a network is a prime example of an organizationally complex system, since it involves numerous nonlinearly interdependent elements, and for many of these, the functional significance is unknown. (See any general immunology text [98] for further discussion of the elements and interactions in the immune network.)

Characteristics of Suppressor Lymphocytes

Immunologic responsiveness to normal self-antigens typically is minimized to reduce autoimmune disease, whereas responsiveness to foreign antigens is enhanced to provide resistance to infectious diseases and resistance to widespread metastasis of tumors. These two conflicting demands of tolerance to self and responsiveness against invasions require tight regulation of a large repertoire of lymphocytes as an essential characteristic of the immune response. Evidence exists that this type of regulation involves suppressor lymphocytes that attenuate the response of effector lymphocytes directed against normal self-antigens (e.g., see [104]).

In addition to maintenance of tolerance, another more subtle, yet perhaps equally important, function of suppressor lymphocytes may be modulation of the immune response to foreign antigens. Although suppression of the immune response to foreign antigens may seem counterintuitive, since invasion should be minimized, regulation of effector lymphocytes by idiotype-stimulated suppressor lymphocytes has been proposed on the basis of two sets of observations (e.g., see [97, 99–101, 103]). First, suppressor lymphocytes specific for a given idiotype are capable of modulating the response of corresponding effector lymphocytes, and second, injection of an antibody into an animal can increase the level of suppressor lymphocytes specific for the idiotype of that antibody. However, due to the exquisite sensitivity of the immune system to perturbation, it has been difficult to prove by laboratory studies alone that suppression of the immune response against experimentally introduced foreign antigens is more than an artifact. Hence, an important motivation for the mathematical analysis reviewed in this paper is determining whether suppressor regulation of the immune response to foreign antigens is functionally significant or just an experimental artifact.

Biological Models and S-System Representation

A schematic representation of suppressor regulation of the immune response to foreign antigens is presented as A in Fig. 5.1. This model is simplified further as B in Fig. 5.1 to focus on interactions involved in suppressor regulation of the differentiation and proliferation of effector lymphocytes (referred to in this paper simply as *suppression*), and also to introduce the standard notation for S-systems. The following conventions have been adopted. The term *source antigen* (Ag_0 or X_0) is used to refer to potentially infectious microorganisms in the environment, to indigenous microflora at permitted sites of the host such as the

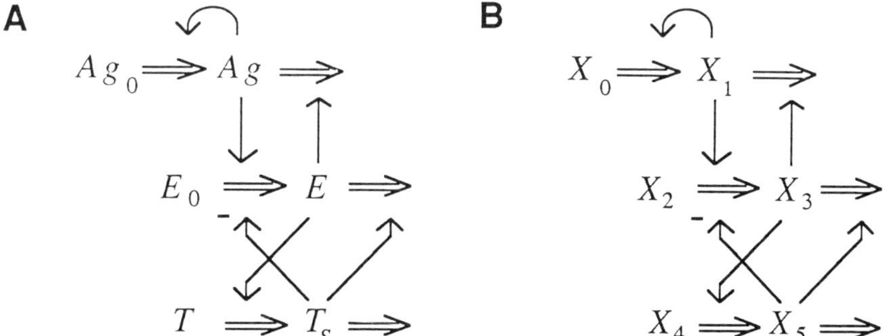

FIGURE 5.1. Schematic Representations of Suppressor Regulation of the Immune Response to Foreign Antigens. Two different symbolic notations are used: A) Conventional Notation, and B) Standard S-System Notation. See text for details.

skin and mucosal membranes, or to sources of tumor cells. The term *systemic antigen* (Ag or X_1) is used to refer to disseminated microorganisms or to tumors that disrupt physiological functions of the host. The term *effector precursor* (E_0 or X_2) is used to refer to B cells or to other immature cells of an effector lineage, and the term *effector lymphocyte* (E or X_3) is used to refer to mature plasma cells or to cytotoxic lymphocytes. The term *suppressor precursor* (T or X_4) is used to refer to immature cells of a suppressor lineage, and the term *suppressor lymphocyte* (T_S or X_5) is used to refer to mature lymphocytes that reduce the level or response of effector lymphocytes. The phrase *immune response to foreign antigens* is used to refer to functionally adaptive secondary immune responses against antigenic determinants associated strictly with invasive microorganims or tumors.

Fig. 5.2 shows the alternative systems that will be compared in this analysis. *System a* represents a system *with suppression*, and *system b* represents a system *without suppression*. All processes other than suppressor regulation of the production of effector lymphocytes are identical in these alternative systems. The nonlinear ordinary differential equations, written in S-system canonical form, representing the dynamics of *system a* are the following:

$$\dot{X}_1 = \alpha_1 X_0^{g_{10}} X_1^{g_{11}} - \beta_1 X_1^{h_{11}} X_3^{h_{13}} \tag{5.1}$$

$$\dot{X}_3 = \alpha_3^a X_2^{g_{32}^a} X_1^{g_{31}^a} X_5^{g_{35}^a} - \beta_3 X_3^{h_{33}} \tag{5.2}$$

$$\dot{X}_5 = \alpha_5 X_4^{g_{54}} X_3^{g_{53}} - \beta_5 X_5^{h_{55}} \tag{5.3}$$

where $\dot{X}_i$ is the change in concentration X_i with time, α_i is the apparent rate constant for net accumulation of X_i, and β_i is the apparent rate constant for net

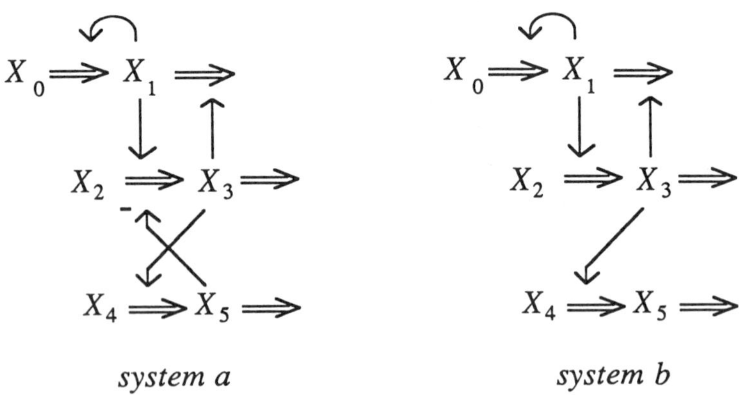

FIGURE 5.2. Immune Networks with (*system a*) and without (*system b*) Suppression. See text for details.

clearance of X_i. The exponential parameters g_{ij} and h_{ij} are apparent kinetic orders representing the effect that X_j has on the rate of net accumulation or clearance of X_i. Stimulatory interactions are represented by positive kinetic orders, inhibitory interactions are represented by negative kinetic orders, and the strength of each interaction is represented by the magnitude of the corresponding kinetic order. The parameter g_{35} represents the only inhibitory interaction in the network and therefore it is the only negative parameter in Eqs. (5.1)–(5.3).

Criteria for Functional Effectiveness

The method of controlled mathematical comparison involves side-by-side evaluation and comparison of systems with and without an interaction of interest on the basis of multiple criteria for functional effectiveness. For this analysis of suppression, the main criterion is that the level of systemic antigen should be minimal, which is analogous to saying that the level of infection or the spread of tumors should be minimal. However, in order to reduce associated metabolic expenditure by the host and immunopathologic side effects produced by effector responses, such as inflammation of tissues after local production of free radicals or precipitation of immune complexes, the effector response also should be minimal. Six independent criteria for functional effectiveness, derived from these predominant benefits and costs are defined below.

Minimum basal levels of systemic antigen and effector lymphocytes. The basal level of systemic antigen is the low-level amount of systemic antigen that persists or the low-level amount of endogenous cross-reacting antigen present as background in the host. The basal level of systemic antigen determines the host's basal level of immunity to that antigen, or, in other words, it determines the basal level of effector lymphocytes and the basal level of effector molecules. For example, if a host is continually exposed to a particular population of microorganisms, such as microflora of the skin and mucosal membranes, then the basal level of systemic antigen is determined by so-called barriers to invasion, by nonspecific mechanisms of clearance, and by the basal level of immunity against the indigenous microoganism. The metabolic costs of an infection are the host nutrients consumed for growth of the microorganism. The metabolic costs associated with the basal level of immunity also are the host nutrients consumed to produce and to maintain a population of antigen-specific effector lymphocytes and to synthesize antibody molecules and other effector factors. In either case, with disruption of physiological functions caused by an infection or by the inflammatory actions of effector molecules, the pathologic costs are relatively obvious. On the basis of both metabolic costs and reduced pathologic effects, both the basal level of systemic antigen and the basal level of effector lymphocytes should be minimal.

Minimum increase in steady-state levels of systemic antigen. The increase in the steady-state level of systemic antigen following challenge is called the *antigen gain*, which is a measure of the severity of an infection. An increase in the steady-state level of systemic antigen can occur in response to any challenge, such as an increase in exposure to source antigen, an increase in rate of invasion by source antigen, an increase in rate of growth of systemic antigen, a decrease in rate of clearance of systemic antigen, a depletion of lymphocyte precursor pools, a decrease in rate of generation of lymphocytes, or an increase in rate of clearance of lymphocytes. Both the metabolic costs and the pathologic effects of an infection are minimized when the immune response minimizes the antigen gain.

Minimum increase in steady-state levels of effector lymphocytes. The increase in the steady-state level of effector lymphocytes following challenge is called the *effector gain*, which is a measure of the metabolic costs and pathological side effects of mounting an immune response. An increase in the steady-state level of effector lymphocytes can occur in response to any challenge, such as an increase in exposure to source antigen, an increase in rate of invasion by source antigen, an increase in rate of growth of systemic antigen, or a decrease in rate of clearance of systemic antigen. Both the metabolic costs and the pathologic side effects of mounting an immune response are minimized when the effector gain is minimized.

Minimum dynamic levels of systemic antigen. The peak level of systemic antigen following a challenge should be minimal. This is a critical characteristic to measure, in addition to the increase in the steady-state level of systemic antigen, because an infection above a certain threshold can produce catastrophic effects like failure of an organ system. Peak levels of systemic antigen are determined from dynamic response curves for systemic antigen. Cumulative exposure to systemic antigen following a challenge also should be minimal. This is important because the severity of an infection is related both to the duration and to the intensity of an invasion. Cumulative exposure to systemic antigen is determined by the integration of dynamic response curves for systemic antigen. Both the metabolic costs and the pathologic effects of an infection are minimized when the immune response minimizes dynamic levels of systemic antigen.

Minimum dynamic levels of effector lymphocytes. Peak levels of effector lymphocytes following challenge should be minimal. The peak level of effector lymphocytes is important to consider, in addition to the increase in the steady-state level of effector lymphocytes, because effector lymphocytes and their products above a certain threshold can cause irreversible damage to host tissues.

Peak levels of effector lymphocytes are determined from dynamic response curves for effector lymphocytes. Cumulative production of effector lymphocytes during an immune response also should be minimal, since this is a measure of the total metabolic cost of mounting an immune response and a measure of the cumulative pathologic side effects that a response has on the host. Both effects are a function of the duration and the intensity of an effector response. Cumulative levels of effector lymphocytes are determined by the integration of dynamic response curves for effector lymphocytes. Both the metabolic costs of mounting an immune response and the pathologic side effects are minimized when dynamic levels of effector lymphocytes are minimized.

Minimum sensitivity to parameter variation. The sensitivity of the immune response to deleterious fluctuations in the environment should be minimal. Sensitivity of a system to natural fluctuations is measured as the percent change in antigen gain with respect to a one-percent change in the value of each parameter of the system and is called *system sensitivity*. Low sensitivity is an indication of how well a system is buffered against potentially harmful fluctuations in the environment. A system with minimum sensitivity can be considered more robust and more reliable than one with higher sensitivity.

Results Using Uncontrolled Mathematical Comparison

The steady-state values for the system represented with suppression in Fig. 5.2 (*system a*) are obtained by setting Eqs. (5.1)–(5.3) equal to zero and solving them (cf. Chapter 2). The implicit steady-state solution for the logarithmic values of X_1, X_3, and X_5 can be written in symbolic form as follows:

$$a_{11}Y_1 - h_{13}Y_3 = B_1 \tag{5.4}$$

$$g_{31}^aY_1 - h_{33}Y_3 + g_{35}^aY_5 = B_3^a \tag{5.5}$$

$$g_{53}Y_3 - h_{55}Y_5 = B_5 \tag{5.6}$$

where $Y_i = \log(X_i)$, $a_{11} = (g_{11} - h_{11})$ and $B_i = \log(\beta_i/\alpha_i) - g_{ii-1}Y_{i-1}$. The values g_{35}^a and a_{11} are negative, and all other parameters are positive, except each B_i can be positive, negative, or zero. It is worth noting that the parameter g_{35}^a represents the strength of suppression. When the determinant is nonzero, Eqs. (5.4)–(5.6) can be solved for explicit values of each Y_i by conventional linear algebraic methods:

$$Y_1^a = [B_1(h_{33}h_{55} - g_{35}^ag_{53}) - B_3^ah_{13}h_{55} - B_5h_{13}g_{35}^a]/D^a \tag{5.7}$$

$$Y_3^a = [B_1g_{31}^ah_{55} - B_3^aa_{11}h_{55} - B_5a_{11}g_{35}^a]/D^a \tag{5.8}$$

$$Y_5^a = [B_1 g_{31}^a g_{53} - B_3^a a_{11} g_{53} - B_5(a_{11} h_{33} - g_{31}^a h_{13})]/D^a \qquad (5.9)$$

$$D^a = a_{11}(h_{33} h_{55} - g_{35}^a g_{53}) - g_{31}^a h_{13} h_{55}$$

These equations will be used to calculate the steady-state response characteristics for the system with suppression (*system a*).

Minimal basal levels of systemic antigen and effector lymphocytes. The first question we can ask with Eqs. (5.1)–(5.3) is, What happens to the basal levels of systemic antigen (X_1) and effector lymphocytes (X_3) when suppression is lost? Figs. 5.3 illustrates what can happen when suppression is lost. In this simulation, the system with suppression starts off in steady state at time zero, and then at time equals one, suppression is lost, converting *system a* into *system b* (i.e., $g_{35}^b = 0$ is substituted for g_{35}^a). The effector lymphocyte concentration begins to rise causing the systemic antigen concentration to fall, and then both levels approach new steady states. The basal level of systemic antigen goes down, but there is a trade off since the basal level of effector lymphocytes goes up. Without knowing if the benefit of reduced levels of systemic antigen is more important than the cost of increased levels of effector lymphocytes, we cannot tell from this result whether having suppression is an advantage or a disadvantage. A different set of parameter values can invert the results shown in Fig. 5.3; that is, the systemic antigen concentration can go up and the effector lymphocyte concentration can go down with loss of suppression. However, once again, without knowing the relative costs and benefits of this type of response, we still cannot tell whether having suppression is an advantage or a disadvantage.

Minimum increase in steady-state levels of systemic antigen. The increase in the steady-state level of systemic antigen (X_1) can be measured as the percent change in systemic antigen concentration with respect to a one-percent change in exposure to source antigen (X_0). This is called the *antigen gain* of the system and is calculated as follows (cf. Chapter 2):

$$L_{10} = (\partial X_1/\partial X_0)(X_0/X_1) = \partial Y_1/\partial Y_0 \qquad (5.10)$$

For the system with suppression (*system a*), the antigen gain can be calculated using Eq. (7):

$$L_{10}^a = -g_{10}(h_{33} h_{55} - g_{35}^a g_{53})/D^a \qquad (5.11)$$

For the system without suppression (*system b*), the antigen gain can be obtained by substituting $g_{35}^b = 0$ for g_{35}^a and D^b for D^a in Eq. (5.11):

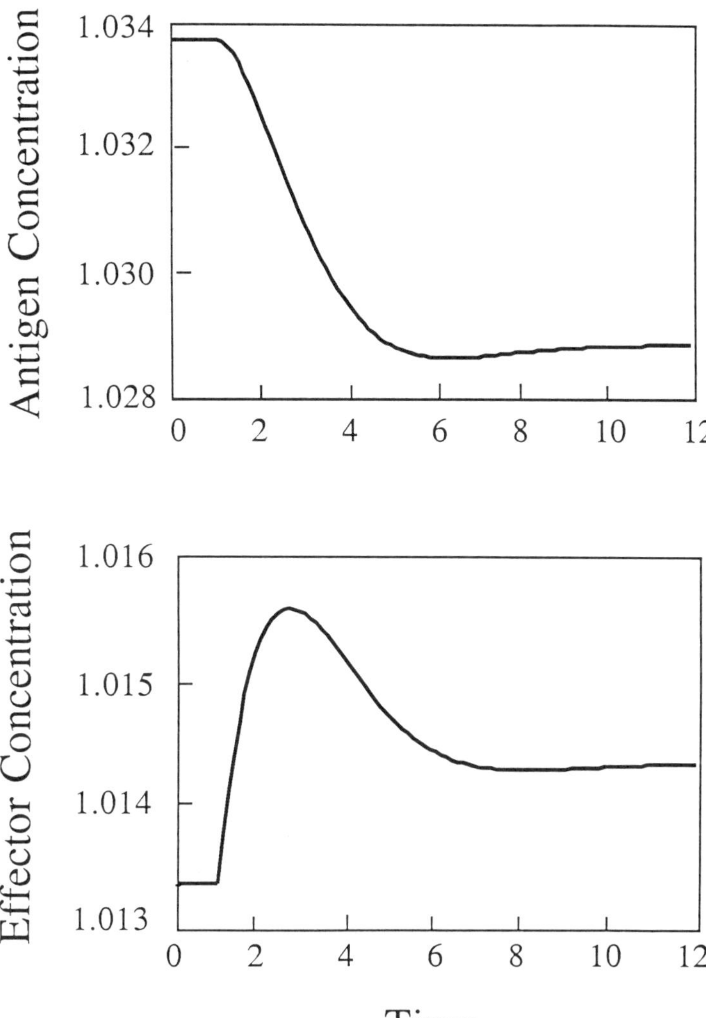

FIGURE 5.3. Change in Basal Levels of Systemic Antigen and Effector Lymphocytes following Loss of Suppression. Parameter values for the S-system with suppression (*system a*) before time equals one are as follows: $\alpha_1 = 2.02$, $\beta_1 = 2.0$, $\alpha_3^a = \beta_3 = \alpha_5 = \beta_5 = X_0 = X_2 = X_4 = g_{10} = g_{32}^a = g_{54} = h_{11} = h_{33} = h_{55} = 1.0$, $g_{11} = 0.9$, $g_{31}^a = g_{53} = h_{13} = 0.5$, $g_{35}^a = -0.5$, and all other g_{ij}'s and h_{ij}'s equal zero. Dynamic solutions were obtained using the program ESSYNS (cf. Chapter 7). At time equals zero, the basal level of systemic antigen is 1.0337, and the basal level of effector lymphocytes is 1.0134. [The basal level of suppressor lymphocytes (not shown) is 1.0067.] At time equals one suppression is lost by substituting $g_{35}^b = 0$ for $g_{35}^a = -0.5$ to change *system a* into *system b*. All other parameters, including the parameters g_{31}^a and g_{31}^b in this uncontrolled comparison, are identical in *systems a* and *b*. After suppression is lost, the system approaches a lower basal level of systemic antigen and a higher basal level of effector lymphocytes.

$$L_{10}^{b} = -g_{10}(h_{33}h_{55})/D^{b}$$
$$D^{b} = a_{11}h_{33}h_{55} - g_{31}^{b}h_{13}h_{55} \tag{5.12}$$

The relative increase in systemic antigen following a challenge can be compared in systems with and without suppression by taking the ratio of the antigen gains in Eqs. (5.11) and (5.12):

$$\begin{aligned} L_{10}^{b}/L_{10}^{a} &= [-g_{10}h_{33}h_{55}/D^{b}][D^{a}/-g_{10}(h_{33}h_{55} - g_{35}^{a}g_{53})] \\ &= \frac{a_{11}h_{33}h_{55}(h_{33}h_{55} - g_{35}^{a}g_{53}) - g_{31}^{a}h_{13}h_{55}(h_{33}h_{55})}{a_{11}h_{33}h_{55}(h_{33}h_{55} - g_{35}^{a}g_{53}) - g_{31}^{b}h_{13}h_{55}(h_{33}h_{55} - g_{35}^{a}g_{53})} \end{aligned} \tag{5.13}$$

This ratio is less than one when $g_{31}^{b} = g_{31}^{a}$, which is the case in this uncontrolled comparison. This type of comparison suggests that the increase in the steady-state level of systemic antigen following a challenge is lower in a system without suppression (*system b*) than it is in a system with suppression (*system a*).

Minimum increase in steady-state levels of effector lymphocytes. The increase in the steady-state level of effector lymphocytes (X_3) can be measured as the percent change in effector lymphocyte concentration with respect to a one-percent change in exposure to source antigen (X_0). This is called the effector gain of the system and it is calculated as follows:

$$L_{30} = (\partial X_3/\partial X_0)(X_0/X_3) = \partial Y_3/\partial Y_0 \tag{5.14}$$

For the system with suppression (*system a*), the effector gain can be calculated by using Eq. (5.8):

$$L_{30}^{a} = -g_{10}g_{31}^{a}h_{55}/D^{a} \tag{5.15}$$

For the system without suppression (*system b*), the effector gain can be obtained by substituting g_{31}^{b} for g_{31}^{a} and D^{b} for D^{a} in Eq. (5.15):

$$L_{30}^{b} = -g_{10}g_{31}^{b}h_{55}/D^{b} \tag{5.16}$$

The relative increase in effector lymphocytes following a challenge can be compared in systems with and without suppression by taking the ratio of the effector gains in Eqs. (5.15) and (5.16):

$$L_{30}^{b}/L_{30}^{a} = (-g_{10}g_{31}^{a}h_{55}/D^{b})(D^{a}/-g_{10}g_{31}^{a}h_{55})$$

$$= \frac{a_{11}(h_{33}h_{55} - g_{35}^{a}g_{53}) - g_{31}^{a}h_{13}h_{55}}{a_{11}(h_{33}h_{55}) - g_{31}^{b}h_{13}h_{55}} \qquad (5.17)$$

This ratio is greater than one when $g_{31}^{b} = g_{31}^{a}$, which is the case in this uncontrolled comparison. This type of comparison suggests that the increase in the steady-state level of effector lymphocytes following a challenge is higher in a system without suppression (*system b*) than it is in a system with suppression (*system a*). This is the opposite of the last result for the increase in the steady-state level of systemtic antigen, which was less in the system without suppression (*system b*). Again, without knowing the importance of decreased levels of systemic antigen relative to increased levels of effector lymphocytes, we cannot tell from these two conflicting results whether suppression is an advantage or a disadvantage.

Minimum dynamic levels of systemic antigen. The dynamic response of systemic antigen following a fourfold increase in exposure to source antigen is given in Fig. 5.4 for systems with and without suppression. The comparison of antigen gains in Eq. (5.13) above showed that the increase in the steady-state level of systemic antigen is lower in the system without suppression (*system b*) than in the system with suppression (*system a*). The dynamic response in Fig. 5.4 also shows that, for this uncontrolled comparison, peak and cumulative levels of systemic antigen also are lower in the system without suppression (*system b*).

Minimum dynamic levels of effector lymphocytes. The dynamic response of effector lymphocytes following a fourfold increase in exposure to source antigen is given in Fig. 5.4 for systems with and without suppression. The comparison of effector gains in Eq. (5.17) above showed that the increase in the steady-state level of effector lymphocytes is lower in the system with suppression (*system a*) than in the system without suppression (*system b*). The dynamic response in Fig. 5.4 also shows that, for this uncontrolled comparison, peak and cumulative levels of effector lymphocytes also are lower in the system with suppression. Once again, this is the opposite of the last result for the dynamic levels of systemic antigen which were less in the system without suppression (*system b*), and we cannot judge on the basis of these two conflicting results whether suppression is an advantage or a disadvantage.

Minimum sensitivity to parameter variation. System sensitivity is the change in the antigen gain of the system with respect to a change in a parameter value.

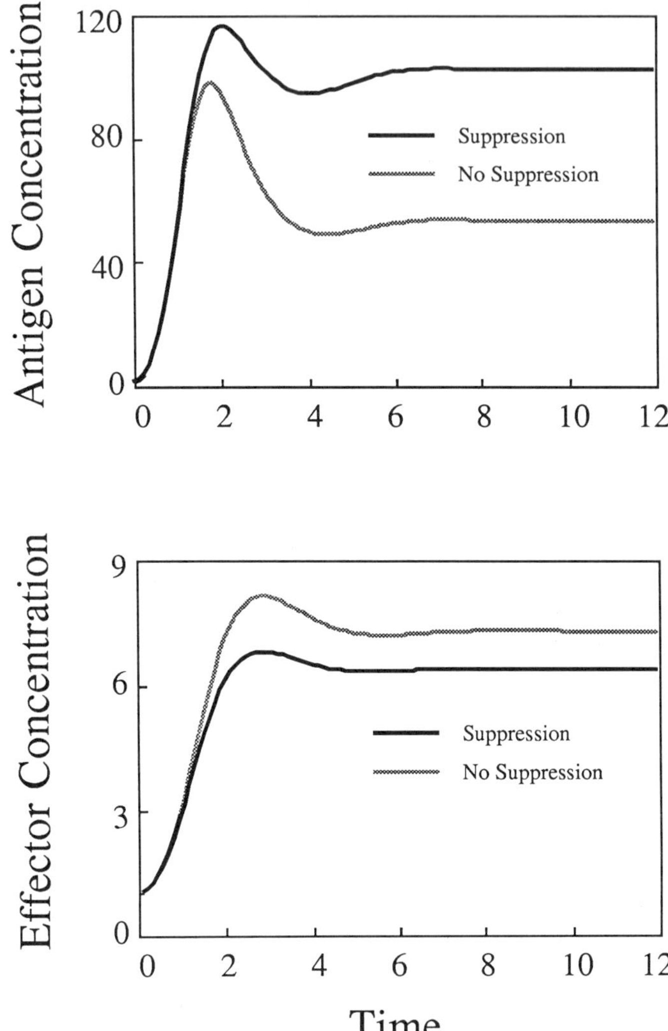

FIGURE 5.4. Response of Alternative Immune Networks following an Increase in Exposure to Source Antigen. This comparison is made without special controls that would eliminate arbitrary changes in steady-state flux and expenditure introduced into the system without suppression (*system b*). Parameter values for each S-system are the same as for the systems with and without suppression in Fig. 5.3, except $\alpha_1 = 2.0$. Concentrations are normalized with respect to the initial steady state, and time is normalized with respect to the pseudo-first-order rate constant for clearance of effector lymphocytes. A fourfold increase in exposure to source antigen is introduced by changing X_0 from a value of 1.0 to a value of 4.0 at time zero.

We will compare the sensitivity of each system to a change in the apparent kinetic order for clearance of effector lymphocytes, h_{33}, which is calculated as follows:

$$S(L_{10}, h_{33}) = (\partial L_{10}/\partial h_{33})(h_{33}/L_{10}) \tag{5.18}$$

The ratio of system sensitivities in *systems a* and *b* is

$$S(L_{10}^b, h_{33})/S(L_{10}^a, h_{33}) = \frac{-a_{11}(h_{33}h_{55} - g_{35}^a g_{53})^2 + g_{31}^a h_{13}h_{55}(h_{33}h_{55} - g_{35}^a g_{53})}{-a_{11}(h_{33}h_{55})^2 + g_{31}^a h_{13}h_{55}(h_{33}h_{55})} \tag{5.19}$$

This ratio is always greater than one, independent of the magnitudes of the parameters. The ratio of system sensitivities to other parameters also is never less than one. This means that the system with suppression (*system a*) is less sensitive to parameter variation than the system without suppression (*system b*), or in other words, suppression helps buffer against potentially harmful fluctuations.

Summary of results using uncontrolled mathematical comparison. Table 5.1 summarizes the results for the uncontrolled mathematical comparison. Loss of suppression in this uncontrolled case results in a decrease in both the steady-state and dynamic levels of systemic antigen, but at a price. In the system without suppression (*system b*), there is a corresponding increase in both the steady-state and dynamic levels of effector lymphocytes and an increase in the sensitivity to fluctuations in the environment. Without knowing the relative importance to assign each of these advantages and disadvantages, it cannot be determined from the uncontrolled mathematical comparison whether suppres-

Table 5.1. Summary of Results for Uncontrolled Mathematical Comparison of Systems With and Without Suppression

Criterion	With Suppression	Without Suppression
Basal Levels	−/+	+/−
Antigen Gain	−	+
Effector Gain	+	−
Dynamic Levels of Antigen	−	+
Dynamic Levels of Effectors	+	−
System Sensitivity	+	−
Overall	?	?

A + sign indicates an advantage, and a − sign indicates a disadvantage relative to the alternative system. A question mark indicates that the overall results are mixed, and without knowing the relative importance of each criterion, the uncontrolled mathematical comparison does not reveal the functional significance of suppression.

sion of the immune response to foreign antigens is functionally significant or is just an experimental artifact.

Results Using Controlled Mathematical Comparison

The system without suppression (*system b*) had some arbitrary differences introduced when suppression was removed in the uncontrolled mathematical comparison. Because these differences were left uncontrolled, they distorted the actual results. Two differences between systems with and without suppression are not unique properties of suppression which can be introduced or altered in other ways. When suppression is removed by setting $g_{35}^{b} = 0$, the basal rate of production of effector lymphocytes is turned up and is accompanied by an increase in the steady-state rate of effector production following a challenge. These two increases in flux mean that more material and energy are expended in *system b* than in *system a*, and differences in expenditure alone change the results.

Since the objective of the analysis is to determine the inherent, irreducible properties of suppression that cannot be attributed to other processes, such as simple changes in flux and expenditure, these arbitrary changes in the background of the system must be controlled. With application of the appropriate controls, the changes in flux and expenditure with removal of suppression can be eliminated so they do not mask unique differences between systems with and without suppression. These types of controlled conditions, with equal steady-state flux and expenditure in both systems, are applied in the following controlled mathematical comparison.

Minimum basal levels of systemic antigen and effector lymphocytes. By making sure the basal rates of effector lymphocyte production are equal in *systems a* and *b*, the basal levels of systemic antigen and effector lymphocytes also become equal. The value of the apparent rate constant for production of effector lymphocytes, α_3^b, that makes the basal rate of effector lymphocyte production equal in *systems a* and *b* can be obtained by equating Y_3^b with Y_3^a and solving for α_3^b or B_3^b. Under these appropriately controlled conditions, with equal basal rates of expenditure on the production of effector lymphocytes, there are no inherent differences between the basal levels of systemic antigen and effector lymphocytes in systems with and without suppression (i.e., $X_1^b = X_1^a$ and $X_3^b = X_3^a$).

Minimum increase in steady-state levels of systemic antigen. The parameter B_3^b determined in the last section, ensures equal basal rates of effector lymphocyte production. However, following a challenge by increased exposure to source antigen, the steady-state rate of production of effector lymphocytes still is higher in the system without suppression. This second difference in expenditure can be avoided by using the appropriate value of g_{31}^{b}. The value of g_{31}^{b} that ensures equal increases in steady-state rates of expenditure on effective lymphocytes is obtained by setting L_{10}^{b} equal to L_{10}^{a} and solving for g_{31}^{b} in terms of the parameters of *system a*. Equating Eqs. (5.11) and (5.12) and solving for g_{31}^{b} yields the following condition that ensures equal increases in steady-state rates of expenditure in systems with and without suppression:

$$g_{31}^{b} = g_{31}^{a} h_{33} h_{55} / (h_{33} h_{55} - g_{35}^{a} g_{53}) \tag{5.20}$$

When this condition is satisfied, there obviously are no inherent differences between the antigen gains in systems with and without suppression (i.e., $L_{10}^{b} = L_{10}^{a}$).

Minimum increase in steady-state levels of effector lymphocytes. By substituting the right-hand side of Eq. (5.20) for the parameter g_{31}^{b} in Eq. (5.16), and then taking the ratio of Eqs. (5.16) and (5.15) the relative magnitude of effector gains in *systems a* and *b* can be compared under the controlled conditions of equal steady-state rates of production of effector lymphocytes:

$$\begin{aligned} L_{30}^{b}/L_{30}^{a} &= D^{a} h_{33} h_{55} / D^{b} (h_{33} h_{55} - g_{35}^{a} g_{53}) \\ &= \frac{[a_{11}(h_{33} h_{55} - g_{35}^{a} g_{53}) - g_{31}^{a} h_{13} h_{55}] h_{33} h_{55}}{[a_{11} h_{33} h_{55} - g_{31}^{a} h_{13} h_{55} h_{33} h_{55} / (h_{33} h_{55} - g_{35}^{a} g_{53})](h_{33} h_{55} - g_{35}^{a} g_{53})} \end{aligned} \tag{5.21}$$

This ratio always has a value of one, independent of the magnitudes of the parameters. Therefore, under the appropriately controlled conditions, there are no inherent differences between effector gains in systems with and without suppression (i.e., $L_{30}^{b} = L_{30}^{a}$).

Minimum dynamic levels of systemic antigen. Fig. 5.5 shows the dynamic response of systems with and without suppression following a fourfold increase in exposure to source antigen. Unlike the uncontrolled comparison (Fig. 5.4), these dynamic responses are under the controlled conditions of equal steady-state rates of effector production. Notice that with equivalent steady-state rates of effector production, the final steady-state levels of systemic antigen are equal, as guaranteed by applying the condition in Eq. (5.20) above. However, the system with suppression has lower peak and cumulative levels of systemic antigen. When different sets of parameter values arc used, there are quantitative changes in peak and cumulative levels of invasion by systemic antigen, but the

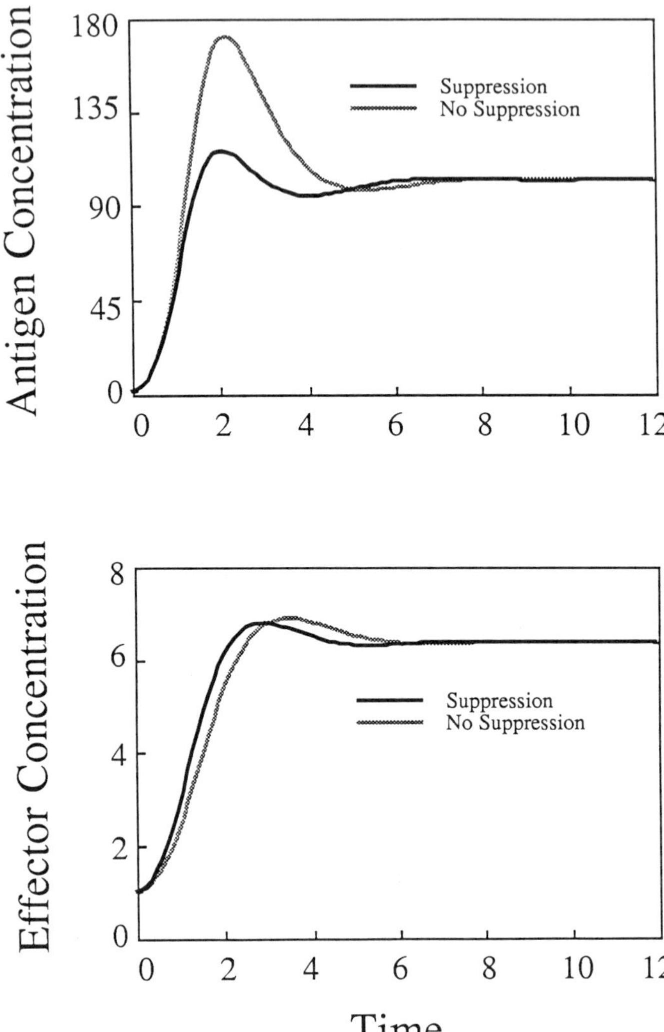

FIGURE 5.5. Response of Alternative Immune Networks following an Increase in Exposure to Source Antigen. This comparison is made with special controls that eliminate arbitrary changes in steady-state flux and expenditure that otherwise are introduced into the system without suppression (*system b*). Parameter values for each S-system are the same as for the systems in Fig. 5.4, except that additional controls are used to ensure equal steady-state flux and expenditure in each system. These special conditions are attained by applying Eq. (5.20) to reset the value of g_{31}^{b} from 0.5 in the uncontrolled comparison to 0.4 in this controlled comparison.

qualitative difference—the system with suppression (*system a*) has lower dynamic levels of invasion than the system without suppression (*system b*)—remains the same. Hence, an inherent functional characteristic of suppression is to lower peak and cumulative levels of invasion by systemic antigen.

Minimum dynamic levels of effector lymphocytes. Fig. 5.5 shows that, even though steady-state levels of effector lymphocytes are the same, the peak and cumulative levels of effector lymphocytes are lower in the system with suppression. Again, when different sets of parameter values are used, there are quantitative changes in peak and cumulative levels of effector lymphocytes, but the qualitative difference—the system with suppression (*system a*) has lower dynamic levels than the system without suppression (*system b*)—remains the same. Hence, an inherent functional characteristic of suppression is to lower peak and cumulative levels of effector lymphocytes.

Minimum sensitivity to parameter variation. By substituting Eq. (20) into Eq. (19), the ratio of system sensitivities for systems with and without suppression can be obtained under controlled conditions:

$$S(L_{10}^{b}, h_{33})/S(L_{10}^{a}, h_{33}) = (h_{33}h_{55} - g_{35}^{a}g_{53})/h_{33}h_{55} \qquad (5.22)$$

This ratio still is always greater than one, independent of the magnitudes of the parameters. The ratio of system sensitivities to other parameters also is never less than one. This means that an inherent property of suppression is to buffer the immune response against potentially harmful fluctuations in the environment.

Summary of results using controlled mathematical comparison. Table 5.2 summarizes the results for the controlled mathematical comparison. Loss of

Table 5.2. Summary of Results for Controlled Mathematical Comparison of Systems With and Without Suppression

Criterion	With Suppression	Without Suppression
Basal Levels	=	=
Antigen Gain	=	=
Effector Gain	=	=
Dynamic Levels of Antigen	+	−
Dynamic Levels of Effectors	+	−
System Sensitivity	+	−
Overall	+	−

An = sign indicates no difference, a + sign indicates an advantage, and a − sign indicates a disadvantage relative to the alternative system. By all criteria, the system with suppression is equal or superior to the system without suppression.

suppression under the appropriately controlled conditions does not change steady-state levels of invasion by systemic antigen and does not change steady-state levels of effector lymphocytes. However, both the dynamic levels of invasion by systemic antigen and the dynamic levels of effector lymphocytes are minimized in a system with suppression. There also is improved buffering against potentially harmful fluctuations in the environment. Now, even without knowing the relative importance to assign each of the criteria for functional effectiveness, the results unambiguously show that an immune system with regulation of the production of effector lymphocytes by suppressor lymphocytes is superior to a system without this form of regulation.

CONCLUSIONS

This analysis has illustrated how the concept of a controlled experiment has been coupled with rigorous techniques from S-systems analysis to yield a new approach called controlled mathematical comparison. It also has shown how the method of controlled mathematical comparison provides unique insight into the functional significance of regulatory interactions in living systems. Another approach to investigating the functional significance of regulatory interactions in living systems might be optimization of alternative mechanisms to determine which performs better on the basis of multiple criteria for functional effectiveness. However, this approach is unnecessary whenever one regulatory mechanism can be shown to be superior on the basis of all the criteria. This type of result, which was the type obtained using the method of controlled mathematical comparison, indicates that even if the optimum members of each class of mechanisms could be found and compared, the mechanism shown to be superior by the method of controlled mathematical comparison still would come out ahead. This sort of general conclusion can be made whenever one mechanism can be shown to be superior to another independent of the magnitudes of the parameters. In other words, the method of controlled mathematical comparison can yield the same kinds of results as optimization and comparison of alternatives, without the serious drawback of having to know the importance and relative weight to assign each of the criteria for functional effectiveness.

What is required for the method of controlled mathematical comparison to work? First, a systematic, analytically tractable nonlinear formalism is required, a formalism that captures essential nonlinear features of a system yet remains workable for side-by-side analytical comparisons. Second, the method of mathematically controlled comparison requires the ability to solve for constraints ensuring that critical characteristics of a system are not changed arbitrarily when a single interaction is altered. Third, robust methods for symbolic steady-state analysis and stability analysis and methods for efficient solution of nonlinear dynamics are required to compare alternative systems on the basis of multiple criteria for functional effectiveness. (For brevity, stability analysis was not included in this review; see [2], [25–26] for examples.) S-systems meet all of these requirements (cf. Chapters 2, 6, and 7) and have been applied successfully to assist in understanding the inherent function of specific regulatory interactions in metabolic pathways, genetic circuits, ecological systems, and immune networks. The method of controlled mathematical comparison using S-systems is based on general mathematical principles and could be extended to reveal the functional significance of individual components in other types of complex nonlinear systems.

SUMMARY

The concept of a control experiment, which is well established in molecular and cellular biology, coupled with rigorous mathematical techniques from S-systems analysis has yielded a new methodology called *controlled mathematical comparison*. This methodology is illustrated and evaluated with an analysis of suppressor regulation of the immune response. Two types of mathematical approaches are part of the evaluation. One involves a mathematical comparison of alternative systems with and without a particular interaction of interest. The other is similar, but it also involves the application of special conditions that minimize irrelevant differences between the alternative systems; this is the method of controlled mathematical comparison. Contrasting the results obtained with controlled and uncontrolled mathematical comparisons shows why experimental designs must be thought out carefully if the functional significance of regulatory interactions in complex systems is to be determined without distortion from background artifacts.

6. Symbolic Analysis of S-Systems with MACSYMA

VOLKER WEINBERGER

An S-system can be solved and analyzed numerically only if all parameter values are known. The program ESSYNS (cf. Chapter 7 and [30, 38]) then yields numerical results for steady states, stability, fluxes, and times courses. However, if there are unknown parameters or if one is interested in the influence of particular parameters on the results (e.g., steady state) in terms of a functional relationship, one has to analyze the system symbolically. Although S-systems are nonlinear and dynamical solutions often cannot be obtained analytically [24], many properties, such as steady states, fluxes, and stability, can be evaluated analytically due to the structural homogeneity and simplicity of S-system differential equations. For instance, symbolic analysis of S-systems contributes essentially to the work of Savageau (e.g. [2, 14, 16, 21]) and Irvine (Chapter 5 and [25–26]). A practical drawback has been that symbolic analysis lacks the support of computer programs and therefore it is time consuming and often monotonous.

Several commercially available computer programs claim to be able to perform all symbolic manipulations necessary for a complete S-system analysis. A powerful, and probably the most prevalent, general-purpose computer algebra system is MACSYMA (Project MAC's SYmbolic MAnipulation System) [110, 106], which was developed by the Mathlab Group of the MIT Laboratory of Computer Science during the seventies and early eighties and is now maintained by Symbolics, Inc. MACSYMA is available for several mainframes and workstations and written in LISP, an artificial intelligence computer language. We used MACSYMA on a SUN-3 workstation, which is almost the smallest possible configuration.

For our purpose of analyzing S-systems, the ideal properties of a symbolic mathematics program would be (1) that all calculations be done by the program in an easy way to save time and allow for symbolic S-systems analysis without attention to mathematical details, and (2) that the results are free of computational errors that frequently occur during extensive paper and pencil calculations. Since we will generally use very basic functions, computational errors are unlikely even though most large and complex computer systems are not perfect (cf. [108] for MACSYMA).

We will examine the usefulness of MACSYMA for steady-state analysis, stability analysis, and calculation of parameter sensitivities and logarithmic gains. We shall give a brief description of MACSYMA in the following section, show how to use the functions developed for the analysis of S-systems in Section 3, and give an example of developing such functions in Section 4. Typical problems and the implications of our work will be summarized and discussed in the final section of this chapter. A disk containing the complete set of functions and further documentation can be obtained from the author.

THE BASICS OF MACSYMA

Although written in LISP, which is the standard example of a functional programming language (a detailed discussion of functional languages may be found in [106]), MACSYMA provides many elements of conventional imperative programming languages like FORTRAN or PASCAL. In particular, values can be assigned to variables and commands will be executed in the usual sequential order. By reducing the number of required parenthesis in comparison to LISP and allowing familiar structures like IF - THEN - ELSE for conditional statements or FOR - FROM - THRU -DO / FOR - IN - DO for loops, MACSYMA also makes programs appear more conventional.

Nevertheless, because of the close relationships between MACSYMA and LISP in structures and concepts, some experience in LISP is a good basis for advanced work with MACSYMA. Both provide an interactive user interface and are based on the evaluation of functions that compute exactly one result. Note that possible results include lists of several values and dummy results like "done." For illustration purposes MACSYMA functions may be compared to functions or subroutines in PASCAL or FORTRAN.

MACSYMA provides many preprogrammed or built-in functions, for example, functions for symbolic calculation of eigenvalues, solutions to systems of linear equations, and solutions to differential equations. Programming in MACSYMA means extending the current repertoire of functions by new (user-defined) functions using the existing functions. There is no conceptual difference between built-in and user-defined functions.

It is not too hard to get started with MACSYMA and use its basic capabilities. However, MACSYMA is much more complex than other programming languages. It provides hundreds of functions and the behavior of these functions is influenced by some hundred options. In many cases the actual result differs in some way from the desired result, since the ''best'' representation of the result often is a matter of taste. We will give an example later in this section [cf. Eq. (d10)].

During a MACSYMA session, one enters a command on successively numbered C-lines (command lines). After execution, the result is usually shown on the corresponding D-line (display line) and the next command may be entered on the following C-line. In general, the format of the two-dimensional output will be quite different from the one-dimensional input. The result (d1), after entering a simple algebraic expression (c1) shall demonstrate this. Throughout this chapter we will accentuate user input and program output in the examples through different typefaces.

```
(c1) x[6] * x[7]^-1 - x[1] * x[4] * x[5]^-1;

                                              x     x  x
                                               6     1  4
(d1)                                          -- - -----
                                              x     x
                                               7     5

(c2)
```

Previous C-lines and D-lines can be referred to at any time later. As a short cut, a percent sign (%) references the most recent D-line. Symbolic values may be assigned to variables, a concept similar to the variables in other programming languages. A colon (:) serves as assignment operator. Each command line has to be terminated either by a semicolon (;) if the result shall be displayed, or with a dollar-sign ($) to omit the D-line. The details and subtleties of MACSYMA are beyond the scope of this chapter, and it is recommended for the beginning user to gain some hands-on experience.

Our intent is not to compute dynamical solutions of S-systems with MACSYMA but to calculate steady-state properties. Nonetheless, it is instructive to study how MACSYMA can be used to solve the first order differential equation for the logistic growth law Eq. (6.1) with the final value K and the growth-rate constant r.

$$\dot{x} = rx - \frac{r}{K}x^2 \tag{6.1}$$

```
(c2) logistic : 'diff(x, t) = r * x - r/k * x^2;

                                                        2
                                  dx               r x
(d2)                              -- = r x -  ----
                                  dt                k

(c3) assume( x > 0, x < k );

(d3)                              [x > 0, k > x]

(c4) solu: ode2( logistic, x, t ), logabs;

                            log(k - x) - log(x)
(d4)                      - ------------------- = t + %c
                                     r

(c5) method;

(d5)                              separable
```

The first command (c2) is to assign the differential equation to the variable *logistic*. In (c3) we provide MACSYMA with more information about x and K. With the next command (c4), the differential equation *logistic* is solved using the built-in solver *ode2*; the resulting implicit solution in assigned to the variable *solu*. The option *logabs* forces MACSYMA to use absolute values when a logarithm is generated by integration. Since K is larger than x, MACSYMA knows that $(K - x)$ is positive. *log* denotes the natural logarithm; *%c* denotes an arbitrary constant. The global variable *method* has been set as a side effect during the execution of the function *ode2* and provides the additional information that our differential equation has been solved by a separation of variables. Side effects are any effects a function has in addition to returning a result, for example changes of global variables or information printed during the execution. They are frequently used to overcome the restriction of returning exactly one result and they make a major contribution to the complexity of MACSYMA.

To obtain an explicit solution $x(t)$ we have to transform the implicit solution (d4) in (c6)–(c8), before MACSYMA is able to solve for x (c9). *lhs* and *rhs* return the left-hand side and the right-hand side of an equation, respectively.

```
(c6) logcontract(solu);

                                   x
                          log(- -----)
                                x - k
(d6)                      ------------ = t + %c
                               r

(c7) % * r;
```

```
                                        x
(d7)                          log(- -----) = r (t + %c)
                                      x - k

(c8) exp( lhs( % )) = exp( rhs( % ));

                                  x        r (t + %c)
(d8)                          - ----- = %e
                                x - k

(c9) solve(%, [x]);

                                           r t + %c r
                                      k %e
(d9)                          [x = ---------------]
                                        r t + %c r
                                      %e           + 1
```

From the human point of view, a simple representation of the result would be Eq. (6.2).

$$x(t) = \frac{K}{1 + e^{-r(t+c)}} \tag{6.2}$$

To approach this form we can factor out r, but MACSYMA still won't divide by the exponential term, leaving us with a representation that most humans probably would consider suboptimal.

```
(c10) factor( first( d9 ));

                                        r (t + %c)
                                    k %e
(d10)                          x = ---------------
                                       r (t + %c)
                                     %e           + 1
```

The latter is a general observation and a major problem: the user and MACSYMA frequently will have different ideas about the simplest form of an expression. If one insists on a particular representation and everything else fails, the last resort always is to do the simplification manually and reenter the result. One may even verify the manipulated result by calculating appropriate differences with MACSYMA.

STEADY-STATE ANALYSIS OF S-SYSTEMS

Several functions have been developed to support the analysis of S-systems. Yet this can be considered only a first step toward automatic analysis. Using our set of functions effectively is a highly interactive process and requires some familiarity with MACSYMA, which is in part due to the very restricted possibilities of programming a comfortable user interface with MACSYMA. The method of choice, which combines the power of symbolic programming with the natural intelligence and resourcefulness of the user, seems to be interactive usage of MACSYMA. In this way we have been able to reproduce for example the analyses of a simple feedback system (cf. [2]) and of Kemper's [107] model for the dynamics of endemic infections [39]. Tables 6.1 and 6.2 give an overview of our self-defined, S-system specific functions and variables.

The starting point for an analysis with the current set of functions always is an S-system entered as an ordered list of right-hand sides of the differential equations. This input could probably be improved, since it is very difficult for MACSYMA to extract S-system parameters like rate constants (α's or β's) or exponents from an algebraic expression. This difficulty characterizes the second major type of problems that could be solved by humans at a glance but sometimes are very hard to solve with MACSYMA: problems involving "seeing." For example, a function for examining the existence of all possible zero steady states could readily be written if MACSYMA would be able to "see" whether a differential equation contains both a production term and a degradation term. Moreover, the two functions *VarName* and *VarIndex* had to be written using internal structures and LISP atoms, just because there was no other way for extracting the name and index of a subscripted variable. On the other hand, tasks like solving a parameterized system of linear equations are no particular problem for MACSYMA, but they might be very time consuming if done manually. The apparently different and complementary skills of the human user and MACSYMA again suggest interactive usage of the package.

The process of analyzing an S-system will be illustrated with a didactic example. Let's consider the S-system Eq. (6.3) representing the simple substrate cycle shown in Fig. 6.1:

$$\begin{aligned}
\dot{X}_1 &= \alpha_1 X_0^{g_{10}} X_2 - \beta_1 X_1 X_2 \\
\dot{X}_2 &= \alpha_2 X_3 - \beta_2 X_2 \\
\dot{X}_3 &= \beta_1 X_1 X_2 - \alpha_2 X_3
\end{aligned} \tag{6.3}$$

As the first step, the system is entered in (c11) as a list of right-hand sides of the differential equations, using a for α and b for β. The result (d11) is rather difficult to survey. The list elements are separated by commas.

```
(c11) sys1: [ a[1] * x[0]^g[10] * x[2] - b[1] * x[1]
            * x[2], a[2] * x[3] - b[2] * x[2],
            b[1] * x[1] * x[2] - a[2] * x[3] ];
```

Table 6.1. Overview of User Functions

Function	Description
ana (*sys*)	steady-state analysis of the S-system *sys*.
calc_flux (*i*, *alphas*, *gmat*, *nvars*)	calculate F-factor (relative flux) *i* at the steady state. Used by function *stabmat2*.
convert_log (*symb*)	convert a symbol of the original system to its counterpart in the logarithmic system (see Table 6.3).
convert_norm (*symb*)	reverse of function *convert_log*.
expo (*sys*)	exponentiate a system *sys* of equations.
get_dv (*vars*, *nvars*, *coef*, *augcoef*, *crank*, *arank*)	user interface to select dependent variables. Includes suggestion. Used by function *ana*.
log_gains (*i*)	calculate logarithmic gains for independent variable x_i.
reconvert (*log_equ*)	resubstitute all original symbol names in *log_equ*.
routh (*matr*)	apply Routh's procedure to a Jacobian *matr* to determine stability of the steady state.
routh2 (*list1*)	apply Routh's procedure to a list of coefficients of the characteristic polynomial of a matrix.
routhphi (*matr*)	calculate a list of the normalized coefficients of the characteristic polynomial of a matrix *matr*.
sensiab (*param*)	calculate parameter sensitivities for *a*'s or *b*'s.
sensigh (*param*)	calculate parameter sensitivities for exponents (*g*, *h*).
solution (*logsys*, *ydepvars*, *varlist*, *arank*, *crank*)	calculate steady state and constraints. Called by *ana*.
stabmat (*sys*)	calculate the Jacobian for relative changes as a function of relative fluxes, independent of the steady state.
stabmat2 (*sys*)	like function *stabmat*; in addition, the relative fluxes at the precalculated steady state are used.
VarIndex (*var*)	returns the index of the subscripted variable *var*.
VarName (*var*)	returns the name of the subscripted variable *var*.

Table 6.2. Overview of Variables Set by User Functions

alphas	list of *a*-values Extracted by functions *stabmat* or *stabmat2*.
betas	list of *b*-values. Extracted by functions *stabmat* or *stabmat2*.
gmat	matrix of exponents g_{ij}. Extracted by functions *stabmat* or *stabmat2*.
hmat	matrix of exponents h_{ij}. Extracted by functions *stabmat* or *stabmat2*.
constraints	list of constraints necessary for the existence of a steady state. Set by *solution*.
csys	original system after substitution of the *constraints*. Set be function *ana*.
steady_state	steady state calculated by *solution*.
steady_rhs	list of right-hand sides of the steady state equations. Set by function *ana*.

$$\text{(d11)} \quad [x_0^{g_{10}} a_1 x_2 - b_1 x_1 x_2,\ a_2 x_3 - b_2 x_2,\ b_1 x_1 x_2 - a_2 x_3]$$

To find the steady state we use the function *ana*. This function first tells us that we have three linear independent equations in logarithmic space but only two dependent system variables (the system is underdetermined). Thus we need to impose one constraint on the remaining parameters in order to make the system consistent and solvable for a unique non-zero steady state.

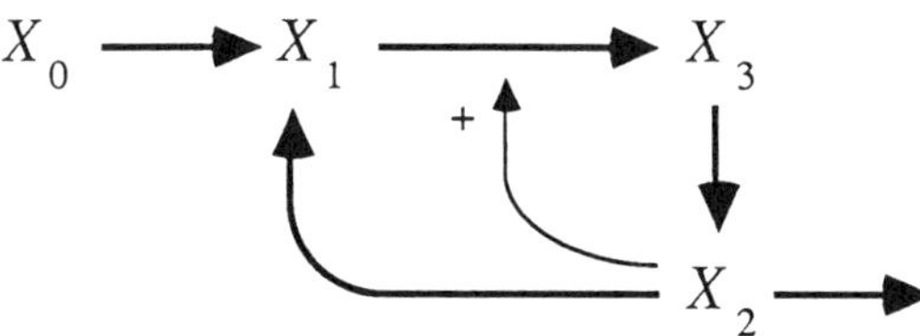

FIGURE 6.1.

```
(c12) ana(sys1);
Number of linear independent log-equations: 3
There is no unique solution for this system.
Number of dependent system variables: 2
Number of constraints required: 1
```

The user is asked to select the dependent variables and the parameters on which the constraints should be imposed. In both cases the program makes a suggestion for a possible choice and checks the user selection for feasibility. If a selection is not feasible, as the choice of parameter b_1 for a constraint in the example, the input will not be accepted. This very plain type of user interface is typical for MACSYMA.

```
Suggestion for dependent variables:  [x , x ]
                                       1   2

Enter []; to confirm  -  or  list of variables ;
[1,3];

Your selection:  [x , x ]
                   1   3

>>>> Selected dependent variables :  [x , x ]
                                       1   3

Looking for a suitable selection of constraints...

suggestion for constraints:  [x ]
                               0

required:  1  constraint(s) on  [x , a , b , a , b , g  ]
                                  0   1   1   2   2   10

Enter []; to confirm  -  or  list of constraints ;
[b[1]];

Your selection:  [b ]
                   1

No (unique) solution with this choice of constraints!

suggestion for constraints:  [x ]
                               0

required:  1  constraint(s) on  [x , a , b , a , b , g  ]
                                  0   1   1   2   2   10

Enter []; to confirm  -  or  list of constraints ;
[a[1]];

Your selection:  [a ]
                   1
```

When a selection is acceptable, MACSYMA calculates the steady state and the constraints and stores these results in some variables (*constraints, steady_state, steady_rhs, csys*) [cf. Table 6.2].

In our example, the result will appear after some seconds. An intuitive interpretation for the constraint in this example is that the production of X_1 must be equal to the degradation of X_2 at steady state. In other words, the total amount of substrate in the cycle needs to be constant.

```
Constraints:

                                          - g
                                             10
                  constraints = [a   = x         b ]
                                  1     0         2

Consistent System stored in CSys (for stability
calculations)!

RHS of Steady State stored in Steady_RHS.

   Steady State:

                       - 1              - 1
(d12)         [x   = b      b , x   = a      b  x ]
                1     1      2   3     2      2  2
```

The stability of a steady state generally can be examined by calculating the eigenvalues of the Jacobian $\boldsymbol{J}_0$ evaluated at steady state. It is stable if all eigenvalues have negative real parts and unstable if there is at least one eigenvalue with a positive real part. Routh's procedure is an elegant way to avoid the calculation of eigenvalues in stability analysis and thus it provides a practicable way to detect the existence of eigenvalues with positive real parts that disclose themselves in sign changes in a sequence (list) of values to be calculated. Adopting the notation of Savageau (cf. Chapter 2 and [2]), the usual input to Routh's procedure are the coefficients ϕ_i of the characteristic polynomial of the Jacobian, which can be calculated by the function *routhphi* and subsequently analyzed with the function *routh2*. The function *routh* combines these two functions.

In stability analyses of S-systems it is common practice to use the Jacobian $\boldsymbol{J}_{R0}$ for relative changes of the variates (cf. Chapter 2 and [2]). $\boldsymbol{J}_{R0}$ can readily be constructed from the exponents of the S-system and the F-factors, which represent relative fluxes at steady state. If no variate is zero at steady state, the validity of the simple relationship $\boldsymbol{J}_0 = \boldsymbol{D}\boldsymbol{J}_{R0}\boldsymbol{D}^{-1}$ can be shown, where $\boldsymbol{D} = diag(X_{i0})$ is the diagonal matrix of the steady state values X_{i0}. Thus $\boldsymbol{J}_0$ and $\boldsymbol{J}_{R0}$ are similar and have equivalent eigenvalues.

In continuation of our example we can now calculate the Jacobian $\boldsymbol{J}_{R0}$ by use of the functions *stabmat* or *stabmat2*. *Stabmat* yields a matrix independent of the actual steady state since the *F*-factors are not calculated (cf. c13, d13). *Stabmat2* evaluates the *F*-factors at the precalculated steady state (stored in *steady_rhs*) and substitutes them into the matrix (cf. c14, d14). Note that the consistent system (stored in *csys*) has to be used since the steady state is only valid under consideration of the constraints.

```
(c13) stabmat(csys);

                [ - flux       0          0     ]
                [       1                       ]
                [                               ]
(d13)           [    0      - flux      flux    ]
                [                 2         2   ]
                [                               ]
                [  flux       flux     - flux   ]
                [      3          3          3  ]

(c14) stabmat2(csys);

                   [ - b  x    0     0   ]
                   [    1  2             ]
                   [                     ]
(d14)              [    0     - b    b   ]
                   [             2    2  ]
                   [                     ]
                   [   a       a   - a   ]
                   [    2       2     2  ]
```

After applying Routh's procedure to the matrix we obtain a list of three normalized values which all need to be positive to guarantee a stable steady state since the first value is always 1 and is not included in the list.

```
(c15) routh(%);

(d15)     [b  x  + b  + a , (b  b  + b  a ) x , 0]
            1  2    2    2    1  2    1  2   2
```

In this simple example one can verify the signs of the values at a glance, knowing that all parameters and variates are positive. It is shown in (c16)–(d17) that MACSYMA also would be able to do this.

```
(c16) assume_pos:true;

(d16)                                   true

(c17) map( sign, d15);

(d17)                             [pos, pos, zero]
```

Likewise one could examine the stability by calculating the eigenvalues directly since this system has only three equations. The output (d18) shows us the eigenvalues and their multiplicities. However, applying Routh's procedure takes significantly less time than calculating eigenvalues.

```
(c18) eigenvalues(d14);

(d18)        [[- b  x , - b  - a , 0], [1, 1, 1]]
                  1  2     2    2
```

Both the result from Routh's procedure and the eigenvalues show that there is no eigenvalue with a positive real part but there is one zero eigenvalue. Thus, according to nonlinear systems theory (cf. [105]), the information from the linearization (Jacobian) is not sufficient to decide about the stability of this system at steady state. Nevertheless, we find our system to be marginally stable at the calculated continuum of steady states since there are only two dependent variables and thus only two independent equations in the consistent system *csys*. The linear dependency

$$\dot{X}_1 + \dot{X}_2 + \dot{X}_3 = 0$$

in the system equations of the consistent system is responsible for the zero eigenvalue and for the one-dimensional manifold of steady states. Moreover, the system is stable in the remaining two dimensions due to the two negative eigenvalues.

For sensitivity analysis and calculation of logarithmic gains let us consider a slightly more complicated example. We examine the simple feedback system (6.4) depicted in Fig. 6.2, which has been analyzed in some detail by Savageau [2: Chap. 9].

$$\begin{aligned} \dot{X}_1 &= \alpha_1 X_0^{g_{10}} X_3^{g_{13}} - \beta_1 X_1^{h_{11}} \\ \dot{X}_2 &= \beta_1 X_1^{h_{11}} \qquad\quad - \beta_2 X_2^{h_{22}} \\ \dot{X}_3 &= \beta_2 X_2^{h_{22}} \qquad\quad - \beta_3 X_3^{h_{33}} \end{aligned} \tag{6.4}$$

Again, the first step is to enter the system and to determine the steady state (Fig. 6.3). In this case we do not need constraints and the steady state is unique.

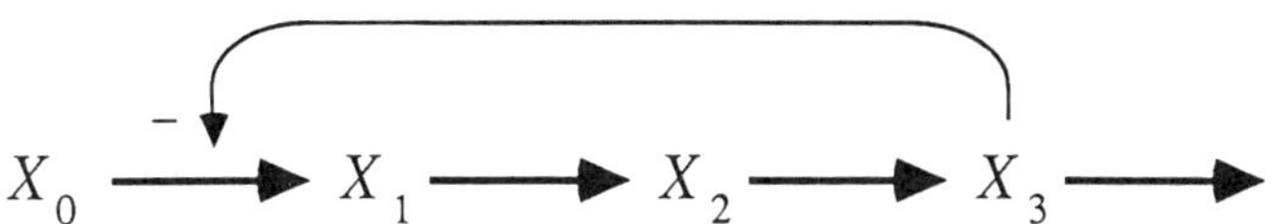

FIGURE 6.2. A Simple Feedback Inhibition System.

```
(c24) ana(feed);

Number of linear independent log-equations:  3

   Constraints:

                                    constraints = []

Consistent System stored in CSys
                  (for stability calculations)!

RHS of Steady State stored in Steady_RHS.

   Steady State:
```

$$(d24)\quad [x_1 = \frac{x_0^{\frac{g_{10}\,h_{33}}{h_{11}\,h_{33} - h_{11}\,g_{13}}}\; a_1^{\frac{h_{33}}{h_{11}\,h_{33} - h_{11}\,g_{13}}}}{b_1^{\frac{1}{h_{11}}}\; b_3^{\frac{g_{13}}{h_{11}\,h_{33} - h_{11}\,g_{13}}}},$$

FIGURE 6.3. First Variable of Steady State of System (4).

Now we can calculate logarithmic gains for changes in the independent variable X_0 and parameter sensitivities (cf. Fig. 6.4). The results correspond to those of Savageau. The sensitivities for the variates X_1 and X_2 and parameter h_{33} (d27 in Fig. 6.4) suggest that results may be quite complex even for relatively simple systems.

While the steady-state values and some parameter sensitivities for this system already are quite involved, the result after applying Routh's procedure fills twelve pages of printout. A small piece is shown in Fig. 6.5. Terms of this complexity usually are beyond the user's insight and appreciation. They can only be used as a basis for further automatic analysis rather than for manual calculations or simplifications. As previously shown in (c17), MACSYMA can help to interpret the results by examination of the signs of the terms after making assumptions about the parameters. Using this approach we were able to determine the signs of two of the three values, both of which turned out to be positive for the feedback inhibition system ($g_{13} < 0$). The sign of the third value however could not be determined in general. This is not surprising since the stability of the steady state might change with changing parameter values at certain bi-

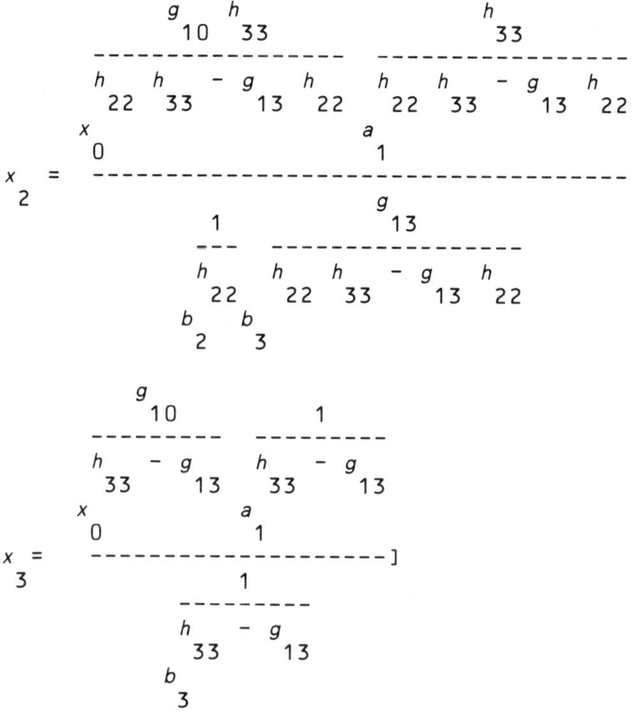

$$x_2 = \frac{x_0^{\frac{g_{10} h_{33}}{h_{22} h_{33} - g_{13} h_{22}}} \, a_1^{\frac{h_{33}}{h_{22} h_{33} - g_{13} h_{22}}}}{b_2^{\frac{1}{h_{22}}} \, b_3^{\frac{g_{13}}{h_{22} h_{33} - g_{13} h_{22}}}},$$

$$x_3 = \frac{x_0^{\frac{g_{10}}{h_{33} - g_{13}}} \, a_1^{\frac{1}{h_{33} - g_{13}}}}{b_3^{\frac{1}{h_{33} - g_{13}}}}]$$

FIGURE 6.3. (cont.) Variables x_2 and x_3 to Steady State of System (4).

furcation points. In these situations one can readily simplify the calculated general result by assuming numerical values for some of the parameters that generate a unique determination of signs. Familiarity with MACSYMA is of major importance during the post-processing phase.

Increasing the complexity in terms of larger numbers of equations or parameters overproportionally increases solution times. Execution times on our workstation are on the order of seconds for usual tasks, however, they increase to minutes when applying Routh's procedure to the Jacobian of the feedback system (6.4) and to hours if one tries to determine the eigenvalues of that matrix. The latter procedure actually failed after some time because our system could not provide enough storage space.

DEVELOPING A USER FUNCTION

To avoid typing the same sequence of commands for each analysis, one might soon want to write a user function. It is usually a good idea to develop the required sequence of commands interactively before combining them into a new

```
(c25) log_gains(0);
```

(d25)
$$\left[\frac{g_{10}\,h_{33}}{h_{11}\,(h_{33} - g_{13})},\ \frac{g_{10}\,h_{33}}{h_{22}\,(h_{33} - g_{13})},\ \frac{g_{10}}{h_{33} - g_{13}}\right]$$

```
(c26) sensiab(a[1]);
```

(d26)
$$\left[\frac{h_{33}}{h_{11}\,(h_{33} - g_{13})},\ \frac{h_{33}}{h_{22}\,(h_{33} - g_{13})},\ \frac{1}{h_{33} - g_{13}}\right]$$

```
(c27) sensigh(h[33]);
```

(d27)
$$[- g_{13}\,h_{33}\,(\log(b_3) - \log(a_1) - g_{10}\log(x_0))$$
$$/((h_{33} - g_{13})\,(g_{13}\log(b_3) + h_{33}\log(b_1)$$
$$- g_{13}\log(b_1) - h_{33}\log(a_1) - g_{10}\,h_{33}\log(x_0))),$$

$$- g_{13}\,h_{33}\,(\log(b_3) - \log(a_1) - g_{10}\log(x_0))$$
$$/((h_{33} - g_{13})\,(g_{13}\log(b_3) + h_{33}\log(b_2)$$
$$- g_{13}\log(b_2) - h_{33}\log(a_1) - g_{10}\,h_{33}\log(x_0))),$$

$$- \frac{h_{33}}{h_{33} - g_{13}}]$$

FIGURE 6.4. Logarithmic Gains and Some Parameter Sensitivities.

function. As an example, we will develop the function *sensiab* for calculation of the parameter sensitivities of *a*'s and *b*'s here. After that, we shall comment on further listings of some of our functions. Unfortunately, MACSYMA does not allow for comments when one enters the function unless the function is

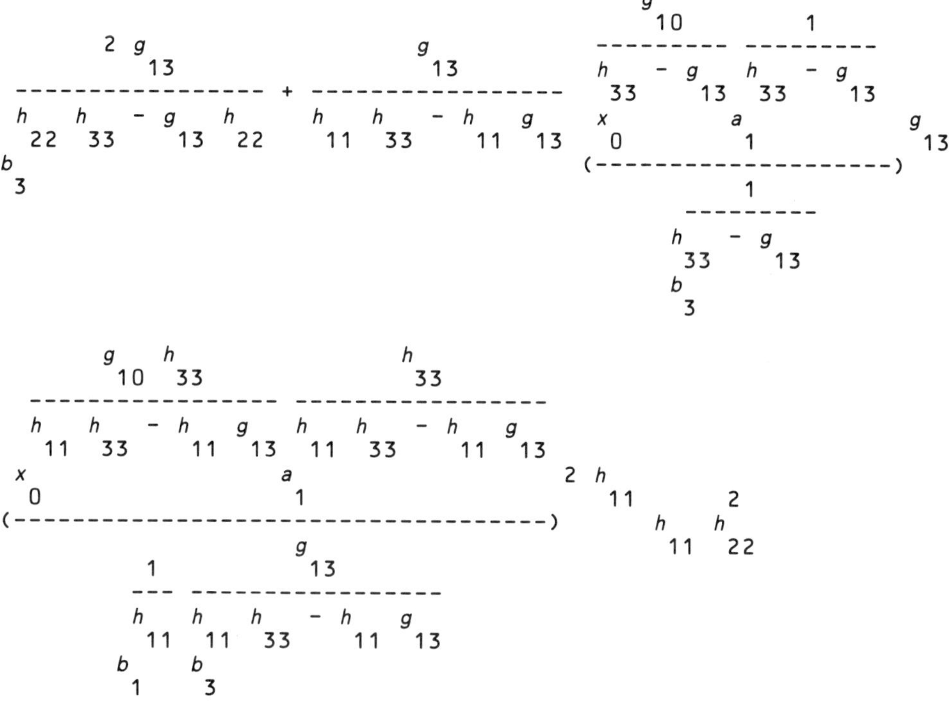

FIGURE 6.5. A Small Part of a Complex Output (see text for details).

batched from a text-edited file. We will therefore add line numbers to the listings where appropriate to allow for references to specific parts of the functions.

The purpose of our function *sensiab* is to calculate the parameter sensitivities $S(X_i, p)$ of all variables X_i with respect to changes in parameter p, which is an apparent rate constant (cf. Chapter 2 and [2: pp. 156–157]). These sensitivities are defined as

$$S(X_i, p) = \frac{\partial \log (X_i)}{\partial \log (p)} = \frac{\partial X_i}{\partial p} \frac{p}{X_i} \tag{6.5}$$

and have to be evaluated at steady state. The basis for the calculation is thus the steady state calculated by the function *ana*. It is stored in the variables *steady_state* and *steady_rhs* (cf. Table 6.2). Because the steady state is in symbolical form, the calculation of the parameter sensitivities is straightforward.

As an example we will calculate $S(X_3, a_1)$ for the feedback system (6.4) introduced in the previous section. Let us assume that the steady state (Fig. 6.3) has already been determined and the right-hand sides are stored in the variable *steady_rhs*, which is a list of three elements: $X_{1,0}$, $X_{2,0}$, $X_{3,0}$. With the command (c30) we extract the steady-state value $X_{3,0}$ from the list *steady_rhs*. With the next command we readily calculate the parameter sensitivity by applying (6.5). The built-in function *diff* is used for symbolic differentiation.

```
(c30) last(steady_RHS);

                        -1                 -1                    -1
       g   (h     - g  )       (h     - g  )      - (h     - g  )
        10   33      13           33     13           33      13
(d30) x                       a                  b
       0                       1                  3

(c31) diff( %, a[1] ) * a[1] / %;

                                       - 1
(d31)                   (h     - g  )
                           33     13
```

In the same way we may obtain the parameter sensitivity $S(X_3, b_3)$:

```
(c32) diff( d30, b[3]) * b[3] / d30;

                                       - 1
(d32)                 - (h     - g  )
                           33     13
```

In the next step we want to obtain the parameter sensitivities for all variables X_i. The result will be a list. The most appropriate iteration for this task is the FOR-element-IN-list-DO form, which applies a command sequentially to all elements in a given list. We need to calculate the parameter sensitivity for each element in the list *steady_rhs*. Since our result also has to be a list we will generate this list by starting from an empty list h and by appending the results successively to that list using the LISP-like function *endcons*. The command sequence for the parameter b_3 is:

```
(c33) h:[];

(d33)                    []

(c34) FOR term IN Steady_RHS DO

        h : endcons( diff(term, b[3]) * b[3] / term, h);

(d34)                         done
```

```
(c35) h;
                                          - 1
(d35) [- g    (h    h    - h    g   )        ,
          13    11   33     11   13

                                            - 1                       - 1
         - g    (h    h    - g    h   )        , - (h    - g   )         ]
            13    22   33     13   22                33     13
```

The result is stored in the global variable h, which is not the return value '*done*' (d34) of the iteration. To see the result we have to evaluate h in (c35). A more compact implementation of our function would employ the LISP-like function *map* to apply a function to each element of a list.

One might want to simplify the result (d35) by factorization:

```
(c36) factor(%);

          - 1                    - 1
(d36) [- h      g    (h    - g   )     ,
          11     13    33     13

                 - 1                 - 1                      - 1
         - g    h      (h    - g   )      , - (h    - g   )      ]
            13   22      33     13              33     13
```

With the latter result (d36), we have reached our goal. We can now combine the necessary steps in the new user function *sensiab*, which has the S-system parameter *param* as a parameter. The *block*-command is necessary to introduce the local variable h. Successive commands are separated by commas.

```
(c37)

sensiab(param) := block([h], h:[],

FOR term IN Steady_RHS DO

 h : endcons(diff(term, param) * param / term, h),

factor(h)

)$
```

To test this function we again calculate the parameter sensitivities $S(X_i, b_3)$, employing the new function *sensiab*. The result (d38) coincides with the previous result (d36). A different representation (d40) can be obtained by changing the option *exptdispflag*.

```
(c38) sensiab(b[3]);

               - 1                  - 1
(d38) [- h        g     (h     - g   )    ,
          11       13     33      13

                      - 1                 - 1                       - 1
              - g     h       (h     - g   )    , - (h     - g   )     ]
                 13    22       33      13              33      13

(c39) exptdispflag:true;

(d39)                             true

(c40) d38;

                 g                       g
                  13                      13                    1
(d40) [- ---------------, - ---------------, - ---------]
         h    (h    - g  )    h    (h    - g  )    h    - g
          11    33     13      22    33     13      33     13

(c41) sensiab(a[1])$
```

Now all other parameter sensitivities can be calculated conveniently. The result of (c41) is shown in Fig. 6.4. Functions like *sensigh* or *log_gains* are similar.

Our next example is the function *routhphi*, which calculates the coefficients ϕ_i of the characteristic polynomial of matrix *matr*.

```
1   Routhphi(matr)   :=   block([cp, res],
2        cp:   ratsimp(expand(charpoly(matr, lambda000))),
3       res:   [],
4       for i from 0 thru length(matr) do
5          res:   cons(ratcoef(cp, lambda000, i), res),
6       res/ first(res) )
```

In the first line we declare the local variables *cp* for the characteristic polynomial and *res* for the result. The characteristic polynomial is calculated in line 2 using the dummy variable *lambda000*. The functions *expand* and *ratsimp* are applied to the result to guarantee a simplified standard form of the polynomial. The coefficients of $lambda000^i$ are then extracted (line 5) starting with $i = 0$ and ending with the highest power of *lambda000*, which is equal to the size of the square matrix *matr*. A list of these coefficients is generated. Finally (line 6), the coefficients are normalized by division to guarantee a positive first coefficient.

The function *ana* is listed in Fig. 6.6 as a more complex example. It is reasonable to use an editor to develop functions of this size. Note that the function *ana* uses two user functions that are even larger: *get_dv* (line 28) to select the

```
1   ana(sys):=block([XVars,YVars, NVars,Coef,AugCoef,CRank,ARank,
2                        XDepVars, YDepVars, NormSys, LogSys, VarList ],
3   logexpand:all,
4   dynamalloc: print,
5   ratsimpexpons: true,
6   NVars: length(sys),
7   XVars:makelist(x[i],i,1,NVars),
8   YVars:makelist(y[i],i,1,NVars),
9   VarList:listofvars(sys),
10  NormSys:map(lambda([x],
11          IF length(x) = 2 THEN log(first(x))-log(-rest(x))
12              ELSE (IF is(x = 0) THEN 0
13                      ELSE return("No steady state."))),sys),
14  LogSys: NormSys,
15  for symbol1 in VarList do
16    block([SName, SIndex], SName : VarName(symbol1),
17       IF not is(SName = false) THEN
18       ( SIndex : VarIndex(symbol1),
19         IF is(Sname = x)
20         THEN LogSys: subst(y[SIndex], log(x[SIndex]), LogSys)
21         ELSE IF is(SName = a)
22         THEN LogSys: subst(la[SIndex], log(a[SIndex]), LogSys)
23         ELSE IF is(SName = b)
24         THEN LogSys: subst(lb[SIndex], log(b[SIndex]), LogSys) )
25       ),
26  AugCoef:augcoefmatrix(LogSys,YVars),ARank:rank(AugCoef),
27  Coef:submatrix(AugCoef, NVars+1),CRank:rank(Coef),
28  XDepVars : get_dv(XVars, NVars, Coef, AugCoef, CRank, ARank),
29  YDepVars : makelist(y[varindex(i)], i, XDepVars),
30  VarList: sort(setdifference(VarList, XVars)),
31  solution(LogSys, YDepVars, VarList, ARank, CRank),
32
33  print("Consistent System stored in CSys
                                (for stability calculations)!"),
34  CSys: radcan(lratsubst(constraints, sys)),
35
36  print("RHS of Steady State stored in Steady_RHS."),
37  VarList: setdifference(XVars, XDepVars),
38  Steady_RHS: Steady_State,
39  FOR elem in VarList DO Steady_RHS: cons(elem = elem, Steady_RHS),
40  Steady_RHS: sort(Steady_RHS),
41  Steady_RHS: map(RHS, Steady_RHS),
42
43  print(" Steady State: "),
44  Steady_State
45  )
```

FIGURE 6.6. Function *ana*.

dependent system variables and *solution* (line 31) to solve for the steady state and the constraints.

Some of the commands of the function *ana* are self-explanatory and we only sketch its key features. Some system options are set in lines 3 to 5. It is reasonable to start with these options since *ana* will usually be the first function used in the analysis of an *S*-system and the options will thereafter remain valid for the entire session. *NVars* is set to the number of equations; *XVars* and *YVars* are lists of the variates X_i and Y_i, respectively.

Table 6.3. Correspondence of Symbols

Logarithmic System	Original System	Comment
y_i	$\log(x_i)$	
la_i	$\log(a_i)$	a_i is used for α_i
lb_i	$\log(b_i)$	b_i is used for β_i

VarList is set in line 9 to the list of all variables and parameters occurring in the system *sys*. In lines 10 to 13 the logarithms of the production terms are set equal to the logarithms of the degradation terms. In this system of equations some logarithmic variables and parameters are then substituted by other symbols in lines 15 to 25, according to Table 6.3. To determine linear dependencies the rank *CRank* of the coefficient matrix *Coef* and the rank *ARank* of the augmented coefficient matrix *AugCoef* are determined in lines 26 and 27. After a valid selection of dependent system variables has been made in *get_dv*, the system will be solved by the function *solution*. The values of the variables *csys* and *steady_rhs* are then determined based on the variables *constraints* and *steady_state*, which have previously been set as side effects of the function *solution*. Finally the function *ana* returns the calculated steady state (line 44).

We conclude our excursion into the development of user functions with a few remarks on writing own functions in LISP, or in MACSYMA using LISP atoms. Symbolics discourages writing functions at the LISP level for several reasons (cf. [110: sec. 4.5]). Most important, it is necessary to depend on internal structures of MACSYMA when one writes functions in LISP. Internal structures are not documented and they may be changed in subsequent releases of MACSYMA. However, we have not found a way of implementing the two very basic functions, *VarName* and *VarIndex*, shown in Fig. 6.7, without depending on internal structures.

```
(c60) VarName ( var ) := IF ?listp(var) AND ?listp( ?car(var) )
                            THEN ?caar(var)
                            ELSE [] $

(c61) varname(x[2]);

(d61)                                x

(c62) VarIndex(var) := IF ?listp(var) AND NOT ?null( ?cdr(var) )
                          THEN ?cadr(var)
                          ELSE [] $

(c63) VarIndex(x[2]);

(d63)                                2
```

FIGURE 6.7. Functions Employing LISP Atoms.

DISCUSSION

We have examined the usefulness of the computer algebra program MACSYMA for the symbolic steady-state analysis of S-systems. Although MACSYMA has proven to be useful we have encountered several problems:

- the "simplest" or desired form of an expression is frequently not found;
- tasks a human can solve by inspection are often difficult to solve with MACSYMA;
- MACSYMA offers no provisions to program an easy and convenient user interface;
- MACSYMA is significantly more complex than other programming languages;
- some functions can only be implemented at the LISP level;
- even simple S-systems may produce an output of several pages that is usually not interpretable by the user without further support of MACSYMA. However, such analyses would probably never be made without the support of computer algebra programs.

Despite these problems, we found that symbolic mathematics provides valuable support in many respects. In particular, the results can be expected to be free of computational errors and they are obtained in a fraction of the time one would need to compute them manually.

Our work suggests that the most effective use of a computer algebra program is the interactive employment of both built-in functions and user functions. This interactive mode allows the user to combine his expertise with the facilities provided by the program. In doing this, Lewis (Chapter 16) has been successful in determining criteria for the birth of limit cycles through Hopf bifurcations in general S-systems with two equations.

This chapter has alluded to the very basic options of S-system analysis with MACSYMA. Further work is needed to improve and examine the possibilities of a fully automated analysis of S-systems and to create functions for the automatic S-system recasting of nonlinear systems (for theory and applications of S-system recasting see Chapters 12–14).

SUMMARY

The chapter examines the usefulness of the computer algebra program MACSYMA for the analysis of S-systems. It provides an introduction to symbolic computing and it shows how MACSYMA can be used for evaluating steady-state features of S-systems. It points out the advantages and the current problems of employing computer algebra for the automatization of symbolic analyses. The desired form of an expression is often not found and some tasks a human can solve at a glance are difficult to solve with MACSYMA. On the other hand, computer algebra can save significant amounts of time and the results can be expected to be free of computational errors. Presently, the most effective method of working with MACSYMA involves an interactive mode in which the user determines the necessary steps of an analysis and lets MACSYMA execute them.

ACKNOWLEDGMENTS

The author would like to express his gratitude to Dr. Eberhard O. Voit for his encouragement and support in the course of this project and for his many helpful comments that contributed to the improvement of this chapter. The work at MUSC during the academic year 1988/89 was made possible by a Fulbright grant.

7. Analysis of Complex Dynamic Networks with ESSYNS

A Program for the Efficient Solution of Nonlinear Differential Equations in S-System Form

DOUGLAS H. IRVINE, EBERHARD O. VOIT, AND MICHAEL A. SAVAGEAU

In the analysis of complex dynamic networks, such as those constituting living systems, information about kinetic parameters for the intact system usually is unavailable, and measuring a complete set of parameters can require a prohibitive amount of time and resources. Even when detailed and accurate information is available, it still can be difficult to make valid generalizations. Only special cases may be represented, and the analytical results may change qualitatively with small errors in the data or with natural fluctuations that alter the response of the system.

Ideally, an analysis of nonlinear dynamic networks can be accomplished with general, symbolic solutions so that valid conclusions can be drawn without confusion from spurious results or from nonrepresentative special cases. The ability to make valid generalizations about nonlinear processes is a major advantage of S-systems (e.g., see [2], [8–9], [21]). Steady-state solutions, conditions for local asymptotic stability, logarithmic gains, and complete sensitivity profiles all can be obtained symbolically using this structured system of nonlinear ordinary differential equations. Often, little or no information about the magnitudes of parameters is required to reveal the functional significance of a particular network interaction (e.g., see [2], [25–26]). The steady-state response and function of nonlinear processes, independent of parameter values, are important features of S-systems attributable to their unique, regular structure.

The most general symbolic solutions for the dynamics of S-systems [24] are for relatively simple sets of one or two differential equations. Although more complex sets may be solved symbolically, this cannot be done with most S-system of three or more equations, and a different approach based on numerical solution is required. An advantage of numerical solution is that it can be applied to networks of virtually any complexity, and each solution is easy to visualize graphically. A disadvantage is that parameter values must be specified; this means that valid generalizations about dynamics remain a challenge.

For complex nonlinear networks, the need to specify parameter values for dynamic solutions appears inescapable, even with S-systems; on the other hand, relying on a narrow set of parameter values is insufficient for general conclusions. Characterizing a more complete repertoire of responses for a network involves numerous solutions with an extended range of parameter values. For a system with p independent parameters, examining each response for n different values of each parameter takes n^p solutions. For example, 59,049 solutions are required to obtain all responses using three different values of ten parameters, an average number for an S-system with three equations. Obviously, making general conclusions about even moderately complex dynamic nonlinear systems demands a robust and efficient numerical method, coupled with utilities to automate solutions over a range of parameter values and utilities to summarize critical features of large sets of data. The successful development of more robust and efficient methods for numerical solution of S-systems has been motivated by a desire to make general conclusions about the response and function of complex dynamic networks. This chapter gives an example of how the regular structure of S-systems makes the numerical methods in ESSYNS more effective [38]. It also provides an overview of complementary analytical methods and support utilities provided in the program [7].

UTILIZING THE REGULAR STRUCTURE OF S-SYSTEMS

Conventional Runge-Kutta and Adams methods approximate local Taylor-series solutions for differential equations. Approximations are used because formulas for higher derivatives change in a problem-dependent manner, and these formulas can be difficult to express. Even when formulas for higher derivatives are available, evaluation for all but the simplest nonlinear ordinary differential equations can be cumbersome and inefficient for solutions by Taylor series.

On the other hand, all S-systems are represented in the following canonical form:

$$\frac{dX_i}{dt} = X_i^{(1)} = \alpha_i \prod_{j=1}^{n} X_j^{g_{ij}} - \beta_i \prod_{j=1}^{n} X_j^{h_{ij}} \qquad i = 1, \ldots, n$$

with n equal to the number of differential equations. Because the basic formulas for higher derivatives are problem-independent, the following recurrence formulas can be used to generate a high-order Taylor-series solution for any S-system in logarithmic coordinates (i.e., $Y_i = \log X_i$):

$$\frac{1}{(m-1)!} \frac{d^m Y_i}{dt^m} = \tilde{Y}_i^{(m)}$$

$$\tilde{Y}_i^{(m)} = \tilde{A}_i^{(m)} - \tilde{B}_i^{(m)}$$

with

$$\tilde{A}_i^{(1)} = \alpha_i \exp\left(\sum_{j=1}^{n} g'_{ij} Y_j\right)$$

$$\tilde{B}_i^{(1)} = \beta_i \exp\left(\sum_{j=1}^{n} h'_{ij} Y_j\right)$$

and for m greater than one

$$\tilde{A}_i^{(m)} = \frac{1}{m-1}\left[\sum_{q=1}^{m-1} \tilde{A}_i^{(m-q)} \tilde{G}_i^{(q)}\right]$$

$$\tilde{B}_i^{(m)} = \frac{1}{m-1}\left[\sum_{q=1}^{m-1} \tilde{B}_i^{(m-q)} \tilde{H}_i^{(q)}\right]$$

$$\tilde{G}_i^{(q)} = \sum_{j=1}^{n} g'_{ij} \tilde{Y}_j^{(q)}$$

$$\tilde{H}_i^{(q)} = \sum_{j=1}^{n} h'_{ij} \tilde{Y}_j^{(q)}$$

where $g'_{ij} = g_{ij} - \delta_{ij}$, $h'_{ij} = h_{ij} - \delta_{ij}$, $\delta_{ii} = 1$, and $\delta_{ij} = 0$ for i not equal j; the tildes are used for consistency with the original notation in [38]. The unique, regular structure of S-systems makes these Taylor coefficients efficient to evaluate, and hence, numerical solution of S-systems by Taylor series has several advantages. Extensive tests [38] show that the variable-order, variable-step-length Taylor-series method in ESSYNS is one-to-two orders of magnitude faster and also is more accurate and more robust than conventional Runge-Kutta and Adams methods. These advantages are attributable to the method's effective use of the regular structure of S-systems. Additional advantages may be realized by capitalizing even more on this structure.

CORE COMMANDS FOR EVALUATION AND SIMULATION OF SYNERGISTIC SYSTEMS

A typical interactive session within ESSYNS involves a definition of parameter values for an S-system, a numerical solution, and a graphical display of results. The user interfaces for these options and other support utilities are outlined below.

Definition of S-System Parameters

The first step in the analysis of S-systems is to define the differential equations by specifying parameter values. When the *Define* command is selected, ESSYNS prompts for the title of the S-system, for the number of dependent variables (i.e., the number of differential equations), for the value of each α_i and β_i, for the value of each g_{ij} and h_{ij}, for the number of independent variables, and for the value of each g_{ij} and h_{ij} associated with an independent variable.

The following S-system represents an immune network with suppressor regulation [25–26]:

$$X_1^{(1)} = 8X_1 - 8X_1X_2^{0.5}$$

$$X_2^{(1)} = X_1^{0.5}\,X_3^{-0.5} - X_2$$

$$X_3^{(1)} = X_2^{0.5} - X_3$$

As an illustration of the *Define* command, the parameter values for the above S-system can be entered in ESSYNS as follows:

```
Command >d

Define a new system

Enter a title for the system, Q to quit, or <CR> for
>Title of Previous System
>Immune Network with Suppression
```

```
Enter the number (1–25) of dependent variables, Q to quit, or ⟨CR⟩ for
8
> 3

Enter alpha 1 to 3, Q to quit, or ⟨CR⟩ for all ones.
> 8 1 1

Enter beta 1 to 3, Q to quit, or ⟨CR⟩ for all ones.
> 8 1 1

Enter the 3 × 3 G matrix one row at a time.
Press ⟨CR⟩ for a row of all zeros, or Q to quit.
> 1 0 0
> .5 0 −.5
> 0 .5 0

Enter the 3 × 3 H matrix one row at a time.
Press ⟨CR⟩ for a row of all zeros, or Q to quit.
> 1 .5 0
> 0 1 0
> 0 0 1

Enter the number (0–22) of independent variables, Q to quit, or ⟨CR⟩
for 0
> 0
```

This information completely specifies the S-system to be used in the subsequent analysis. The parameter values can be displayed with the *View* command, or they can be changed with the *Modify* command. The S-system can be saved in a file with the *Write* command and can be reactivated later with the *Load* command. Steady-state values, eigenvalues, logarithmic gains, and complete sensitivity profiles can be calculated with the *Steady-State* command. Each of these support utilities is selected with a single-letter command, and prompts are given for all additional input.

Numerical Solution

The next step in the analysis of dynamics is to solve the differential equations numerically by running the current S-system. When the *Run* command is selected, ESSYNS prompts for the title of the run, for the timing information (i.e., the initial time, final time, and report interval), for the initial value of each dependent variable (i.e., the first point in each solution), and for the value of each independent variable. Then the Taylor-series method is applied to solve the S-system, and the solution is stored in memory. A typical run with a moderately complex, nonstiff S-system requires only a few seconds on a personal computer.

As an illustration of the *Run* command, the S-system defined in the preceding section can be solved numerically by providing the following information:

```
Command > r

Run current system

Enter a title for the data, Q to quit, or ⟨CR⟩ for
> Title of Previous Run
> Fourfold Step of Immune Network

Enter the initial time, final time, and increment, Q to quit, or ⟨CR⟩ for
> 0 10 .1
> 0 12 .12

Enter 3 initial values, Q to quit, or ⟨CR⟩ for
> 1 1 1
> 1/1024  1/16  1/4

| 23 steps | 101 points | 100.0% complete | order 9 | 1.42 seconds |

Command >
```

Solutions can be saved in a file with the *Output* command and can be reactivated later with the *Input* command. Columns of data in a solution can be transformed with any valid combination of mathematical operations in the *X-Y Transformations* command.

Graphics Display

The next step in the analysis of dynamics is to visualize characteristic responses by displaying solutions graphically. The *Time Course* command displays all variables or any combination of variables in the current solution as a function of time. As an illustration of the *Time Course* command, a graph of the solution obtained with the *Run* command in the preceding section can be displayed by entering the following information:

```
Command > t

Time course display of current data
```

Enter P to plot data, L to list data, or Q to quit > **P**

Enter the number (1–3) of variables to plot, Q to quit, or ⟨CR⟩ for 3
> **3**

Enter 3 indexes, Q to quit, or ⟨CR⟩ for
> 1 2 3
> **1 2 3**

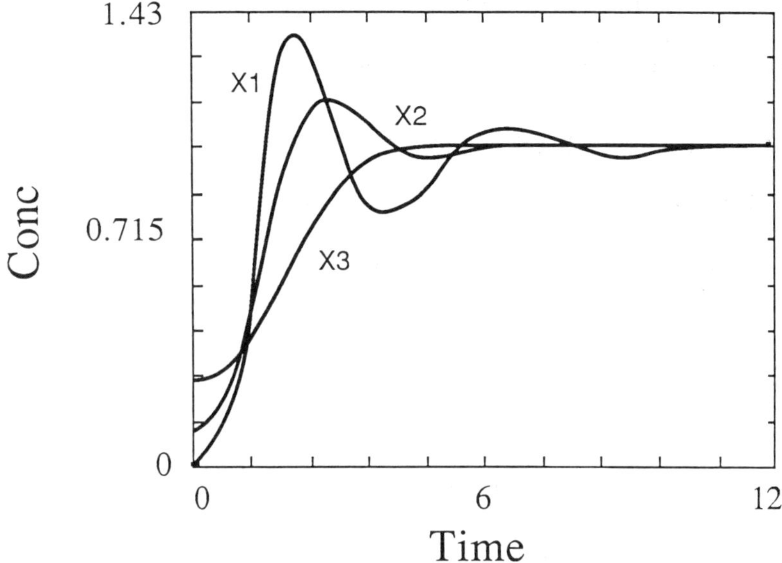

There are three additional commands for plotting graphs: the *Compare* command displays multiple time course plots on the same axis, the *Phase Plane* command displays plots as a function of variables other than time, and the *3-Dimensional* command displays plots in three dimensions. *Compare* plots, *Phase-Plane* plots, and *3-Dimensional* plots can contain data from one or more files. Appropriate scales for the axes are set automatically to include all points, or each scale can be set manually to magnify particular portions of a graph. Graphs can be printed on dot-matrix printers, or tables containing data points in a graph can be saved in a file for use with presentation-quality graphics programs.

Programmed Operation

Many analyses can bc accomplished in one interactive session, but others require numerous simulations that in combination can take a long time. To avoid unnecessary waiting for individual solutions, an entire session can be programmed in a macro command file, and the analysis can be completed without additional interactive input. A new command file for programmed operation can be created or an existing file can be edited with the *New Command File* option. The syntax for commands in a file is identical to that for commands in an interactive session. Programmed operation with a command file is activated by the *Use Command File* option. Data from a programmed analysis with a command file can be saved on a disk for subsequent viewing, or graphs can be printed automatically during the analysis. Also, summaries of particular statistics for each solution can be printed automatically. With a programmed operation, repetitious parts of an extensive analysis become not only less tedious but also less prone to inadvertent errors in input.

ADDITIONAL SOURCES OF INFORMATION

This chapter has reviewed briefly the motivation behind new, more robust and efficient numerical methods for the solution of S-systems and it has given one example of how the regular structure of S-systems makes these methods more effective. An overview of complementary analytical methods and support utilities in ESSYNS also was given. Details about the development, implementation, and testing of the variable-order, variable-step Taylor-series method in ESSYNS are available in [38], and complete descriptions of the analytical methods and support utilities are available in *The User's Guide to ESSYNS* [7].

ESSYNS runs on MS-DOS personal computers with 512 kilobytes or more random access memory (RAM) and it is available on two 360-kilobyte, 5.25-inch disks or one 720-kilobyte, 3.5-inch disk. A math coprocessor (80 × 87) is optional but it is used automatically by ESSYNS when it is available for faster computations and graphics. Extended memory (EMS) also is optional but it is used automatically by ESSYNS when it is available for faster execution of commands. Several graphic video standards are supported including Hercules, CGA, EGA, VGA, ATT, PC3270, and IBM8514; multicolor graphics are supported for EGA, VGA, and IBM8514 systems with color monitors. ESSYNS Version 2, and *The User's Guide to ESSYNS* are available upon request.

SUMMARY

The desire to make general conclusions about the response and function of complex dynamic networks has led to the development of more robust and efficient methods for the numerical solution of S-systems. Advantages over conventional methods like Runge-Kutta and Adams have been realized by developing a radically different approach, based on the direct calculation of Taylor series, to capitalize on the canonical structure of all differential equations expressed in the S-system form. A variable-order, variable-step Taylor-series method has been implemented in the program ESSYNS, which stands for *Evaluation and Simulation of SYNergistic Systems*. An example of how the regular structure of S-systems makes the numerical methods in ESSYNS more effective is given along with an overview of complementary analytical methods and support utilities provided in the program.

Part II

Applications

8. Completely Uncoupled or Perfectly Coupled Circuits for Inducible Gene Regulation

MICHAEL A. SAVAGEAU AND PETER J. SANDS

The molecular description of genetic regulatory mechanisms has advanced rapidly during the past two decades. There is now a plethora of designs and molecular components by which these designs are realized. The richness of molecular description has become an embarrassment because no comparable theory exists to give meaning to these various designs for gene regulation. At present there is no ability to predict the molecular design for a system that has yet to be characterized or to predict the functional implications of those that already have been described in molecular terms. In this fundamental sense, there is little understanding of genetic regulatory mechanisms.

To review the experimental developments in gene regulation we shall begin with the influential work of Jacob and Monod [122]. Although the first and best characterized system was the lactose operon in the enteric bacterium *Escherichia coli*, a number of other systems appear to fit the same model [127]. A schematic representation of this model is shown in Fig. 8.1a. For the purposes of this chapter we shall emphasize two of its features. First, the regulator in this system is a negative-acting protein that normally blocks expression of the adjacent structural genes in the absence of inducer. This is referred to as a negative mode of regulation. Second, the level of the regulator itself does not

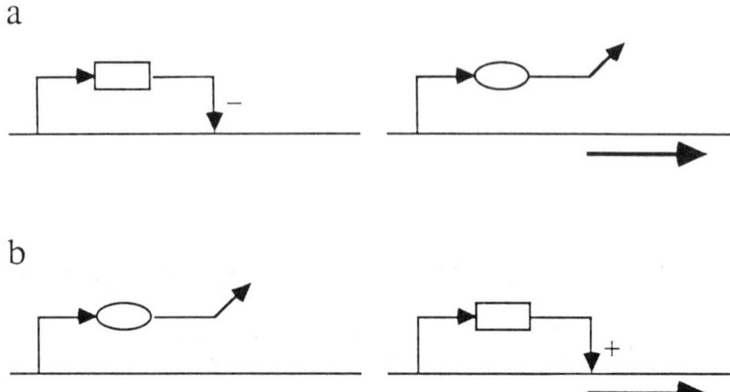

FIGURE 8.1. Two alternative modes of gene regulation. (a) Negative mode (e.g., the *lac* operon) and (b) positive mode (e.g., the *mal* operon).

vary during induction/deinduction of β-galactosidase activity. This is referred to as the classical circuit of gene regulation [13] (see Fig. 8.2a).

It was soon realized that regulatory proteins could be activators rather than repressors. The work of Englesberg and his colleagues [118] showed that the arabinose catabolic system in *E. coli* is regulated primarily by a positive-acting protein that normally requires inducer to activate transcription. Other systems also fit this model [117, 130]. A schematic model of this positive mode of regulation is shown in Fig. 8.1b.

Somewhat later it was found that the level of the regulatory protein was itself

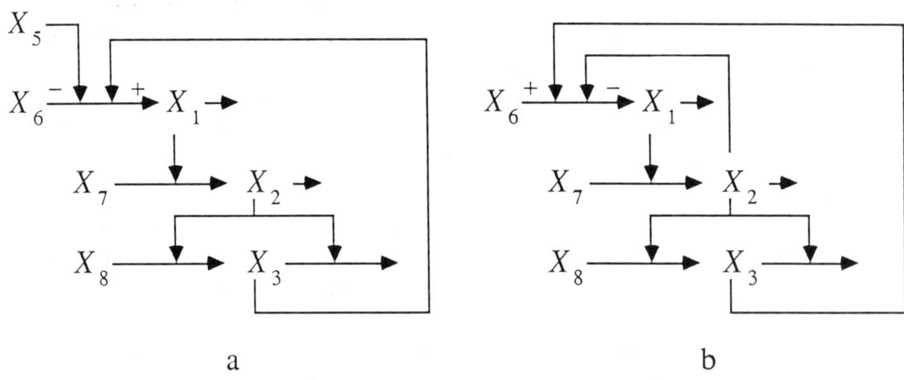

FIGURE 8.2. Classical vs. autogenous circuits of gene regulation. (a) Classical circuit in which the level of regulator (X_5) is invariant during the induction/deinduction of X_2. The expression of X_2 and X_5 is completely uncoupled. (b) Autogenous circuit in which the level of regulator ($X_5 = X_2$ in this case) varies coordinately with that of X_2 during induction/deinduction. The expression of X_2 and X_4 is perfectly coupled.

regulated in the case of the histidine catabolic system in *Salmonella typhimurium* [139]. The structural gene for the regulatory protein was shown to lie within a transcriptional unit that was itself subject to negative control by the repressor protein. This type of regulation, in which a regulatory protein is involved directly in regulating expression of its own structural gene, has been called autogenous regulation [119]. A schematic model of the autogenous circuit of regulation is shown in Fig. 8.2b. Thus, neither negative control by repressor protein nor the classical circuit of regulation that characterizes the lactose operon represents general features of gene regulation.

More recently it has been discovered that regulatory genes may be regulated but yet they have their structural genes located in transcriptional units separate from the other structural genes under their control [115–116, 126, 129, 131]. For example, the regulator of the D-serine deaminase system in *E. coli* is a positive-acting protein modulating the expression of the other regulated products (effectors) in the system. The regulator, which is located in a separate transcriptional unit, also acts as a negative element in the regulation of its own structural gene. Furthermore, during induction/deinduction the levels of regulator and effector proteins vary by different amounts [126]. This is in contrast to classical regulation, in which the regulator level does not change, and autogenous regulation in which the regulator and the effector proteins vary in a coordinate fashion. A schematic model of this type of coupled regulatory circuit is shown in Fig. 8.3.

There are a number of variations on the above themes, which represent a large number of systems that are well described at the molecular level. Although, as we have indicated above, there is no general theory that provides

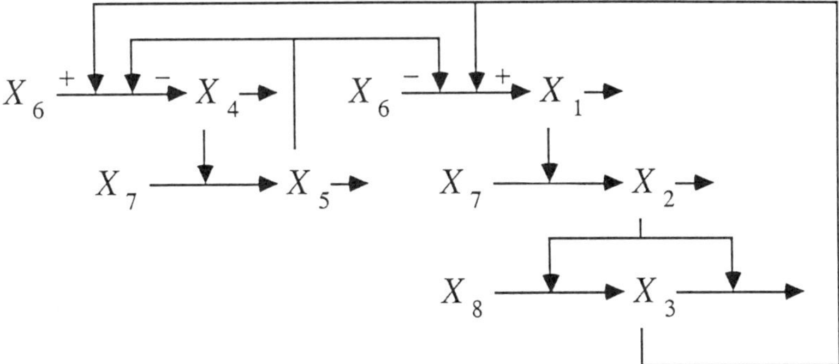

FIGURE 8.3. Coupled circuits of gene regulation. In this circuit the expression of X_2 and X_5 is coupled in a more general fashion that may be neither uncoupled nor perfectly coupled. However, under appropriate assumptions regarding the parameter values in this coupled circuit, one can represent the special cases of completely uncoupled and perfectly coupled circuits shown in Figure 8.2. See text for discussion.

fundamental understanding to these various designs, elements of such a theory are beginning to appear. The following is a brief review of one of these theoretical developments.

An understanding of classical vs. autogenous circuits of gene regulation has been developed from the comparative analysis of these alternative designs [2, 13, 17, 134]. According to these analyses, the autogenous circuit of regulation is superior to the classical circuit in inducible systems with a negative mode of regulation and in repressible systems with a positive mode of regulation. The converse is true in inducible systems with a positive mode of regulation and in repressible systems with a negative mode of regulation. On the basis of several criteria for functional effectiveness, this theory also predicts a negative mode of regulation for autonomous circuits of regulation. Autonomous circuits are special cases of autogenous circuits in which there are no regulatory modifiers other than the regulators themselves [2, 134].

These predictions and the experimental data for autonomous circuits are in agreement [134]. The predictions with regard to autogenous vs. classical regulation are supported in those cases for which the experimental data clearly indicate classical or autogenous circuits of regulation, the notable exception being the lactose operon [134]. However, as noted above, other systems have neither the classical nor the autogenous circuit of regulation. In these cases, the regulator is located in a distinct but coupled regulatory circuit that represents a new class of circuitry for which extension of the theory is required.

It is our goal to develop such a theory. In the next section we introduce a model of coupled regulatory circuits that includes classical and autogenous circuits as special cases. In the methods section we give criteria for functional effectiveness by which alternative designs can be compared, and we describe the method of mathematically controlled comparisons by which we shall analyze the relative merits of alternative designs for gene regulation. In this chapter, we focus on the special cases of complete uncoupling and perfect coupling. The general case of coupled gene circuits will be presented elsewhere.

MODEL

The model of a coupled regulatory circuit shown in Fig. 8.3 follows the conventions established elsewhere (e.g., see Chapter 2 and [2]), but we shall briefly outline some of the salient points. The horizontal arrows represent synthetic or degradative processes whereas the vertical arrows identify catalytic or modifier influences that affect the process in question. The X's represent chemical concentrations with the subscripts indicating the identity and/or location of the chemical species. The independent variables (X_6, X_7, X_8) are determined by the environment, which may include the remainder of the cellular metabolic network as well as the extracellular milieu, or they may be determined directly by the experimentalist during an experiment *in vitro*. The dependent variables (X_1 through X_5) have their values determined in turn by the values of independent variables. The variables X_6 and X_7, representing precursors for mRNA and protein synthesis, are themselves the products of regulated biosynthetic pathways and hence normally will have relatively invariant values. By contrast, X_8, representing the substrate of an inducible catabolic system, is subject to much larger variation under natural conditions.

There also are a number of processes and variables that are represented implicitly. Several "microscopic" processes occur much faster than the temporally dominant processes that are emphasized in Fig. 8.3. These "microscopic" processes determine the parameter values associated with the dominant processes (e.g., see [2:Chapter 15]), but they need not be considered explicitly. Similarly, a number of variables, such as the concentrations of RNA polymerase, ribosomes, associated factors and cofactors, are important elements in the dominant processes. However, for our purposes these concentrations may be considered relatively invariant or available in excess so that they need not be considered explicitly. Should it become desirable to include any of these processes or variables explicitly, this is readily accomplished within this formalism by adding the appropriate X's and arrows.

Equations

The interactions in such systems are, in general, highly nonlinear, characteristic of saturable, multiplicative, synergistic phenomena. As we have shown elsewhere, such phenomena can be appropriately represented as a "synergistic system" of equations [3–4, 19] based on the general power-law formalism developed earlier (cf. Chapters 1, 2, and [1, 9, 12]). In the synergistic-system representation, there is one equation for each dependent variable. Each equation consists of a difference between the net rate of synthesis and the net rate of degradation. These net rate terms in turn consist of a product of power-law functions, one for each reactant or modifier that affects the process in question.

For the coupled regulatory circuit in Fig. 8.3 the synergistic system is

$$\begin{aligned}
\dot{X}_1 &= \alpha_1 X_6^{g_{16}} X_3^{g_{13}} X_5^{g_{15}} - \beta_1 X_1^{h_{11}} \\
\dot{X}_2 &= \alpha_2 X_7^{g_{27}} X_1^{g_{21}} - \beta_2 X_2^{h_{22}} \\
\dot{X}_3 &= \alpha_3 X_8^{g_{38}} X_2^{g_{32}} - \beta_3 X_2^{h_{32}} X_3^{h_{33}} \\
\dot{X}_4 &= \alpha_4 X_6^{g_{46}} X_3^{g_{43}} X_5^{g_{45}} - \beta_4 X_4^{h_{44}} \\
\dot{X}_5 &= \alpha_5 X_7^{g_{57}} X_4^{g_{54}} - \beta_5 X_5^{h_{55}}
\end{aligned} \tag{8.1}$$

where $\dot{X}$ is the time derivative of X, α_i and β_i are rate constants, and the g_{ij} and h_{ij} are kinetic orders.

All the parameters associated with the model in Fig. 8.3 are positive except the regulatory parameters—g_{13}, g_{15}, g_{43} and g_{45}—which may be positive, negative, or zero. A positive value for a regulatory parameter indicates that an increase in the concentration of the corresponding regulatory protein (activator) or regulatory modifier (inducer) causes an increase in the rate of the affected process. A negative value indicates that an increase in the concentration of the corresponding regulatory protein (repressor) or regulatory modifier (corepressor) causes a decrease in the rate of the affected process.

Parameter Values

We shall base our numerical solutions on parameter values obtained for the *lac* operon of *E. coli* operating over the regulatable region of its induction characteristic [79]. There are two principal advantages of this choice: (1) it provides us with a relatively complete and accurate set of parameter values, and (2) it provides us with an appropriate context within which one might hope to account for the classical circuit of the *lac* system. This is one of the goals of our extended circuit theory.

1. h_{11} has been determined by examining the kinetic order for the degradation of radioactively labeled *lac* mRNA. The results show that this is a first-order process [113, 125]. Thus, $h_{11} = 1$.
2. β_1 was determined in these same studies to have a value of 0.347 min^{-1}.
3. h_{22} also represents a first-order process, the dilution of stable β-galactosidase molecules by exponential growth. Thus, $h_{22} = 1$.
4. β_2 for cells growing exponentially with a doubling time of 60 min. will have a value of $(\ln 2)/60 = 0.116\ min^{-1}$. This is simply the nonspecific effect of dilution. The contribution from specific degradation of β-galactosidase, which has a half-life of at least 8 hours and perhaps much greater [123, 132], is negligible.
5. $g_{21} = 1$ because translation in steady state is proportional to the amount of specific mRNA.
6. $g_{32} = h_{32} = 1$ because the activities of β-galactosidase are proportional to the amount of protein [120].
7. $h_{33} = 1$ for the following reasons. The K_m is smaller and the V_m is greater for the intermediate allolactose than for the substrate lactose [120]. Since both these reactants are catabolized by the same enzyme (X_2), the level of allolactose can be expected to be much lower than that of lactose during steady state. Even if lactose were present at saturating concentrations, allolactose would have a concentration of no more than its K_m value.

Hence, the kinetic order with respect to substrate for allolactose degradation will be near unity [9, 12].

8. $\beta_3 = 0.00223$ *cell* · *molecule*$^{-1}$ · *min*$^{-1}$. This is determined from V_m/K_m by using the values reported by [120], a molecular weight for β-galactosidase of 540,000 Da, and a cellular volume for *E. coli* that yields the conversion factor 10^{-7} *mM* = 1 *molecule* · *cell*$^{-1}$.
9. $g_{13} = 1$, as determined from filter-binding studies *in vitro* [114]. Although we cannot use this value for coupled circuits, for which the influence of alternative values for regulatory parameters is the object of study, we will have occasion to use this value when representing classical (completely uncoupled) circuits, such as the *lac* operon.

We have not found estimates for the half-lives of repressor mRNA and protein. We will assume these values are the same as those for β-galactosidase. This choice has the advantage that it will allow our model to account for the special case of an autogenous circuit, which also is a goal of our extended theory. When the concentration variables are normalized with respect to their basal steady-state values (see [2]), in particular, we note that the basal steady-state level of β-galactosidase is about 5 *molecules* · *cell*$^{-1}$ [122]. Eq. (8.1) then can be rewritten as

$$\begin{aligned}
\dot{X}_1 &= 0.347\,[X_3^{g_{13}} X_5^{g_{15}} - X_1] \\
\dot{X}_2 &= 0.0116\,[X_1 - X_2] \\
\dot{X}_3 &= 0.0112\,[X_8 X_2 - X_2 X_3] \\
\dot{X}_4 &= 0.347\,[X_3^{g_{43}} X_5^{g_{45}} - X_4] \\
\dot{X}_5 &= 0.0116\,[X_4 - X_5]
\end{aligned} \tag{8.2}$$

The value of g_{38} has been given a value of unity for convenience. Its true value is unknown, but this is unimportant for our purposes. It is sufficient to note that the factor $X_8^{g_{38}}$ represents the induction factor; for induction by a factor of say 100 over the basal level $X_8^{g_{38}}$ must increase by a factor of 100. With $g_{38} = 1$, the induction factor is equal to the normalized concentration of the substrate X_8.

METHODS

For the comparison of alternative systems one requires criteria for functional effectiveness with which to judge alternative circuits of regulation, methods of mathematical and computer analysis with which to quantitate these criteria, and systematic methods for making "well-controlled" comparisons (see Chapter 5 and [5]).

Criteria for the functional effectiveness of inducible catabolic systems have been discussed in detail elsewhere ([2:Chapter 15], [13, 134]). These are summarized in Table 8.1. Other criteria might be added to this list, but based upon the current understanding of cellular physiology, these are likely to represent the major criteria, and thus they are reasonable requirements for inducible systems to function effectively.

Mathematical and computer methods of analysis used to quantitate the above criteria are as follows: Criteria that reflect the steady-state performance of the system are quantitated by the calculation of logarithmic gains, $L_{ij} = \partial(\log X_i)/\partial(\log X_j)$, and parameter sensitivities, $S(X, p) = (\partial X/\partial p)(p/X)$ [1, 2, 9, 12, 34]. Local stability is determined by the conventional Routh-Hurwitz method. The degree of stability is quantitated by the minimum "distance" from the critical boundaries for stability [2], which are given by the last two Routh-Hurwitz criteria for stability. Temporal responsiveness is determined by the time required for the system's dynamic response to a step perturbation to be within $\pm 5\%$ of its final steady-state value. The dynamic responses are obtained using ESSYNS, a program designed specifically for evaluation and simulation of synergistic systems (cf. Chapter 7 and [7, 38, 121]).

In comparing two alternative designs there is the potential for extraneous differences to confuse the results. This is particularly true for complex systems whose integrated behaviors are not intuitively obvious. A strategy for dealing mathematically with such comparisons has been developed; it involves ideal controlled comparisons. First, we require that all processes internal to the two systems that can be made identical have the same set of parameter values. This is called *internal equivalence* ([1], [2:Chapters 9 and 14]). Those processes that necessarily differ will in general have parameter values that are unrelated. In order to reduce the element of arbitrariness in the choice of values for these parameters, we require constraints for *external equivalence* ([1], [2:Chapters 9 and 14], [25]). External equivalence means that the two alternative designs

Table 8.1. Criteria for Functional Effectiveness of Inducible Catabolic Systems[a]

1. Sharp threshold in the concentration of substrate necessary for induction
2. Ability to make the most product available to the organism from a given supra-threshold increment in substrate
3. Insensitivity to perturbations in the system's component parts
4. Degree of stability
5. Temporal responsiveness

[a]These criteria apply to inducible systems of other types, although the interpretation of the criteria may be slightly different (e.g., see [134]).

produce the same behavior from a perspective external to the systems. For example, they should have the same steady-state induction characteristic. Once the two systems have been made maximally equivalent from both the internal and external perspectives, any remaining differences in their behavior can be attributed to inherent differences in their designs. This method of mathematically controlled comparison has allowed some rather general conclusions to be drawn regarding alternative designs of molecular and cellular networks (for a review, see [5]; see also Chapter 5).

CLASSICAL VS. AUTOGENOUS CIRCUITS

The autogenous circuit of gene regulation may be considered a special case of the coupled circuit in which the regulator and effector circuits have the same set of parameter values. When $\alpha_1 = \alpha_4$, $\alpha_2 = \alpha_5$, $\beta_1 = \beta_4$, $\beta_2 = \beta_5$, $g_{13} = g_{43}$, $g_{15} = g_{45}$, $g_{16} = g_{46}$, $g_{21} = g_{54}$, $g_{27} = g_{57}$, $h_{11} = h_{44}$, and $h_{22} = h_{55}$, the regulator and effector circuits are perfectly coupled and $X_1 = X_4$ and $X_2 = X_5$. Under these conditions Eq. (8.1) simplify to the following:

$$\begin{aligned}
\dot{X}_1 &= \alpha_1 X_6^{g_{16}} X_2^{g_{12}} X_3^{g_{13}} - \beta_1 X_1^{h_{11}} \\
\dot{X}_2 &= \alpha_2 X_7^{g_{27}} X_1^{g_{21}} - \beta_2 X_2^{h_{22}} \\
\dot{X}_3 &= \alpha_3 X_8^{g_{38}} X_2^{g_{32}} - \beta_3 X_2^{h_{32}} X_3^{h_{33}} \qquad (8.3)
\end{aligned}$$

The explicit symbolic solution for the system in steady state is obtained by setting the derivatives to zero, taking logarithms, and solving the resulting set of linear algebraic equations.

$$\begin{aligned}
y_1 &= [B_1 h_{22} h_{33} + B_2(a_{32} g_{13} + g_{12} h_{33}) + B_3 g_{13} h_{22}]/\Delta \\
y_2 &= [B_1 g_{21} h_{33} + B_2 h_{11} h_{33} + B_3 g_{13} g_{21}]/\Delta \\
y_3 &= [B_1 a_{32} g_{21} + B_2 a_{32} h_{11} + B_3(h_{11} h_{22} - g_{12} g_{21})]/\Delta \qquad (8.4)
\end{aligned}$$

where $y_i = \log X_i$, $B_1 = b_1 - g_{16} y_6$, $B_2 = b_2 - g_{27} y_7$, $B_3 = b_3 - g_{38} y_8$, $b_i = \log(\beta_i/\alpha_i)$, $a_{32} = g_{32} - h_{32}$, and $\Delta = -[h_{33}(h_{11}h_{22} - g_{12}g_{21}) - a_{32}g_{13}g_{21}]$.

Although the steady-state behavior is given completely by the solution in Eq. (8.4), it is convenient to display the influence of each independent variable (X_6, X_7, X_8) and each parameter separately.

The influence of substrate concentration X_8 on effector mRNA concentration X_2 is given by the logarithmic gain L_{28}, which also is the slope of the steady-state induction characteristic. By simply differentiating y_2 with respect to y_8 in Eq. (8.4)

$$L_{28} = \partial(\log X_2)/\partial(\log X_8) = \partial y_2/\partial y_8 = (\partial y_2/\partial B_3)(\partial B_3/\partial y_8)$$
$$= -g_{13}g_{21}g_{38}/\Delta \tag{8.5}$$

The percentage change in X_2 is L_{28} times the percentage change in X_8.

The influence of a rate constant, such as mRNA half-life β_2, on effector mRNA concentration X_2 is given by the rate-constant sensitivity $S(X_2, \beta_2)$. This is obtained by simply differentiating y_2 with respect to B_2 in Eq. (8.4).

$$S(X_2, \beta_2) = \partial(\log X_2)/\partial(\log \beta_2) = \partial y_2/\partial b_2 = \partial y_2/\partial B_2$$
$$= h_{11}h_{33}/\Delta \tag{8.6}$$

The percentage change in X_2 is $S(X_2, \beta_2)$ times the percentage change in β_2.

The influence of a kinetic order, such as that for the synthesis of X_3 with respect to the enzyme X_2 (i.e., g_{32}), on effector mRNA concentration X_2 is given by the kinetic-order sensitivity $S(X_2, g_{32})$. Again, by suitable differentiation of the explicit solution in Eq. (8.4),

$$S(X_2, g_{32}) = \partial(\log X_2)/\partial(\log g_{32}) = (\partial X_2/\partial g_{32})(g_{32}/X_2)$$
$$= -g_{13}g_{21}g_{32}y_{20}/\Delta \tag{8.7}$$

The percentage change in X_2 is $S(X_2, g_{32})$ times the percentage change in g_{32}. With these three types of calculations one can examine every possible influence separately.

The classical circuit of gene regulation may be considered a special case of the coupled circuit when, in addition to the above conditions on the parameters, the parameter $g_{12} = 0$. This also corresponds to coupled circuits in which $g_{15} = g_{43} = 0$, and the regulator and effector circuits are completely uncoupled.

Internal Equivalence

For the comparison of classical and autogenous circuits we require internal equivalence. The same set of parameter values will apply for both alternatives, except in the case of mRNA synthesis. The values of the parameters α_1^p, g_{12}^p, g_{13}^p and g_{16}^p, which characterize regulated transcription in the perfectly coupled autogenous circuit, will in general differ from those of α_1^u, g_{13}^u and g_{16}^u, which characterize regulated transcription in the completely uncoupled classical circuit. This is because mutational alteration of a single process will in general affect all the parameter values that characterize that process. Hence, we distinguish with superscripts (p = perfectly coupled, u = uncoupled) those parameters that are unique to each alternative circuit. All unsuperscripted parameters have the same value in each circuit. This specification of internal equivalence does not ensure that the two circuits will be equivalent from an external perspective.

External Equivalence

For external equivalence we require the two circuits to be as nearly identical as possible with regard to their systemic behavior. In this particular comparison, three constraints on systemic behavior can be used to fix the values of the three parameters—α_1^u, g_{13}^u, and g_{16}^u—unconstrained by internal equivalence. The two circuits are then maximally equivalent, and there are no unconstrained parameter values.

First, we require the two circuits to have the same slope for their induction characteristics, i.e., $L_{28}^p = L_{28}^u$. The slope L_{28}^p is given in Eq. (8.5), L_{28}^u can be obtained by setting $g_{12} = 0$ in Eq. (8.5), and equating the two expressions gives

$$g_{13}^u = g_{13}^p h_{11} h_{22}/(h_{11} h_{22} - g_{12}^p g_{21}). \qquad (8.8)$$

Second, we require that the two circuits respond equally to fluctuations in the levels of nucleotide precursors (X_6). Thus, from Eq. (8.3)

$$L_{26}^p = \partial y_2/\partial y_6 = -g_{16}^p g_{21} h_{33}/\Delta^p$$

$$L_{26}^u = -g_{16}^u g_{21} h_{33}/\Delta^u$$

Equating these two systemic properties gives

$$g_{16}^u = g_{16}^p h_{11} h_{22}/(h_{11} h_{22} - g_{12}^p g_{21}) \qquad (8.9)$$

Finally, we require the two circuits to have the same basal level of expression and displacement in their induction characteristics. At the threshold level of substrate for induction both circuits will have the same level of effector enzyme; that is, $X_2^p = X_2^u$ (or $y_2^p = y_2^u$). Equating the appropriate versions of Eq. (8.4) yields

$$B_1^u = \left[B_1^p + (g_{12}^p/h_{22})B_2\right] h_{11} h_{22}/(h_{11} h_{22} - g_{12}^p g_{21}) \qquad (8.10)$$

Solving explicitly for the value of α_1^u shows

$$\alpha_1^u = \beta_1\left[(\alpha_1^p/\beta_1)(\alpha_2/\beta_2)^{g_{12}/h_{22}} X_{70}^{g_{12}g_{27}/h_{22}}\right]^{h_{11}h_{22}/(h_{11}h_{22} - g_{12}g_{21})} \qquad (8.11)$$

where X_{70} is the nominal level of amino acid precursors.

One can view the autogenous circuit as given; the unsuperscripted parameters of the classical circuit have the same values as the autogenous circuit. The three superscripted (u) parameters have their values determined by the constraints for external equivalence [Eqs. (8.8), (8.9) and (8.11)] and they are uniquely specified in terms of the given independent concentrations and parameter values of the autogenous circuit.

Having established a suitable basis for comparison, one can then compare the behavior of the two circuits with respect to all possible influences. In this way, one can determine the irreducible differences between the two circuits.

Steady-State Comparisons

There are three dependent variables (X_1, X_2, X_3); thus, there are nine possible influences (logarithmic gains) due to the three "environmental" or independent variables (X_6, X_7, X_8). The comparison of these nine logarithmic gains is summarized in Table 8.2. The ratios for two of the logarithmic gains (L_{26} and L_{28}) are unity because of the constraints for external equivalence. Four other ratios also are unity, and three have values that differ from unity. The interpretation of these latter ratios will be given shortly, after considering the other influences.

There are six rate constants with eighteen possible influences (rate-constant sensitivities). The comparison of the two circuits on the basis of these sensitivities is summarized in Table 8.3. There are six sensitivities, associated with the rate constants for inducer synthesis and degradation, that are identical for the two circuits. The remaining twelve rate-constant sensitivities differ for the alternative circuits.

There are eleven kinetic orders with thirty-three possible influences (kinetic-order sensitivities), that are compared in Table 8.4. Again, the sensitivities

Table 8.2. Percentage Change in a Dependent Variable in Response to a One-Percent Change in an Independent Variable for the Completely Uncoupled (u) Circuit Relative to that for the Perfectly Coupled (p) Circuit of Gene Regulation[a]

	Independent variables		
Dependent variables	X_6	X_7	X_8
X_1	1	B^b	1
X_2	1	A	1
X_3	1	A	1

[a]The percentage change in a dependent variable with respect to a one-percent change in an independent variable [e.g., $L_{27} = \partial(\log X_2)/\partial(\log X_7) = dy_2/dy_7$] is calculated for the completely uncoupled (u) and perfectly coupled (p) circuits of gene regulation, the constraints for external equivalence are imposed, and the ratio of the two logarithmic gains (L_{27}^u/L_{27}^p) is taken.

[b]$A = 1 - (g_{12}g_{21})/(h_{11}h_{22})$ and $B = a_{32}g_{13}/(a_{32}g_{13} + g_{12}h_{33})$

Table 8.3. Percentage Change in a Dependent Variable in Response to a One-Percent Change in a Rate Constant Parameter for the Completely Uncoupled (u) Circuit Relative to that for the Perfectly Coupled (p) Circuit of Gene Regulation[a]

Dependent Variables	Parameters					
	α_1	β_1	α_2	β_2	α_3	β_3
X_1	A[b]	A	B	B	1	1
X_2	A	A	A	A	1	1
X_3	A	A	A	A	1	1

[a]The percentage change in a dependent variable in response to a one-percent change in a rate constant parameter [e.g., $S(X_2, \beta_1) = (dX_2/d\beta_1)(\beta_1/X_2)$] is calculated for the completely uncoupled (u) and perfectly coupled (p) circuits of gene regulation, the constraints for external equivalence are imposed, and the ratio of the two sensitivities (S^u/S^p) is taken.

[b]$A = 1 - (g_{12}g_{21})/(h_{11}h_{22})$ and $B = a_{32}g_{13}/(a_{32}g_{13} + g_{12}h_{33})$

associated with the parameters for inducer synthesis and degradation are the same for both circuits. Those associated with the parameters for synthesis and degradation of enzyme and mRNA differ for the alternative circuits. The sensitivities with respect to changes in the regulatory parameter g_{12} are always zero for the uncoupled circuit because g_{12}^u is zero by definition; those with respect to g_{13} are the same for both circuits because g_{13}^u is proportional to g_{13}^p [see Eq. (8.8)].

For systems with a positive mode of regulation $g_{12} > 0$ (and $g_{12} < h_{11}h_{22}/g_{21}$ for stability, as we shall see in the next section), and twenty-one of the influences are always less in the classical circuit than in the autogenous circuit ($A < 1$ in Tables 8.2–8.4). Six additional influences also are less in the classical

Table 8.4. Percentage Change in a Dependent Variable in Response to a One-Percent Change in a Kinetic Order Parameter for the Completely Uncoupled (u) Circuit Relative to that for the Perfectly Coupled (p) Circuit of Gene Regulation[a]

Dependent Variables	Parameters										
	g_{16}	g_{12}	g_{13}	h_{11}	g_{27}	g_{21}	h_{22}	g_{38}	g_{32}	h_{32}	h_{33}
X_1	1	0	1	A[b]	B	B	B	1	1	1	1
X_2	1	0	1	A	A	A	A	1	1	1	1
X_3	1	0	1	A	A	A	A	1	1	1	1

[a]The percentage change in a dependent variable with respect to a one-percent change in a kinetic order parameter [e.g., $S(X_2, h_{11}) = (dX_2/dh_{11})(h_{11}/X_2)$] is calculated for the completely uncoupled (u) and perfectly coupled (p) circuits of gene regulation, the constraints for external equivalence are imposed, and the ratio of the two sensitivities (S^u/S^p) is taken.

[b]$A = 1 - (g_{12}g_{21})/(h_{11}h_{22})$ and $B = a_{32}g_{13}/(a_{32}g_{13} + g_{12}h_{33})$

circuit than in the autogenous circuit ($B < 1$ in Tables 8.2–8.4) provided $a_{32} > 0$ or small in magnitude, as is the case for the *lac* operon and many other inducible systems (see [2: Chapter 15] for further discussion). Thus, with a positive mode of regulation, the influences on the uncoupled classical circuit are always less than or equal to those on the otherwise equivalent perfectly coupled autogenous circuit. According to the third criterion for functional effectiveness (Table 8.1), the classical circuit is the superior design.

For systems with a negative mode of regulation, $g_{12} < 0$ and twenty-one of the influences are always greater in the classical circuit than in the autogenous circuit ($A > 1$). Six additional influences will differ depending upon the value of the parameter a_{32}. If $a_{32} > -g_{12}h_{33}/2g_{13}$, then these influences are greater in the classical circuit than in the autogenous circuit ($|B| > 1$). However, if $a_{32} < -g_{12}h_{33}/2g_{13}$, then these same influences are less in the classical circuit ($|B| < 1$). Thus, with the negative mode of regulation, forty-eight of the possible sixty influences on the autogenous circuit are always less than or equal to those on the otherwise equivalent classical circuit. Furthermore, since these influences include the physiologically most important ones, one can still conclude that the autogenous circuit is the superior design according to the third criterion for functional effectiveness.

Stability Comparisons

The local stability of the autogenous circuit is determined in the conventional manner from the linearized version of Eq. (8.3) [2, 17, 124]. The characteristic equation is

$$\begin{aligned}\lambda^3 + \varphi_1\lambda^2 + \varphi_2\lambda + \varphi_3 \\ &= \lambda^3 + (F_1h_{11} + F_2h_{22} + F_3h_{33})\lambda^2 \\ &\quad + \left[F_1F_2(h_{11}h_{22} - g_{12}g_{21}) + F_1F_3h_{11}h_{33} + F_2F_3h_{22}h_{33}\right]\lambda \\ &\quad + F_1F_2F_3\left[h_{33}(h_{11}h_{22} - g_{12}g_{21}) - a_{32}g_{13}g_{21}\right] = 0 \qquad (8.12)\end{aligned}$$

where the F_i are positive quantities equal to the pseudo first-order rate constants for the degradation of X_i. Since $\varphi_1 > 0$, the conditions for stability reduce to

$$\varphi_1\varphi_2 - \varphi_3 > 0 \quad \text{and} \quad \varphi_3 > 0$$

In the two-dimensional space of regulatory parameters (g_{12}, g_{13}), these conditions imply two straight-line boundaries for stability:

$$g_{13} = h_{11}h_{22}h_{33}/(a_{32}g_{21}) - (h_{33}/a_{32})g_{12} \qquad (8.13)$$

and

$$g_{13} = -[h_{11}h_{22}h_{33}/(a_{32}g_{21})][2 + (F_1h_{11}/F_2h_{22}) + (F_2h_{22}/F_1h_{11}) + (F_1h_{11}/F_3h_{33}) + (F_3h_{33}/F_1h_{11}) + (F_2h_{22}/F_3h_{33}) + (F_3h_{33}/F_2h_{22})] + [(F_1h_{11} + F_2h_{22})/F_3a_{32}]g_{12} \quad (8.14)$$

The values of the regulatory parameters g_{12} and g_{13} must lie to the left of these two lines if the corresponding system is stable. In the limiting case with $a_{32} = 0$, the lines become vertical and the stable region is all values to the left of $g_{12} = h_{11}h_{22}/g_{21}$ (see Fig. 8.4).

To compare classical and autogenous circuits in Fig. 8.4, one should note that equivalent systems are represented by points along a line with a negative slope passing through the g_{12} axis at a value of $h_{11}h_{22}/g_{21}$. The equation for this line of equivalence is Eq. (8.8). The classical circuits are represented along

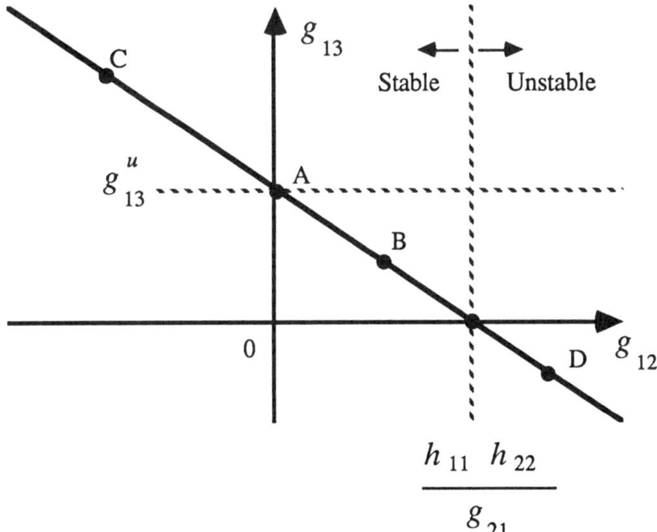

FIGURE 8.4. The two-dimensional space of regulatory parameters (g_{12}, g_{13}) for uncoupled classical circuits and perfectly coupled autogenous circuits. $g_{12} = h_{11}h_{22}/g_{21}$ is the boundary of local instability with $a_{23} = 0$. The inclined line of equivalence is the locus of points representing systems with a given steady-state induction characteristic. The specific examples indicated by the points A (classical), B (positive mode, autogenous circuit), C (negative mode, autogenous circuit), and D (unstable circuit) are compared in Table 8.5 and Fig. 8.5.

the vertical axis ($g_{12} = 0$). Autogenous circuits with a positive mode of regulation are represented along the line of equivalence with positive g_{12} values. In Fig. 8.4 one sees clearly that increasing the strength of the autogenous contribution to the regulation reduces the distance from the boundary of stability, the classical circuit has the greatest degree of stability, and hence, the classical circuit is superior according to criterion 4. Conversely, for circuits with a negative mode of regulation ($g_{12} \leq 0$), increasing the strength of the autogenous contribution to the regulation increases the degree of stability, and the comparable classical circuit has the least degree of stability. Thus, the classical circuit is now inferior according to criterion 4.

From this diagram one can draw the following qualitative conclusions regarding the pattern of regulatory interactions. For completely uncoupled classical circuits there is one relevant regulatory parameter, g_{13}. Only positive values of g_{13} represent physically realizable inducible systems; negative values of g_{13} correspond to repressible systems. For perfectly coupled autogenous circuits there are two relevant regulatory parameters: g_{12} and g_{13}. Of the four possible patterns for regulatory interactions, only two are physically realizable as inducible systems: $g_{12} > 0$ and $g_{13} > 0$, $g_{12} < 0$ and $g_{13} > 0$. Systems with $g_{12} > 0$ and $g_{13} < 0$ are unstable (or repressible) systems, and those with $g_{12} < 0$ and $g_{13} < 0$ are repressible systems. The fraction of realizable patterns is thus one-half for both classical and autogenous circuits. These results are summarized in Table 8.5 and they have been contrasted with those for more generally coupled circuits in which the fraction of realizable patterns is greater [92].

Table 8.5. Qualitative Regulatory Patterns for Completely Uncoupled (u) and Perfectly Coupled (p) Circuits of Gene Regulation[a]

Regulatory Parameters		
g_{13}	g_{12}	Examples
Completely uncoupled (Classical) circuit		
+	0[b]	*A*
−	0	Impossible (Repressible)
Perfectly coupled (Autogenous) circuit		
+	+	*B*
+	−	*C*
−	+	*D* (Unstable)
−	−	Impossible (Repressible)

[a]See Fig. 8.4 and text for discussion.
[b]$g_{12} = 0$ for classical circuits.

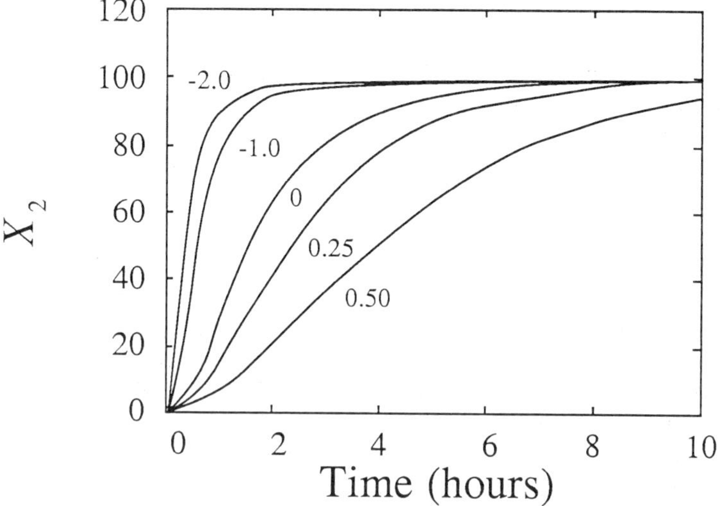

FIGURE 8.5. Temporal responsiveness of induction. Before $t = 0$ the systems are in their basal state. At $t = 0$, the concentration of substrate is increased to a level that yields full induction. The response of effector proteins (X_2) is represented as a function of time. The systems are those represented by a series of points along a given line of equivalence, $g_{13} = 1 - g_{12}$. The number associated with each system response is the value of g_{12}.

Temporal Response Comparisons

The comparisons of autogenous and classical circuits thus far have been general and give us results that are independent of particular numerical parameter values. Indeed, this is one of the principal advantages of the power-law formalism. However, because Eq. (8.3) is nonlinear and has no general dynamic solution in terms of elementary functions, one cannot expect general symbolic results when it comes to comparisons involving temporal responsiveness. For these one must solve the differential equations numerically with specific choices for the parameter values. The parameter values we shall assume are those given earlier for the *lac* operon. Eq. (8.3) then becomes

$$\dot{X}_1 = 0.347[X_2^{g_{12}} X_3^{g_{13}} - X_1]$$

$$\dot{X}_2 = 0.0116[X_1 - X_2]$$

$$\dot{X}_3 = 0.0112[X_8 X_2 - X_2 X_3] \qquad (8.15)$$

The constraint given by Eq. (8.8), which ensures that the systems being compared all have the same steady-state induction characteristic, becomes

$$g_{13}^{p} = 1 - g_{12}^{p} \tag{8.16}$$

since $h_{11} = h_{22} = g_{13}^{u} = 1$ (see *Parameter Values* section).

Consider a series of systems represented by points progressing to the left along a line of equivalence [Eq. (8.16)], as shown in Fig. 8.4. The induction of each system is simulated by a numerical solution of the appropriate equations (8.15), and the resulting temporal response of enzyme (X_2) is plotted in Fig. 8.5.

For systems with a positive mode of gene regulation ($g_{12} \geq 0$), the classical circuit ($g_{12} = 0$) has the best temporal response. As the strength of the autogenous contribution to the regulation (g_{13}) is increased, the response time becomes progressively worse. The conclusions are reversed for systems with a negative mode of gene regulation ($g_{12} \leq 0$). The classical circuit ($g_{12} = 0$) now has the worst response time; increasing the strength of the autogenous contribution improves the response time, although there are diminishing returns in improvement as the value of g_{12} becomes more negative.

DISCUSSION

The results in the previous section allow us to compare the two alternative circuits quantitatively on the basis of the five criteria for functional effectiveness listed in Table 8.1.

The uncoupled classical circuit and the perfectly coupled autogenous circuit are equivalent on the basis of the first two criteria. They have the same sharpness of threshold for induction and the same ability to convert suprathreshold increments of substrate into product because they have the same steady-state induction characteristic. This is ensured by the constraints for external equivalence (Eqs. [8.8] and [8.11]).

The classical circuit is superior to the autogenous circuit according to criterion 3 (insensitivity of the system to perturbations in its parameter values) when the circuits have a positive mode of regulation. The responses of the classical circuit to the sixty possible influences are always less than or equal to those of the equivalent autogenous circuit. These conclusions are largely reversed when their mode of regulation is negative. The responses of the autogenous circuit to fifty-four of the sixty possible influences are always less than or equal to those of the equivalent classical circuit (Tables 8.2–8.4).

The classical circuit is superior to the autogenous circuit on the basis of criterion 4 (the degree of local stability) when the circuits have a positive mode of regulation, but it is inferior to the autogenous circuit when they have a negative mode of regulation (Fig. 8.4).

The classical circuit is superior to the autogenous circuit on the basis of criterion 5 (temporal responsiveness) when the circuits have a positive mode of regulation, but again it is inferior to the autogenous circuit when they have a negative mode of regulation (Fig. 8.5).

Thus, if there is a positive mode of regulation, then the completely uncoupled classical circuit will be superior or equal to the equivalent perfectly coupled autogenous circuit on the basis of all five criteria for functional effectiveness (Table 8.6). If there is a negative mode of regulation, then the autogenous circuit will be superior or equal to the equivalent classical circuit on the basis of criteria 1, 2, 4, and 5. It may be inferior to the classical circuit on the basis of a subset of criterion 3.

As indicated in the introduction, the previous theory of classical vs. autogenous regulation in inducible systems satisfactorily accounted for the classical circuit with a positive mode of regulation and some autogenous circuits with the negative mode of regulation. However, it failed to account for the *lac* operon, which appears to have a classical circuit but a negative mode of regulation.

There are several hypotheses that might explain this notable failure. First, the experimental data characterizing the *lac* operon might be incomplete, and when critical measurements are made, one will find that the *lac* operon is autogenously regulated. This is unlikely. Even though one can criticize the measurements of the *lac* repressor during induction/deinduction, and perhaps more refined measurements would show some variation in the repressor levels, the available evidence suggests that it is much closer to being classically regulated than autogenously regulated.

Second, earlier analyses considered classical vs. autogenous circuits with product as the natural inducer, and classical circuits with substrate, intermediate, or product as the natural inducer. However, autogenous circuits with intermediate as the natural inducer were not examined. Perhaps this combination has some negative consequences that would account for the classical circuit of the *lac* operon, which has the intermediate allolactose as its natural inducer. The results in this paper show that this is not the case. Our results, summarized

Table 8.6. Summary of Comparisons

Criterion	Mode of Molecular Regulation: *Positive*	*Negative*
1	Uncoupled =[a] Perfectly coupled	Perfectly coupled = Uncoupled
2	Uncoupled = Perfectly coupled	Perfectly coupled = Uncoupled
3	Uncoupled ≥ Perfectly coupled	Perfectly coupled ≥ Uncoupled[b]
4	Uncoupled > Perfectly coupled	Perfectly coupled > Uncoupled
5	Uncoupled > Perfectly coupled	Perfectly coupled > Uncoupled

[a]= implies equivalent, > implies superior, and ≥ implies superior or equivalent.
[b]54 out of 60 cases, and can be 60 out of 60. See text for details.

summarized at the beginning of the Discussion, confirm the earlier analyses and extend them to the specific comparison of classical vs. autogenous circuits with the intermediate as the natural inducer.

Third, the earlier analyses were not carried out with the particular parameter values of the *lac* operon, and this could be critical for the results obtained. Most of our comparisons are made in a general fashion that yields symbolic results independent of particular parameter values, and, as indicated in the previous paragraph, the earlier results of this nature have been confirmed and extended in this paper. The only results that could be affected by specific numerical values are those having to do with temporal responsiveness. However, the particular values previously assumed and those of the *lac* operon used in this paper yield the same qualitative results and not the reversal of effects that would be required to explain the classical circuit of the *lac* operon. Thus, the specific values of the *lac* parameters do not lead to the prediction of a classical circuit.

From these considerations one might conclude that *lac* is simply an exception. It is always possible to attribute exceptions to historical accidents [79]. From the low demand for its expression [112] one might argue that this system is under weak selective pressure. In such circumstances, a less than optimal design would be more likely to persist. Alternatively, *lac* might have neither a classical nor an autogenous circuit and thus it falls outside the scope of our analyses to date. In such a case, *lac* may be understood only in terms of a more general model, such as that in Fig. 8.3. Indeed, there is some evidence suggesting that this may be the case (Savageau, in preparation).

SUMMARY

In addition to the well-established classical and autogenous circuits of inducible gene regulation, more general types of coupled circuits have recently been described in enteric bacteria. The current circuit theory of gene regulation accounts for most classical and autogenous examples; the notable exception is the well-studied lactose (*lac*) operon of *Escherichia coli*. However, this theory now must be expanded to account for the new examples of coupled circuits of gene regulation. In this chapter we describe fundamental regulatory concepts, a general model of coupled regulatory circuits, and a methodology for making well-controlled comparisons of alternative gene circuits. With this general framework for an expanded circuit theory, we confirm and extend earlier results regarding the special cases of completely uncoupled (classical) and perfectly coupled (autogenous) circuits. The results show that several hypotheses concerning the anomalous case of the *lac* operon now can be discounted.

ACKNOWLEDGMENTS

We thank D. H. Irvine for helpful comments. MAS also wishes to thank P. Benyon for making it possible to spend a sabbatical year at the Division of Computing Research, and he acknowledges the partial support of U.S. Public Health Service Grant GM 30054 from the National Institutes of Health.

9. GTP-Dependent Kinetic Proofreading

An S-System Model of the Reaction of Poly(U)-Programmed Ribosomes with a Leucyl-tRNA$_2$ · Elongation Factor Tu · GTP Complex

MASAHIRO OKAMOTO, JUN-JI YOSHII, AND KATSUYA HAYASHI

The estimated overall misreading probability of the incorporation of amino acids into proteins has been estimated to be between 1 in 10^4 [142] and 3 in 10^4 [145]. At the same time, Grosjean et al. [143] have reported relative stability in the lifetime of complexes between two complementary tRNAs, and nearly so, of their anticodons. According to them, lifetimes of correct complexes (base triplet interactions) are usually greater than 100 msec, while lifetimes of incorrect complexes usually are between 1 msec and 10 msec. That is, the lifetime of correct complexes typically is between one and two orders of magnitude longer than that of incorrect complexes with mismatching bases. These data show that the error frequency may be estimated as approximately 1 in 10^2, which is significantly different from 1 or 3 in 10^4. Consequently, the ribosome must have a special mechanism that amplifies the small difference in lifetime to produce a greater discrimination between correct and incorrect base pairings.

Hopfield [144] was the first to propose a special mechanism for such an amplification and he termed it *kinetic proofreading*. In his model, the hydrolysis of GTP associated with the reaction is postulated to allow an independent check on the selected aminoacyl-tRNA before the product is made and released by the ribosome. This GTPase reaction is claimed to be essentially irreversible. Thompson and Stone [151] have experimentally demonstrated the existence of a GTP-dependent proofreading step in *Escherichia coli*. In a series of papers [140–141, 148–150], they have determined several kinetic parameters for the reaction of poly(U)-programmed ribosomes in *E. coli* with a near-cognate ternary complex consisting of Leu-$tRNA_2$, elongation factor Tu, and GTP. However, a complete characterization of such a specific proofreading mechanism has never been presented.

In this investigation and elsewhere [147], we have evaluated experimental measurements of kinetic parameters, related to the interaction between poly(U)-programmed ribosomes and ternary complexes, consisting of polypeptide elongation factor Tu, aminoacyl-tRNA (Phe-tRNA or Leu-$tRNA_2$) and GTP. We have incorporated them into the currently best-established mathematical model for such a system. The model was subjected to a detailed steady-state analysis that allowed us to predict numerical values of all kinetic parameters and steady-state concentrations. These qualitative predictions are of particular interest since some of the parameter values have not yet been measured experimentally or are difficult to determine. Furthermore, we have examined the dynamic behavior of the intact system by numerically solving the kinetic equations that characterize the system. These equations are represented in S-system form [cf. 71–72, 146].

MODEL

The mechanism that best describes the overall process of GTP-dependent kinetic proofreading in aminoacyl-tRNA-recognition on ribosomes is shown in Diagram 9.1 (see also [140, 148, 151]). RS represents an mRNA programmed ribosome; the rejection of aminoacyl-tRNA (aa-tRNA) from the ribosome creates a new intermediate; and the rate limiting break-down step of this process involves the hydrolysis of GTP. Our goal is to investigate the single proofreading reaction of poly(U)-programmed ribosomes with the near-cognate ternary complex that consists of leucyl-$tRNA_2$, elongation factor Tu (EF-Tu), and GTP.

But first let's examine Fig. 9.1, which illustrates the general branching structure of this proofreading mechanism. There are four independent fluxes: *a*, *b*, *c*, and *d*. *a* is the net flux representing GTP-hydrolysis of correctly matched aa-tRNA (phenylalanyl-tRNA); *b* is the net flux representing production of correctly coded amino acid (phenylalanine); *c* is the net flux representing hydrolysis of incorrectly matched aa-tRNA (leucyl-$tRNA_2$); and *d* is the net flux rep-

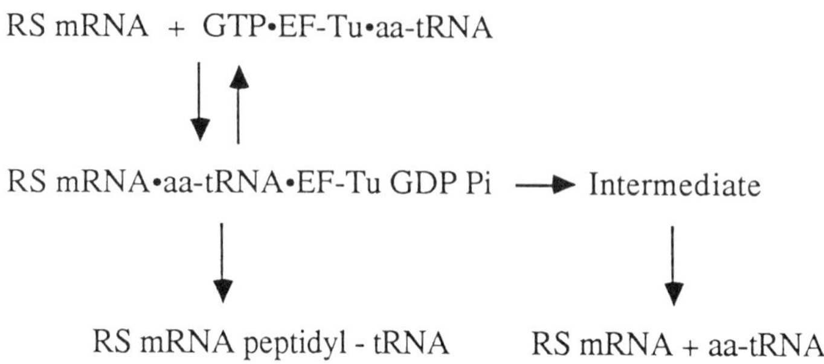

DIAGRAM 9.1. Schematic Representation of Kinetic Proofreading. RS mRNA represents the ribosome programmed with mRNA.

resenting production of incorrectly coded amino acid (leucine). The macroscopic behavior of this system can be represented with a set of eight macroscopic parameters defined as follows:

1. initial discrimination ratio: $I_1 = (a + b)/(c + d)$
2. post discrimination ratio: $I_2 = b/d$
3. proofreading discrimination ratio: $P = (c/d)/(a/b)$
4. net forward flux: $F = b + d$
5. net recycling flux: $R = a + c$
6. net error: $E = d/(b + d)$
7. cost of proofreading: $C = R/F = (a + c)/(b + d)$
8. waste-energy ratio: $WE = a/c$

Fig. 9.2 shows the specific kinetic proofreading mechanism of the reaction of poly(U)-programmed ribosomes with a leucyl-tRNA$_2$ · EFTu · GTP complex corresponding to the general branching model in Fig. 9.1. The binary complex of EF-Tu and GTP reacts with aa-tRNA (Phe-tRNA = X_3, Leu-tRNA$_2$ = X_4) to form the ternary complex of aa-tRNA · EF-Tu · GTP (X_5 or X_6), followed by the reversible binding (X_7 or X_8) of the ternary complex to the ribosome (X_2), programmed with poly(U)-mRNA and by the irreversible formation of an obligatory high-energy intermediate (X_9 or X_{10}). The parameters k_9 and k_{10} represent rate constants for peptidyl transfer or for peptide formation. The variables X_{11} and X_{12} represent intermediates of the branched pathway, that is, the rejecting reaction from the ribosome that has been termed *proofreading* or *editing*. The parameters k_1 and k_2 represent rate constants for the GTP-hydrolyzed proofreading reaction that yields free aa-tRNA and free ribosome.

The mechanism is assumed to operate in *E. coli* cells that grow exponentially in a minimal glucose medium with an average doubling time of 40 min. The method of estimating numerical values of kinetic parameters and steady-state

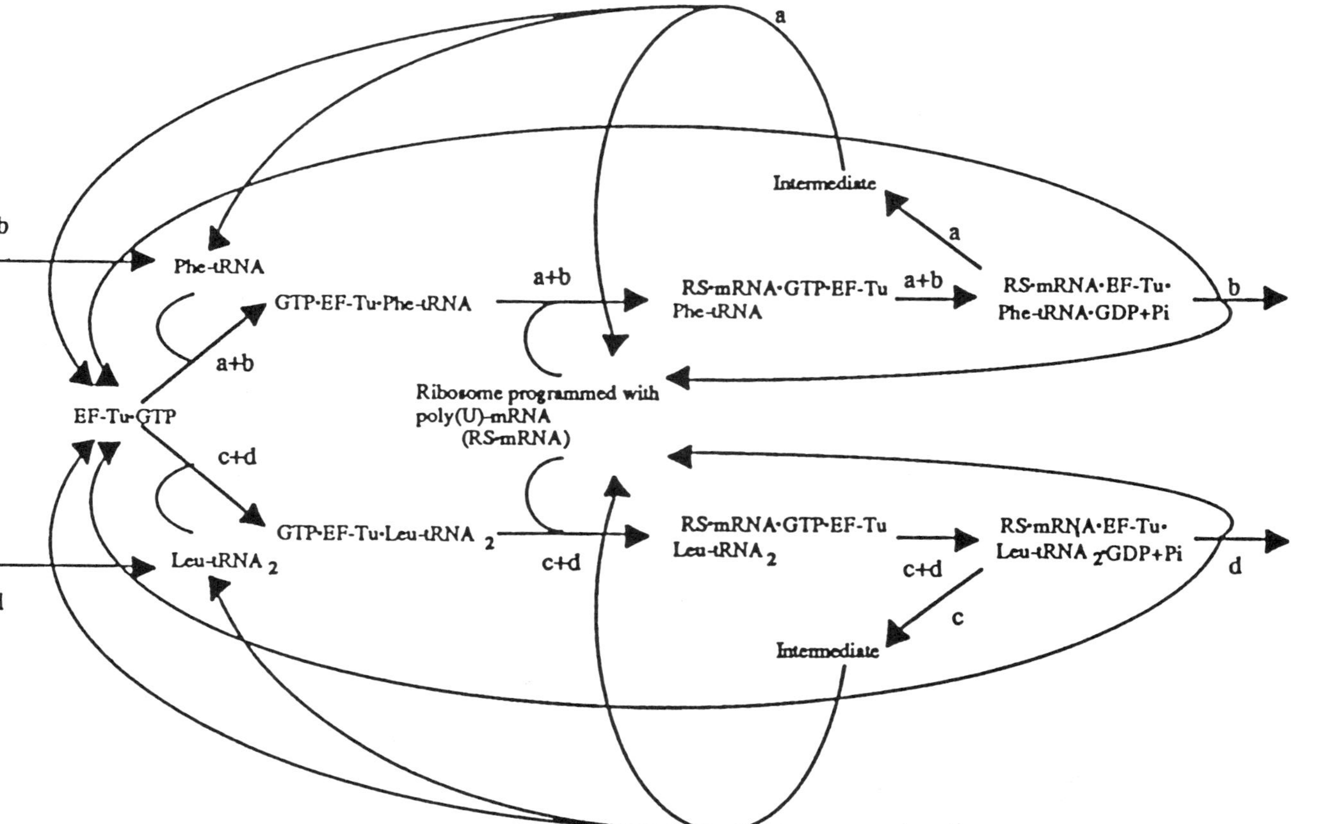

FIGURE 9.1. Kinetic proofreading model of poly (U)-programmed ribosomes with Leu-tRNA$_2$ · EF-Tu · GTP complex. Small letters (a through d) represent the flux at the corresponding step.

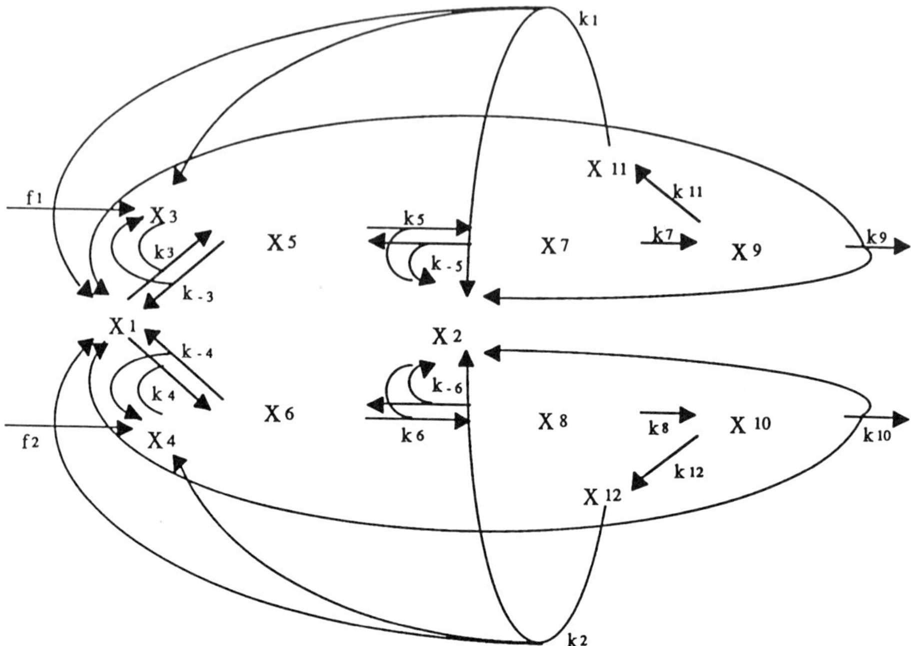

FIGURE 9.2. Mathematical model of Fig. 9.1. The k_i represent the rate constant at the corresponding step.

concentrations is described elsewhere [147]; results are summarized in Table 9.1. From these values, one can estimate the values for the eight macroscopic parameters described above: $I_1 = 1147$, $I_2 = 9090$, $P = 103.0$, $F = 11.29$ μMs^{-1}, $R = 0.8314$ μMs^{-1}, $E = 1.1 \times 10^{-4}$, $C = 0.07363$ and $WE = 88.25$. These values define the nominal values of the system parameters.

Method of Numerical Solution

The kinetic equations (fundamental equations) governing the system in Fig. 9.2 can be written as:

$$\dot{X}_1 = (k_1X_{11} + k_2X_{12} + k_9X_9 + k_{10}X_{10} + k_{-3}X_5 + k_{-4}X_6) - (k_3X_1X_3 + k_4X_1X_4)$$

$$\dot{X}_2 = (k_1X_{11} + k_2X_{12} + k_9X_9 + k_{10}X_{10} + k_{-5}X_7 + k_{-6}X_8) - (k_5X_2X_5 + k_6X_2X_6)$$

$$\dot{X}_3 = (f_1 + k_1X_{11} + k_{-3}X_5) - k_3X_1X_3$$

$$
\begin{aligned}
\dot{X}_4 &= (f_2 + k_2X_{12} + k_{-4}X_6) - k_4X_1X_4 \\
\dot{X}_5 &= (k_3X_1X_3 + k_{-5}X_7) - (k_{-3}X_5 + k_5X_2X_5) \\
\dot{X}_6 &= (k_4X_1X_4 + k_{-6}X_8) - (k_{-4}X_6 + k_6X_2X_6) \\
\dot{X}_7 &= k_5X_2X_5 - (k_{-5} + k_7)\,X_7 \\
\dot{X}_8 &= k_6X_2X_6 - (k_{-6} + k_8)\,X_8 \\
\dot{X}_9 &= k_7X_7 - (k_9 + k_{11})\,X_9 \\
\dot{X}_{10} &= k_8X_8 - (k_{10} + k_{12})\,X_{10} \\
\dot{X}_{11} &= k_{11}X_9 - k_1X_{11} \\
\dot{X}_{12} &= k_{12}X_{10} - k_2X_{12} \qquad (9.1)
\end{aligned}
$$

Table 9.1. Standard Values at Steady State

k_1	$= 0.12$	(sec^{-1})
k_2	$= 3.2$	(sec^{-1})
k_3	$= 1.0$	$(\text{M}^{-1}\text{sec}^{-1})$
k_{-3}	$= 2.0 \times 10^{-8}$	(sec^{-1})
k_4	$= 0.7128816 \times 10^{-3}$	$(\text{M}^{-1}\text{sec}^{-1})$
k_{-4}	$= 0.1425763 \times 10^{-10}$	(sec^{-1})
k_5	$= 4.0 \times 10^{7}$	$(\text{M}^{-1}\text{sec}^{-1})$
k_{-5}	$= 1.76 \times 10^{-2}$	(sec^{-1})
k_6	$= 3.2 \times 10^{6}$	$(\text{M}^{-1}\text{sec}^{-1})$
k_{-6}	$= 48$	(sec^{-1})
k_7	$= 160$	(sec^{-1})
k_8	$= 32$	(sec^{-1})
k_9	$= 6.4$	(sec^{-1})
k_{10}	$= 6.4$	(sec^{-1})
k_{11}	$= 0.4660194$	(sec^{-1})
k_{12}	$= 48$	(sec^{-1})
f_1	$= 0.1129 \times 10^{-4}$	$(\text{M} \cdot \text{sec}^{-1})$
f_2	$= 0.1242037 \times 10^{-8}$	$(\text{M} \cdot \text{sec}^{-1})$
X_{1S}	$= 0.2247744 \times 10^{-3}$	(M)
X_{2S}	$= 0.9806075 \times 10^{-5}$	(M)
X_{3S}	$= 0.5388554 \times 10^{-1}$	(M)
X_{4S}	$= 0.6588540 \times 10^{-1}$	(M)
X_{5S}	$= 0.3088244 \times 10^{-7}$	(M)
X_{6S}	$= 0.8411010 \times 10^{-9}$	(M)
X_{7S}	$= 0.7570055 \times 10^{-7}$	(M)
X_{8S}	$= 0.3299160 \times 10^{-9}$	(M)
X_{9S}	$= 0.1764063 \times 10^{-5}$	(M)
X_{10S}	$= 0.1940683 \times 10^{-9}$	(M)
X_{11S}	$= 0.6850728 \times 10^{-5}$	(M)
X_{12S}	$= 0.2911023 \times 10^{-8}$	(M)

Although these simultaneous differential equations can be solved directly by using available numerical routines, the cost is prohibitive because the system is very "stiff". Okamoto and Savageau have developed an alternative approach that allows us to obtain numerical solutions of such a large-scale stiff system in a convenient way [71–72, 146]. We have adopted this method here. The first step is to convert Eq. (9.1) into corresponding S-system equations [2–3, 29]:

$$
\begin{aligned}
\dot{X}_1 &= \alpha_1 X_5^{g_{1,5}} X_6^{g_{1,6}} X_9^{g_{1,9}} X_{10}^{g_{1,10}} X_{11}^{g_{1,11}} X_{12}^{g_{12}} - \beta_1 X_1^{h_{1,1}} X_3^{h_{1,3}} X_4^{h_{1,4}} \\
\dot{X}_2 &= \alpha_2 X_7^{g_{2,7}} X_8^{g_{2,8}} X_9^{g_{2,9}} X_{10}^{g_{2,10}} X_{11}^{g_{2,11}} X_{12}^{g_{2,12}} - \beta_2 X_2^{h_{2,2}} X_5^{h_{2,5}} X_6^{h_{2,6}} \\
\dot{X}_3 &= \alpha_3 X_5^{g_{3,5}} X_{11}^{g_{3,11}} - \beta_3 X_1^{h_{3,1}} X_3^{h_{3,3}} \\
\dot{X}_4 &= \alpha_4 X_6^{g_{4,6}} X_{12}^{g_{4,12}} - \beta_4 X_1^{h_{4,1}} X_4^{h_{4,4}} \\
\dot{X}_5 &= \alpha_5 X_1^{g_{5,1}} X_3^{g_{5,3}} X_7^{g_{5,7}} - \beta_5 X_2^{h_{5,2}} X_5^{h_{5,5}} \\
\dot{X}_6 &= \alpha_6 X_1^{g_{6,1}} X_4^{g_{6,4}} X_8^{g_{6,8}} - \beta_6 X_2^{h_{6,2}} X_6^{h_{6,6}} \\
\dot{X}_7 &= \alpha_7 X_2^{g_{7,2}} X_5^{g_{7,5}} - \beta_7 X_7^{h_{7,7}} \\
\dot{X}_8 &= \alpha_8 X_2^{g_{8,2}} X_6^{g_{8,6}} - \beta_8 X_8^{h_{8,8}} \\
\dot{X}_9 &= \alpha_9 X_7^{g_{9,7}} - \beta_9 X_9^{h_{9,9}} \\
\dot{X}_{10} &= \alpha_{10} X_8^{g_{10,8}} - \beta_{10} X_{10}^{h_{10,10}} \\
\dot{X}_{11} &= \alpha_{11} X_9^{g_{11,9}} - \beta_{11} X_{11}^{h_{11,11}} \\
\dot{X}_{12} &= \alpha_{12} X_{10}^{g_{12,10}} - \beta_{12} X_{12}^{h_{12,12}}
\end{aligned} \tag{9.2}
$$

Numerical values for the apparent kinetic orders ($g_{i,j}$ and $h_{i,j}$) and the apparent rate constants (α_i and β_i) are determined from the experimental data shown in Table 9.1. As an illustration, let's determine the parameter $g_{1,5}$ in Eq. (9.2). According to the power-law formalism (cf. Chapter 2), one first approximates the sum F_1^+ of positive terms in the first equation with a product of power-law functions, and then computes $g_{1,5}$ as the relative derivative with respect to X_5 at an operating point of our choice:

$$
\begin{aligned}
F_1^+ &= k_1 X_{11} + k_2 X_{12} + k_9 X_9 + k_{10} X_{10} + k_{-3} X_5 + k_{-4} X_6 \\
&\approx \alpha_1 X_5^{g_{1,5}} X_6^{g_{1,6}} X_9^{g_{1,9}} X_{10}^{g_{1,10}} X_{11}^{g_{1,11}} X_{12}^{g_{1,12}}
\end{aligned}
$$

Thus, the parameter $g_{1,5}$ is

$$g_{1,5} = (\partial F_1^+/\partial X_5)(X_5/F_1^+)|_{(op)}$$
$$= k_{-3}X_5(op)/[k_1 X_{11}(op) + k_2 X_{12}(op) + k_9 X_9(op) + k_{10}X_{10}(op) + k_{-3}X_5(op) + k_{-4}X_6(op)] \quad (9.3)$$

where $X_i(op)$ is the *in situ* concentration of metabolite X_i at the operating point. Since the S-system representation of F_1^+ is $\alpha_1 \Pi X_j^{g_{1,j}}$, the apparent rate constant α_1 can be estimated when all parameters $g_{1,j}$ are given (cf. Chapter 2).

The second step of our analysis was explicit evaluation of the relaxation times for the reactant variables in Eq. (9.1). We identified those with the shortest relaxation time and assumed the corresponding equations in system (9.2) to be in a quasi-steady state. The latter assumption allowed us to convert the differential equations in question into a corresponding subset of *algebraic* equations, leaving a smaller set of differential equations that was less stiff than the original set. The algebraic quasi-steady-state equations are solved simultaneously with the differential equations.

The third step is incorporation of constraints among the variables. Since not all of the fundamental equations [Eq. (9.1)] are independent, for example,

$$\dot{X}_1 = -(\dot{X}_5 + \dot{X}_6 + \dot{X}_7 + \dot{X}_8 + \dot{X}_9 + \dot{X}_{10} + \dot{X}_{11} + \dot{X}_{12})$$
$$\dot{X}_2 = -(\dot{X}_7 + \dot{X}_8 + \dot{X}_9 + \dot{X}_{10} + \dot{X}_{11} + \dot{X}_{12})$$

the total concentration of EF-Tu (denoted as X_E) and the total concentration of ribosome (denoted as X_T) may be considered constant throughout the period of time during which we have examined the dynamic behavior:

$$X_E = X_1 + X_5 + X_6 + X_7 + X_8 + X_9 + X_{10} + X_{11} + X_{12}$$
$$X_T = X_2 + X_7 + X_8 + X_9 + X_{10} + X_{11} + X_{12}$$

RESULTS

Effect of Speed on Accuracy

Throughout our experiments, we used 1.223 as the ratio of the concentrations of Leu-tRNA$_2$ and Phe-tRNA; that is, $X_{4S}/X_{3S} = 1.223$ (cf. Table 9.1). Any effects due to changes in these concentrations will be parallel and thus will eliminate effects due to changes in competition for both aa-tRNAs. Because the concentrations of both aa-tRNAs are varied, the steady-state flux through the system, defined as F (see *Model* section), changes as shown in Fig. 9.3. *OP* represents the standard operating point whose numerical values are given in

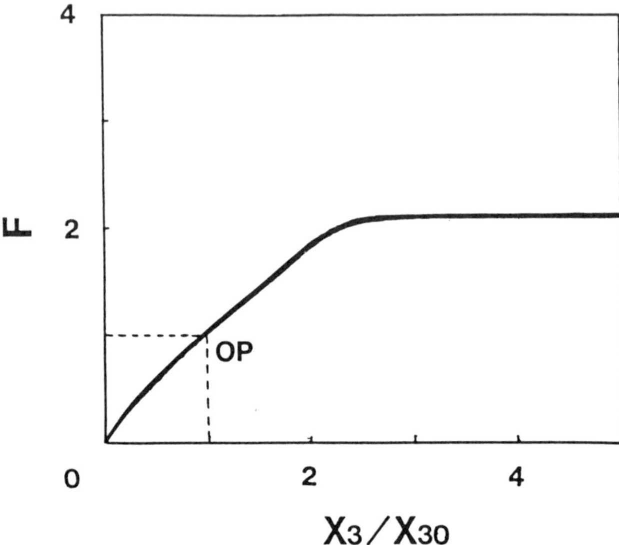

FIGURE 9.3. Flux through the system as a function of Phe-tRNA concentration. The net forward flux (F) is normalized with respect to its nominal value ($F_0 = 11.29\ \mu Ms^{-1}$). The ratio of X_4/X_3 is maintained at 1.223 while the concentrations are varied in parallel. OP denotes the operating point. X_{30} is equal to X_{3S}.

Table 9.1. Fig. 9.3 shows that the system normally operates in the midrange of the rate-vs.-concentration curve. Since the net forward flux is not saturated around the operating point, we can fruitfully go on to examine the dynamic response to changes in rate.

(A) Effects of changes in concentrations of aa-tRNAs. The effects on eight macroscopic parameters in the system that result from doubling or halving the concentrations of both aa-tRNAs are shown in Fig. 9.4. The ordinate in each panel represents the relative change from the nominal value. As shown in Fig. 9.4, the four macroscopic parameters P, F, C, and WE change rapidly in response to perturbations while the others gradually approach their respective steady-state values. With an increase in aa-tRNAs, the proofreading discrimination ratio P increases rapidly and then decreases to its preperturbation steady-state level. The waste-energy ratio WE decreases and then approaches its preperturbation level. The net recycling flux R increases and asymptotically approaches a new steady state.

These results show that during fast transient periods (< 2 sec), GTP hydrolyzes more for proofreading of Leu-$tRNA_2$ (k_2-reaction step) immediately after the perturbation than it does at the k_1-reaction step. Within 2 sec, WE decreases, in spite of the increase in P and R. Since the net forward flux F increases more

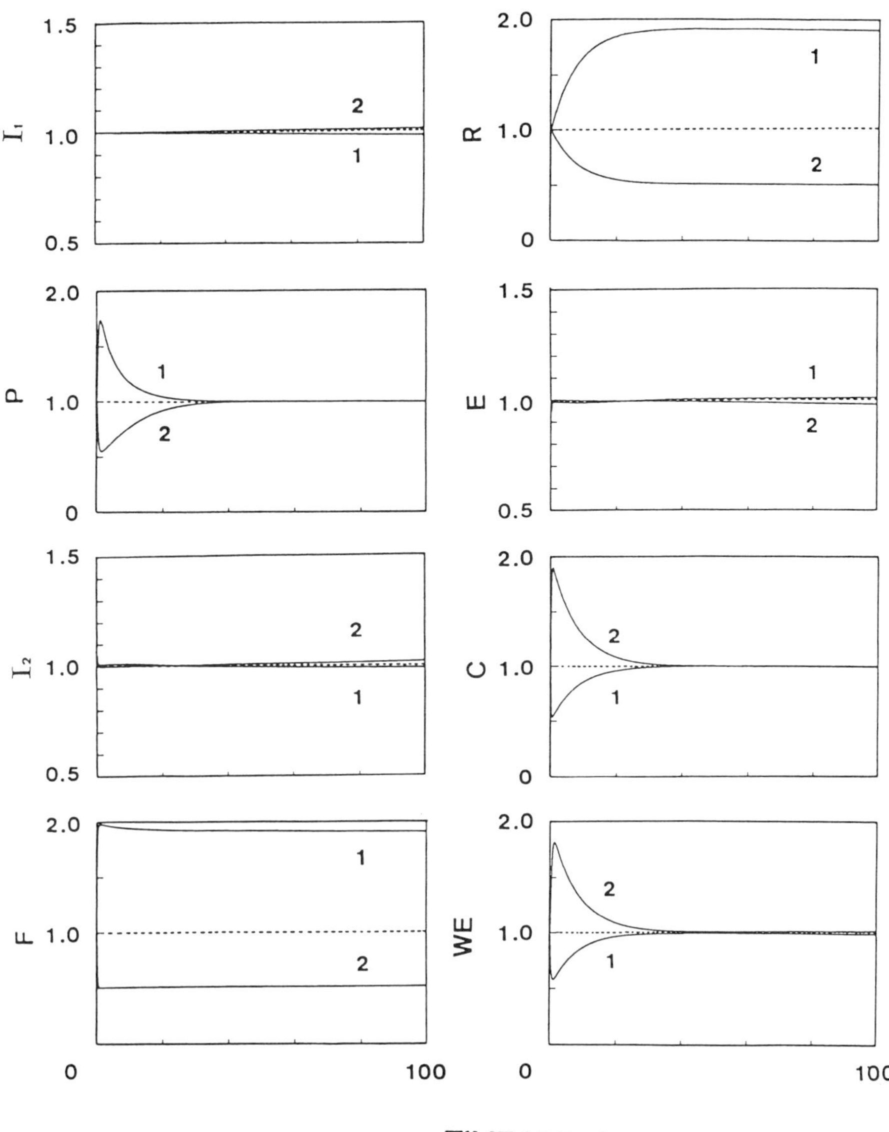

FIGURE 9.4. Effects of speed on accuracy: Changes in speed produced by alterations in aminoacyl-tRNA concentrations that leave aminoacyl-tRNA competition unaffected. Initial concentrations of Phe-tRNA (X_3) and Leu-tRNA$_2$ (X_4): increased (curve 1) by doubling their values at the operating point (OP in Fig. 9.3); decreased (curve 2) by halving their values at the operating point. The broken line indicates the case of the nominal operating point.

rapidly than R and reaches a new steady-state value at the same level as R, the cost of proofreading C initially decreases and then returns to its preperturbation steady-state level. In spite of a 100% increase in aa-tRNAs, the initial discrimination ratio I_1 and the postdiscrimination ratio I_2 hardly change, and remain within a 2% range about their nominal values. Because I_1, P, and C are almost unaffected by a 100% increase in aa-tRNAs, the net error E stays at its nominal value. The responses to a decrease in aa-tRNAs are nearly mirror images of the responses to an increase.

(B) Effects of changes in demand for product. Flow through the system also can be influenced by the demand for product which is reflected mathematically in a change of the numerical values of rate constants k_9 and k_{10}. This change may correspond to a fluctuation of rate of peptidyl transfer or of peptide formation. We examined the dynamic behavior of eight macroscopic parameters in response to a 140% increase or a 57% decrease of rate constants k_9 and k_{10} (where $k_9 = k_{10}$). The results are shown in Fig. 9.5. Upon a 57% decrease in the values of k_9 and k_{10} (curve 2 at each panel), the value of P increases rapidly and then decreases to its preperturbation level. Both R and WE increase in value and asymptotically approach their new steady states. A comparison of curve 2 of panels P and R in Fig. 9.5 and curve 1 of the corresponding panels in Fig. 9.4 shows that their dynamic responses are very similar to each other, but the response of WE is quite different (see curve 2 of panels P and R in Fig. 9.5 and curve 1 in Fig. 9.4).

The saturation pattern of WE in Fig. 9.5 shows that GTP hydrolyzes more at the k_1-reaction step (an unnecessary proofreading step) than at the k_2-reaction step (an indispensable proofreading step). At the cost of this waste of energy, however, the postdiscrimination ratio I_2 doubles and the net error E decreases to half the preperturbation value; that is, the use of energy results in a 50% decrease in error. Furthermore, the net forward flux F almost returns to its preperturbation value. These results show that in response to a significant reduction (57%) in k_9 and k_{10}, the system can produce protein that is twice as accurate. The produced amount, however, is nearly unaffected.

Effect of Competition on Accuracy

In a second series of experiments, we have varied the ratio of noncognate (X_4) to cognate (X_3) aa-tRNA in such a way that the net forward flux F is unaltered. Table 9.2 shows the values of X_3 and X_4 that approximately double or halve their ratio while maintaining the same nominal value for the net forward flux (F). The dynamic responses to a sudden increase or decrease in the ratio of noncognate to cognate aa-tRNA are shown in Fig. 9.6. With a decrease in the ratio X_4/X_3 (curve 1 in each panel), I_1 increases precipitously and reaches a

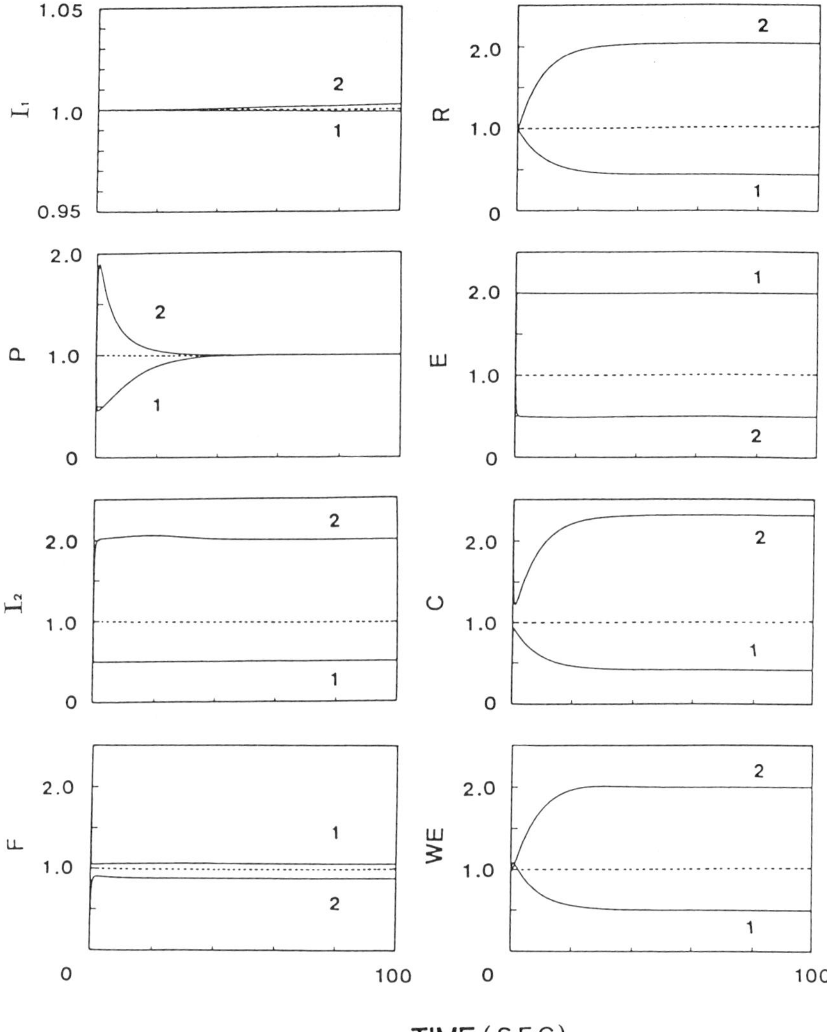

FIGURE 9.5. Effects of speed on accuracy: Changes in speed produced by alterations in demand for product that leave aminoacyl-tRNA competition unaffected ($X_4/X_3 = 1.223$). Rate constants of k_9 and k_{10} are: increased by 140% increase in both values at time zero (curve 1); decreased by 57% decrease in both values (curve 2). The broken line indicates the case of the nominal operating point.

new steady state. In spite of the increase in the values of I_1 and I_2, the net recycling flux R hardly changes from its preperturbation level.

These results show that a decrease in net error E may derive from an increase

Table 9.2. Aminoacyl-tRNA Concentrations that Maintain the Same Net Forward Flux Through the System

Aminoacyl-tRNA Concentration [M]			
Cognate (X_3)	Noncognate (X_4)	Ratio X_4/X_3	Net forward Flux (F) [M · SEC^{-1}]
0.5388840×10^{-1}	0.3294271×10^{-1}	0.6113	0.1129124×10^{-4}
0.5388554×10^{-1}†	0.6588540×10^{-1}†	1.2227	0.1129124×10^{-4}
0.5388050×10^{-1}	0.1317708	2.4456	0.1129124×10^{-4}

†corresponds to that at standard point (operating point)

in the initial discrimination ratio (I_1), and not from a proofreading mechanism, such as the hydrolysis of GTP for the proofreading of noncognate aa-tRNA.

CONCLUSION

In an integrated biochemical system, most of the data come from kinetic experiments *in vitro* or from parameter estimation with a corresponding mathematical model. Often the parameters differ greatly in their numerical values, and as a consequence, the corresponding differential equations are very stiff. Furthermore, the number of equations typically is very large. Instead of numerically solving such a large set of equations directly, it has turned out to be advantageous to use S-systems that derive as power-law approximations of the original equations. In our example, we found that this approach is faster, and thus less costly, and absolute errors are maintained within 2%.

In previous papers [71–72, 146], we have integrated all of the data relating to isoleucyl-tRNA synthetase and its ATP-dependent proofreading of valyl-tRNA^{Ile} into a conventional Michaelis-Menten model whose dynamic behavior then was determined by numerical solution of the relevant kinetic equations. In the same manner as described here, the effect of speed on accuracy was examined; that is, (1) we maintained the ratio of noncognate to cognate amino acid concentration at nominal value while changing the concentrations in parallel, and (2) we changed the rate constants for the demand for product. The results showed that speed and accuracy are inversely related: the higher the value of F (net forward flux), the lower the accuracy (the higher the error E).

Fig. 9.4 of this chapter clearly shows that doubling or halving aa-tRNA concentrations (X_3 and X_4) only slightly influences the net error, while Fig. 9.5 demonstrates that the error is decreased by 50% when the rate constants k_9 and k_{10} are reduced (by 57%). That is, accuracy is much more influenced by changes in demand for product (k_9 and k_{10}). Furthermore, this 57% reduction of k_9 and

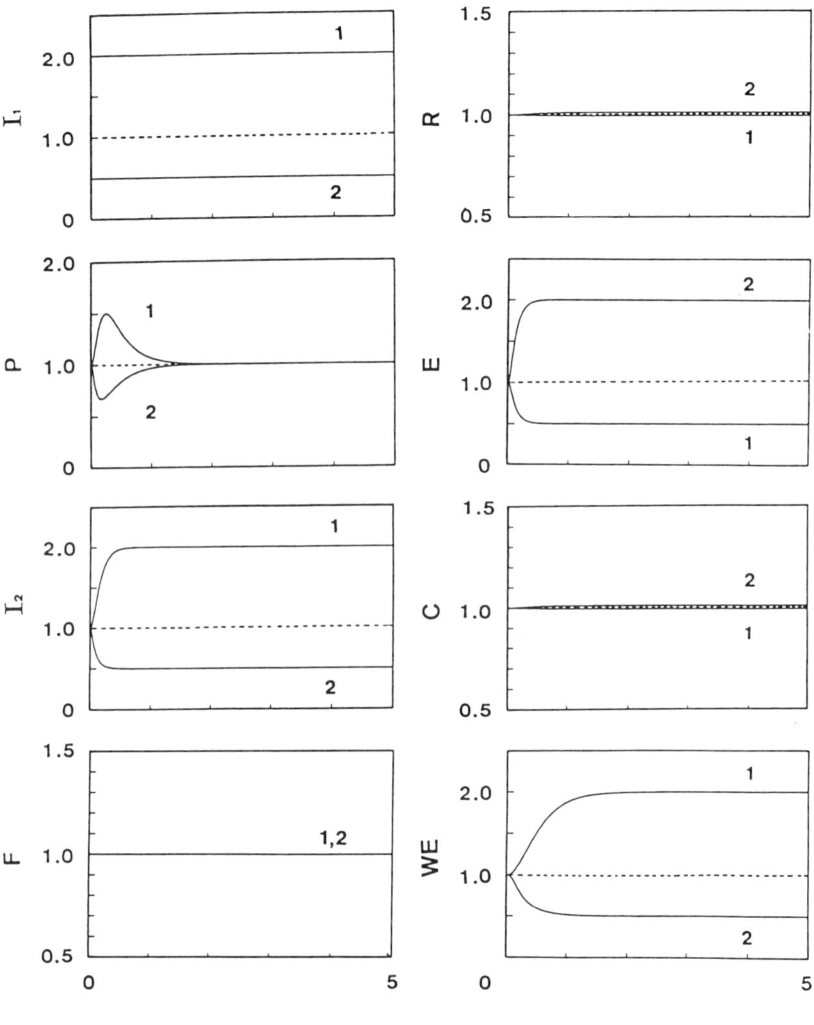

FIGURE 9.6. Effects of aminoacyl-tRNA competition on accuracy with the net forward flux fixed at steady state ($F_0 = 11.29\ \mu Ms^{-1}$). Curve 1: $X_3 = 53.89\ \mu M$, $X_4 = 32.94\ \mu M$, $X_4/X_3 = 0.611$. Curve 2: $X_3 = 53.88\ \mu M$, $X_4 = 131.77\ \mu M$, $X_4/X_3 = 2.446$. The broken line indicates the case that $X_3 = 53.89\ \mu M$, $X_4 = 65.89\ \mu M$ and $X_4/X_3 = 1.223$.

k_{10} did not significantly impair productivity: the system produced nearly the same amount of doubly accurate protein with twice the amount of GTP-hydrolysis (see curve 2 in Fig. 9.5).

SUMMARY

Experimental measurements of kinetic parameters, related to the interaction between poly(U)-programmed ribosomes and ternary complexes, consisting of polypeptide elongation factor Tu, aminoacyl-tRNA (Phe-tRNA or Leu-$tRNA_2$) and GTP, have been incorporated into the currently best-established kinetic proofreading model. The dynamic behavior of this model has been determined with numerical solutions of kinetic equations that are characterized by products of power-law functions (S-system form). The main results for data of *Escherichia coli* are the following: (1) a 57% reduction of the rate constant for peptidyl transfer or for peptide formation evokes the production of a nearly constant amount of doubly accurate protein with twice the amount of GTP-hydrolysis; and (2) doubling or halving the concentrations of cognate aminoacyl-tRNA (Phe-tRNA) and near-cognate aminoacyl-tRNA (Leu-$tRNA_2$), while holding the concentration-ratio of both aminoacyl-tRNAs constant, has less influence on the accuracy of protein, as measured by the change in net error.

ACKNOWLEDGMENT

We thank Drs. Michael A. Savageau, Eberhard O. Voit, and Douglas Irvine for their helpful suggestions and constructive criticism throughout this work.

10. An Analysis of Forestry Data

JONI A. TORSELLA AND AHMAD M. BIN RAZALI

The decision of when to harvest a forest in order to obtain maximum wood yield or maximum tree size has long been the concern of silvicultural economists. While experience with forests of different species and under different conditions is the prominent guide for such a decision, mathematical modeling provides an additional powerful tool for understanding and analyzing the growth and the dynamics of a forest. Many mathematical approaches could be of help, but the most promising appears to be modeling with systems of differential equations. Several differential equation models have been proposed in the literature (cf. [33] and [152]). However, they all were based on assumptions that seem too restrictive. In particular, some of these models ignore essential dynamic interactions between the system constituents, whereas others use *ad hoc* functional representations that are difficult to justify.

In this chapter, we describe a model for the growth and dynamics of even-aged forests of *Loblolly* pine (*Pinus taeda*) that is based on S-systems. First we discuss the traditional modeling approaches and explain why S-systems are an attractive modeling tool. Then we show how the S-system model can be applied to actual data. Specifically, we discuss various strategies for estimating the parameters of an S-system model from growth data.

EXPERIMENTAL OBSERVATIONS

The experimental data we will analyze were provided by the Southeastern Forest Experiment Station of the USDA Forest Service. They arise from a large study intended to yield insight into the dynamics of even-aged stands of *Loblolly* pine (*Pinus taeda*). The specific goal of this study is to develop models that reliably predict the effects of density, fertilizer, rainfall, thinning, and the reduction of competition from weed species on the growth of trees. Our aim in this project is to develop and evaluate S-system models that represent the growth of *Loblolly* pine trees in relationship to varying planting densities.

Fig. 10.1 shows a graphical representation of a typical data set. Plotted are the number of plants and their average size; both variables are represented in logarithmic coordinates. It is evident that, as the trees grow, the tree population declines in number. Furthermore, even though this representation does not show time explicitly, the trees grow faster when the tree number is small. Thus, by varying the initial planting density one can affect total yield and average tree size. This average tree size is important since only big trees can be used for lumber while small trees are of pulp quality at best.

Our data set is rather large and has almost no missing data. Each set is accompanied by appropriate control groups. The data consist of a number of variables, among which the following are the most important for our purpose:

Treatment number. The treatment number refers to the number of trees per acre established at age three. Specifically, Treatment 1-1,000, Treatment 2-2,000, Treatment 3-4,000, Treatment 4-8,000, and Treatment 5-16,000.

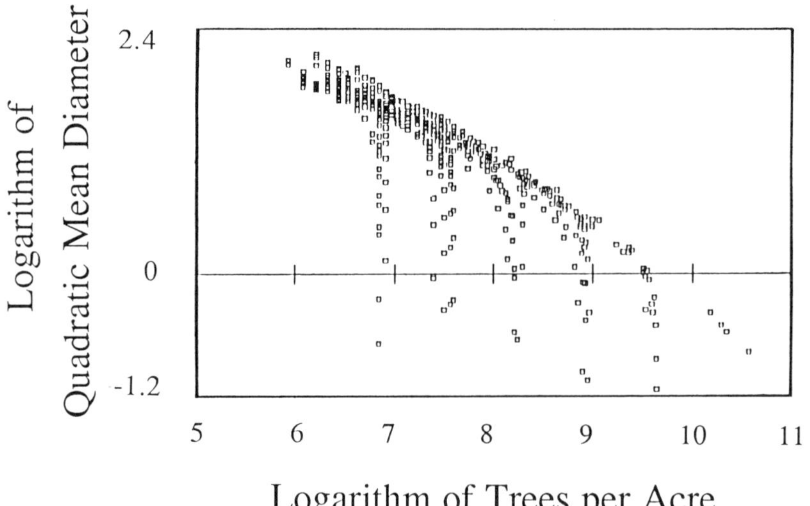

FIGURE 10.1. Trajectorial plot of observed tree growth data.

Block. The block represents the replication number. There are four blocks per treatment.

Age. Age ranges from three to twenty-six years. All plots are even-aged; that is, all trees start at the same age and trees that die are not replaced. Thus density decreases with age (time) and all trees in a plot have the same age.

Various measures of size. There are several measures of size. They include height, diameter at breast height (1.3 m), crown length, and crown width.

Trees per acre (density). As specified above, the number of trees per acre defines treatment. The measurement plot is 0.02 acre. But to avoid edge effects there is an area of trees surrounding the actual measurement area so that the gross area is 0.10 acre. In some instances, we have converted density to growing space. These two quantities are mathematically equivalent, with growing space being just a reciprocal function of density. In a sense, growing space is intuitively more appealing because it is a measure based on an individual tree rather than on the whole population of trees.

REPRESENTATION OF FOREST DYNAMICS

Traditional Modeling Approaches

Essentially three major approaches can be pursued in modeling a system. The first approach is linear approximation. It enjoys great popularity since most parts of linear analysis are straightforward. In particular, methods such as Laplace transformation exist for solving linear systems, and calculation of steady states is easy. Since the representing equations are always linear, alternative linear models differ only in their parameter values but not in their functional form. The major drawback is that linear approximations are sufficiently accurate only over a limited range of variation in the independent variable. Thus linearization is appropriate in systems that are close to linear over a large range of variation or that can be designed to operate within a restricted range. In contrast to technical systems, natural biological systems typically do not satisfy either of these two criteria. Most biological systems exhibit intrinsically nonlinear characteristics, such as synergism and saturation, and they are usually given to us and cannot be designed by us.

The second approach is nonlinear approximation. The range of validity for nonlinear representations of biological systems often is much wider than it is for linear representations. Furthermore, common nonlinear behaviors such as stable limit cycles can be captured with nonlinear functions but not often with linear approximations. The major drawback of nonlinear representations is that one usually has no *a priori* information about the most appropriate functional form. So not only do systems differ in parameter values, but they also differ in their structure. In addition, mathematical analysis is considerably more complicated. Dynamical solutions of nonlinear differential equations are usually not known, and even steady-state analysis is difficult; even if a steady state exists, it is often difficult to find.

One could consider computer simulation as a third approach. In fact, it is really not an approach on its own; it is a method that can be used to evaluate linear or nonlinear models. Specifically, one must know the functional form of the system before one can perform computer simulations. Once the model is specified, computer simulations are relatively easy to execute. However, they only supply fragmentary knowledge about the system and they never offer the generality and stringency of mathematical deduction. Furthermore, as the number of parameters in a system increases, the time and cost of an exhaustive evaluation become prohibitive.

S-System Approach

The initial modeling step is to formulate S-system equations for density or growing space plus one measure of size. Other measures can be represented as auxiliary variables that are expressed as algebraic functions rather than as differential equations. Often, these algebraic functions are, at least approximately, simple allometric relationships (Dr. William Harms, pers. comm; see also [19]).

The two-variable S-system representation is obviously a simplification because the growth dynamics of tree populations depends on many influences. These could be included in the S-system at the cost of additional variables and equations. However, all processes occurring much faster than macroscopic growth, such as biomolecular reactions, and all processes occurring much slower than macroscopic growth, such as evolutionary changes, are implicitly incorporated in the S-system. On the relevant time scale, the fast processes are in a quasi steady state and the slow processes are quasi constant during the period of observation [19].

Without further restrictions, a two-variable S-system can exhibit a variety of behaviors such as oscillations, limit cycles, and growth without bound. Some of these behaviors are obviously not realistic either because they are physically impossible or because the experimental design limits such behaviors. For example, it is physically impossible for height to increase without bound; the number of plants cannot increase because the experiment is performed in a way that the stand remains even-aged.

The following observations and controls can be translated into constraints on parameter values (cf. [33]):

1. The average size increases monotonically.
2. The number of plants decreases monotonically since we are concerned with a pure stand of even-aged plants.
3. The system is stable with the average size reaching a maximum and the number of plants reaching the carrying capacity of the site.
4. For an initial period following establishment there is no self-thinning of the tree population and the average tree size increases.
5. The relationship between average size and number of plants approaches the power-law relationship $S = aN^{-b}$. This "limiting relationship" is a theoretical concept that provides an upper bound for plant growth in stands with declining densities. It has been observed in a great variety of perennial plant species (reviewed in [33]). However, it only gives a limit for old plant stands whereas it is of no use in newly established plant communities. Obviously, these young stands are of importance, and this implies that the simple power-law function is insufficient for a full representation of forest dynamics.

When these constraints are incorporated in the general two-variable S-system model, the number of unknown parameters can be reduced from twelve to ten. In addition, the constraints place bounds on some of the parameter values. These reductions in degrees of freedom result in substantial savings in computer time when it comes to estimating optimal parameter values. (Note: In [33], the number of essential parameters could be reduced further because it was assumed that all observations could be scaled with respect to steady-state values and a suitable time scale. Although these assumptions are useful for theoretical considerations, they cannot be used here because the appropriate scaling factors are unknown.)

SIMULATION RESULTS

Preliminary work with the model involved simulating the dynamics of plant stands with different initial sizes and densities. The purpose of this preliminary work was to assure that the S-system model with incorporated constraints would still be able to capture those features of tree population growth that are considered essential to foresters. Three major observations were made: (1) the two-variable S-system generates dynamic results (tree size vs. time, density vs. time) that are in qualitative agreement with observed data; (2) the model trajectories of tree size vs. density qualitatively reflect trajectories from actual data (see Figs. 10.1 and 10.2); and (3) the model simulations clearly show strong interdependence between average tree size and density (cf. [33]). Specifically, for an arbitrary, fixed set of parameter values and the same initial value for size, variation in the initial value for density results in differently shaped growth

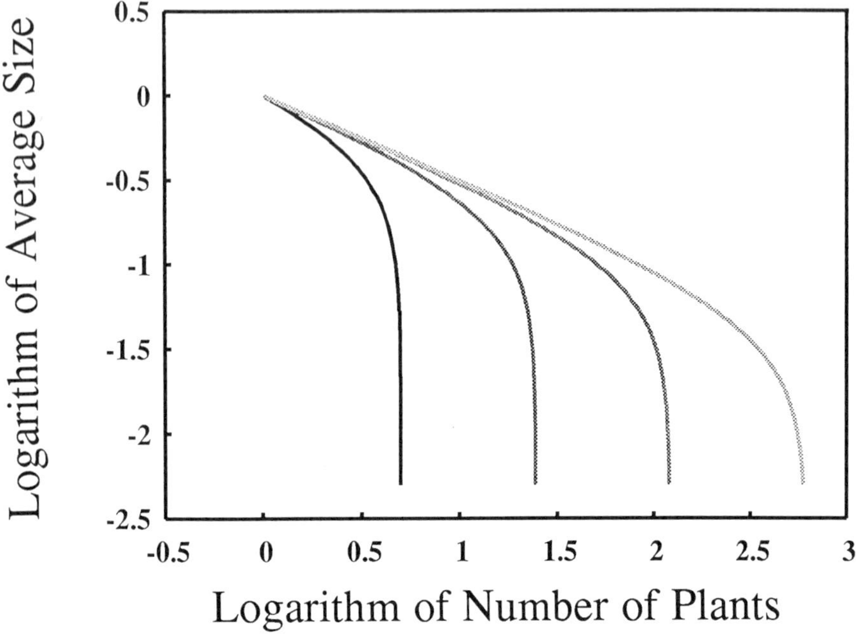

FIGURE 10.2. Trajectorial plot of simulated tree growth data.

curves (cf. Fig. 10.3). If these curves are compared without consideration of the initial tree density (which is not explicitly visible), one might conclude that the four plant stands grew according to different growth functions or according to a general model with different parameter values. Similarly, for an arbitrary, fixed set of parameter values and the same initial value for density, variation in the initial tree size results in density curves of different appearance (cf. Fig. 10.4). Again, this variation is due to the dependence on tree size which is only explicitly visible in the model equations themselves. The important conclusion is that it is necessary to analyze tree size and density simultaneously in order to reveal the true dynamics of the system.

PARAMETER ESTIMATION

Given all parameters of an S-system, numerical solutions can be used to explore the dynamic behavior of biological systems. However, very often experimentalists are interested in the opposite procedure, namely, the determination of parameter values from dynamical or steady-state responses. This determination requires methods of parameter estimation, such as linear or nonlinear regression.

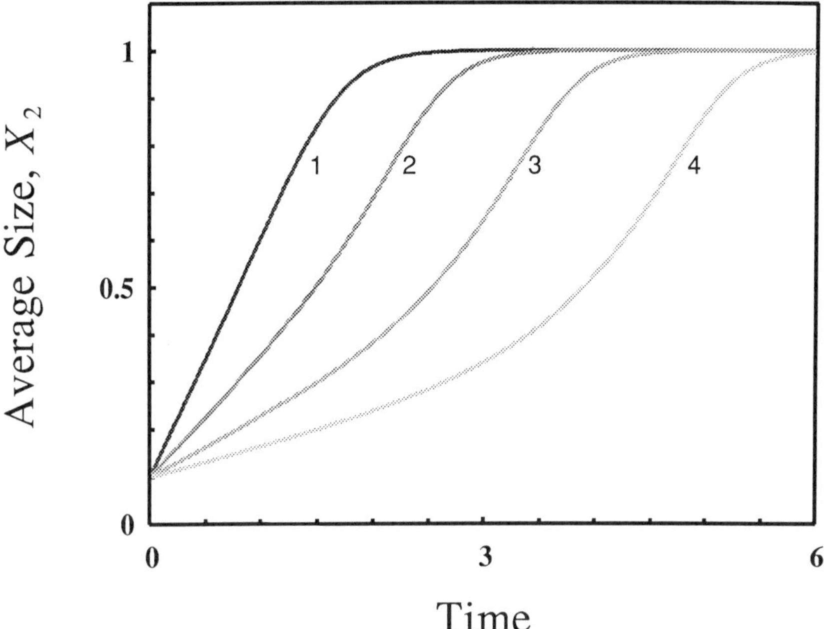

FIGURE 10.3. Tree growth in plots with different densities. (See text for details.)

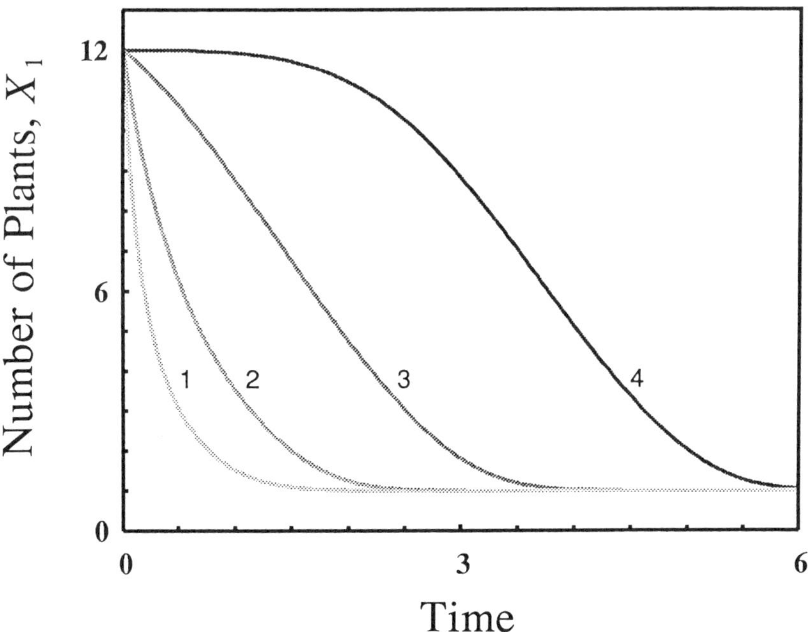

FIGURE 10.4. Decline in density in plots with different tree sizes. (See text for details.)

Linear regression is appropriate when system responses close to a steady state are analyzed, which often is true in biochemical applications ([2]; see also Chapter 2). In the case of steady-state responses, one can furthermore employ methods of sensitivity and logarithmic gain analysis. These methods place constraints on the system parameters and reduce the number of parameter values that must be estimated; thus, the number of degrees of freedom in the search for the full set of optimal values is reduced (cf. Chapters 2 and 4).

The task of estimating parameter values from dynamical data is incomparably more difficult. It is a nonlinear regression problem that is typically very complicated because it involves a global observation of the system, not just an observation of the system at steady state. Specifically, the logarithmic transformation that reduces steady states of S-systems into linear systems is no longer applicable. There has been almost no experience with the problem of parameter estimation for dynamical systems with S-systems (for exceptions, see [22, 31] and Chapter 11).

Nonlinear Regression in S-System Differential Equations

Our preliminary work involved parameter estimation using BMDP's 1988 version of the nonlinear regression program AR. The AR routine estimates parameters of a system of differential equations by using an iterative pseudo-Gauss-Newton algorithm. It computes numerical solutions to the system of differential equations by using a modified fourth and fifth order Runge-Kutta-Fehlberg routine. BMDP has nice features such as automatic computation of the correlation matrix among the parameter values, the ability to specify lower and upper bounds for the parameters, and plots of predicted values, observed values, and residuals. From the analysis of simulated as well as actual data we have gained insight into the functioning of the parameter estimation software. Three results are noteworthy:

1. The parameter estimating algorithm is sensitive to start values for the parameters that have to be specified to initiate the algorithm. For instance, we used as start values the exact parameter values that were known because we worked with simulated data. The parameter estimating algorithm, however, did not always recognize these as the best parameters, but it started extensive searches, sometimes ending up with suboptimal values. We also specified the start values to be 10% above the exact values, 10% below the exact values, or in combinations thereof. In some instances the parameter estimating algorithm terminated with different sets of "best" parameter values. In particular, the parameters of the β-terms were difficult to estimate. This lack of robustness is partly due to the fact that the S-system model is very flexible and it can generate very similar

graphs with quite different combinations of parameter values. In addition, the present data show only the initial growth phase of the *Loblolly* pine stand whereas almost no information is available about the retardation in growth. Since the missing part of the growth curve is dominated by the β-terms of the S-system, the parameters of the β-terms are difficult to estimate.

2. The parameter estimating algorithm is sensitive to the convergence and error criteria that one selects. For example, the algorithm uses certain criteria to decide the relative error of integration that it will tolerate, the absolute error of integration that it will tolerate, and whether or not a solution has been reached. By changing these criteria very different parameter estimates can be obtained.
3. The scales for which the variables are measured influence the parameter estimation algorithm. The parameter estimator works at minimizing the absolute error and, as a consequence, growing space is greatly misrepresented. In response to this problem, we transformed the data so that both variables range between 0 and 1. This was easily accomplished by working with growing space and *reciprocal* height instead of growing space and height itself.

As an alternative to BMDP, we have investigated another software package that is under development at the National Biomedical Simulation Resource at Duke University Medical Center. This package consists of a *S*imulation *Co*ntrol *P*rogram, SCoP, and a parameter estimating program, SCoPfit. SCoP is designed to model either steady-state or time-dependent physical, chemical, or biological systems. SCoPfit is used for estimating those parameter values that produce the closest fit between the computed behavior of the model system and the reference data or experimental observations. Initially we used SCoP for time-dependent systems given in the form of ordinary first-order differential equations. This setup was very similar to that in BMDP's AR program. SCoP allows the modeler to select the integration method for solving the set of differential equations. We tried several of the integrators including fourth-order Runge-Kutta, Euler, and Adeuler methods. The method that seemed most appropriate for our particular problem was the Adeuler subroutine which basically combines Euler's method with adjustable time steps.

Nonlinear Regression in Algebraic S-System Equations

In light of these problems, we transformed the estimation problem into a problem of simultaneous nonlinear algebraic equations rather than first-order ordinary differential equations. This was accomplished by "hand estimating" the slopes of the curves representing growing space vs. time and height vs. time and yielded values for $\dot{X}_1$ and $\dot{X}_2$. Substituting these numbers into the left-hand sides of the differential equations at n time points, we obtained a set of $2n$ algebraic equations in which time was no longer explicit.

It is evident that the procedure for estimating slopes is prone to inaccuracies. Nevertheless, it has been applied successfully in the past (cf. [22]), and preliminary work with *Loblolly* pine data indicates that it may be sufficient at least for obtaining reasonable initial estimates that then can be used in the direct estimation procedures described above. Furthermore, the estimation of $\dot{X}_1$ and $\dot{X}_2$ can be made much more accurate and objective with nonparametric data smoothing techniques; such techniques are currently being developed in our department in collaboration with Dr. A. Georgiev.

The newly derived set of data (X_1, $\dot{X}_1$, X_2, $\dot{X}_2$ at n time points) was analyzed with another routine of the SCoP package that solves simultaneous nonlinear algebraic equations using the Newton-Raphson method. Depending on the start values for the parameters, the program did, in a few cases, meet the convergence criteria. However, the final results depended on the chosen start values.

Nonlinear Regression in a Half-System Model

The currently available *Loblolly* growth data represent trees of ages up to twenty-six years. During this phase, growth in height and diameter seems not to slow down, and the growth curves do not yet reach the inflection point. The same is true for the decrease in density. The monotonicity in rates suggests the use of a *Half-system* instead of the full S-system. This representation contains considerably fewer parameters that can be expected to make the estimation process more reliable. The Half-system can be interpreted as a general system description in which all influences that increase or decrease a system variable X_i are aggregated in a single function which then is approximated with one product of power-law functions (cf. Chapter 2). The general Half-system for increasing diameter at breast height ($DBH = X_1$) and decreasing density (trees per acre; $TPA = X_2$) is

$$\dot{X}_1 = \alpha_1 X_1^{g_{11}} X_2^{g_{12}}$$

$$\dot{X}_2 = -\beta_2 X_1^{h_{21}} X_2^{h_{22}}$$

(cf. Chapters 2 and 12). The reduction to a Half-system offers new avenues of parameter estimation.

Parameter estimation in state space. Given measurements of X_1, $\dot{X}_1$, X_2, and $\dot{X}_2$ at n time points, the set of Half-system equations can be transformed into a linear system by taking the logarithms of both sides of the equations. The resulting system

$$\log(\dot{X}_1) = \log(\alpha_1) + g_{11}\log(X_1) + g_{12}\log(X_2)$$
$$\log(-\dot{X}_2) = \log(\beta_2) + h_{21}\log(X_1) + h_{22}\log(X_2)$$

is linear with respect to the unknown parameters g_{ij}, h_{ij}, $\log(\alpha_1)$ and $\log(\beta_2)$, and the problem of nonlinear parameter estimation becomes a problem of multiple linear regression. Furthermore, the equations are independent of each other and thus can be solved separately with programs like BMDP 1R. Parameter estimates are obtained within seconds as compared to hours or days necessary with the previous methods; they are summarized in Table 10.1.

Although the slopes of diameter and density were estimated by using a ruler and a protractor, the model with the estimated parameters produced close fits to the observed data. Fig. 10.5 shows one set of measured data in comparison with model solutions based on estimated parameter values.

Even though the data fit may be considered sufficiently accurate, the results in Table 10.1 are not entirely satisfactory because the estimated parameter values within the same treatment show no resemblance at all, even in cases where the corresponding growth curves are very similar. This disparity can be explained with redundancies among the parameters. In other words, different combinations of parameter values can produce almost identical solution curves of the Half-system over the observed range. To show some of these redundancies, we computed regression lines for different pairs of parameters. The results are given in Table 10.2.

Table 10.1.

Tmt	Blk	$\log\hat{\alpha}_1$	$\hat{g}_{11}$	$\hat{g}_{12}$	$\log\hat{\beta}_2$	$\hat{h}_{21}$	$\hat{h}_{22}$
1	1	4.8040	−1.0700	−0.6000	−7.5430	3.0500	0.7880
1	2	−3.3330	−0.5163	0.4695	−1.6996	1.1222	0.4899
1	3	−1.7264	−0.9709	0.2548	16.6300	2.0568	−2.6645
1	4	−12.0320	0.5882	1.6539	−25.1890	1.6965	4.1427
2	1	−3.4020	−0.5178	0.4309	−13.5760	0.9600	2.2510
2	2	−1.3383	−0.4767	0.1086	−10.6550	1.8318	1.7665
2	3	6.4480	−0.8940	−0.9308	6.0806	0.7608	−0.4440
2	4	1.1540	−0.4641	−0.2328	−8.6682	2.0003	1.3886
3	1	5.0450	−0.9338	−0.6718	−11.3062	2.3961	1.7273
3	2	−2.9665	−0.1389	0.2631	−28.2210	3.2065	3.9659
3	3	−2.1806	−0.2334	0.1354	−10.1490	1.7747	1.7429
3	4	−1.6370	0.0026	0.0560	−13.9840	2.0428	2.2021
4	1	−1.7078	−0.0040	0.0346	−20.9330	3.3784	2.9303
4	2	−1.7031	0.0209	0.0440	−25.8990	4.0583	3.4204
4	3	−0.1928	0.2000	−0.1593	−18.6780	3.1534	2.7102
4	4	−1.1390	0.0000	0.0000	−12.3150	1.8638	2.1063
5	1	−2.3933	0.0201	0.1253	−27.2330	3.8850	3.6094
5	2	1.9610	−0.5215	−0.3380	−16.1090	2.4884	2.4697
5	3	−2.5893	0.0435	0.1403	−12.7240	1.8352	2.1078
5	4	0.2548	−0.0633	−0.1737	−13.8250	2.0923	2.2148

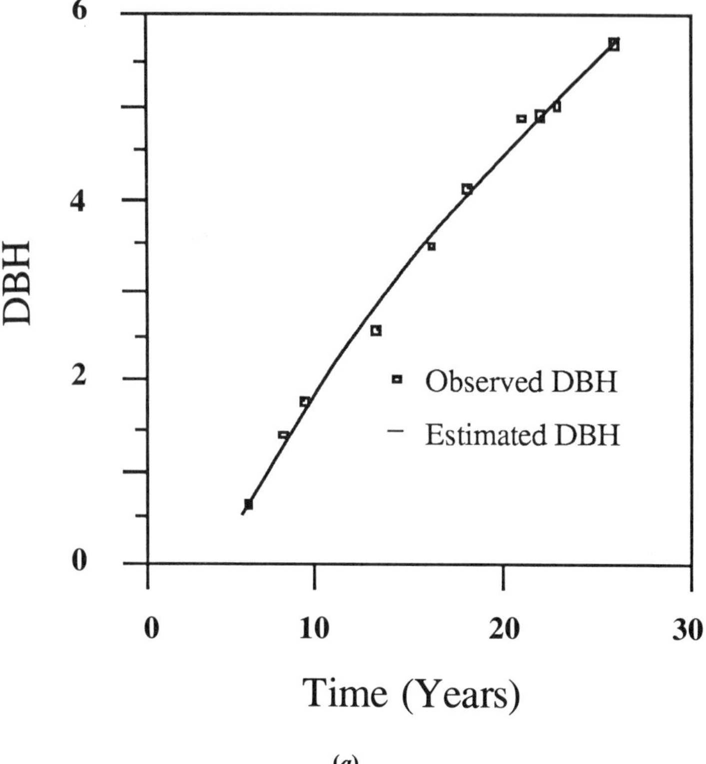

(*a*)

FIGURE 10.5. Observed and fitted tree sizes and densities.

Four p-values of less than 0.001 and three R^2 values greater than 0.95 demonstrate good linear relationships between some of the parameter values (cf. Table 10.2). In cases where R^2 is greater than 0.95, the relationship is almost perfect since more than 95% of the variation in the dependent variable is explained by the independent variable. Graphing the various pairs of parameters confirms these linear relationships between log (α_1) and g_{12}, and between log (β_2) and h_{22}. The interpretation of these results is that the rates of growth and of decrease in density are both related to the influence of tree density but not to the influence of the current tree size.

The linear relationships between the logarithms of the rate constants and the exponents associated with density imply that even though the linear regression process determines a unique optimal solution, other pairs of parameters on the regression lines yield "almost optimal" solutions. This implies that even very small noise in the experimental data may significantly affect the numerical outcome of the linear regression.

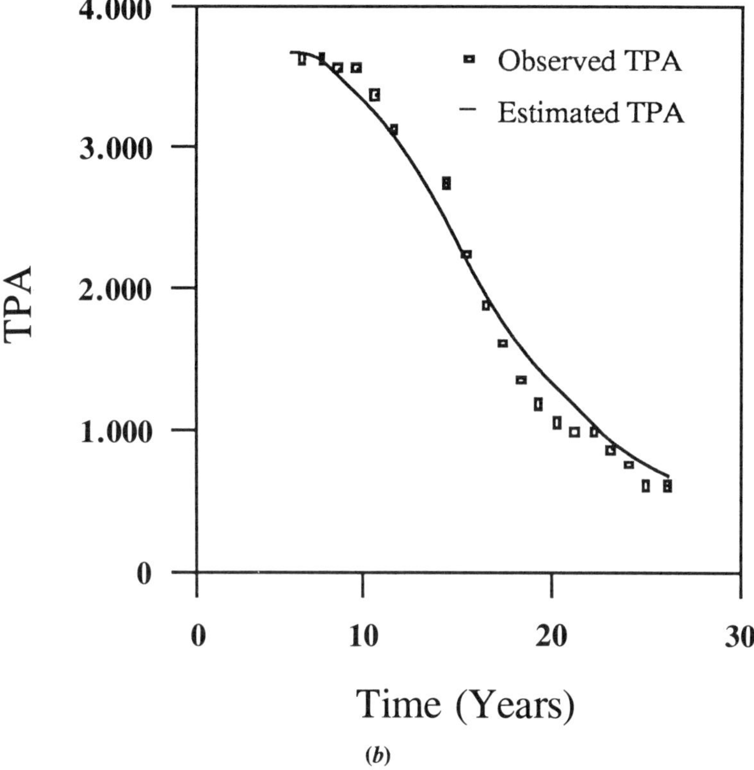

(*b*)

FIGURE 10.5. (*Continued*)

Trajectorial solution. To evaluate further the redundancies between some of the parameters, we studied the trajectories of the system. These are given as relationships between *DBH* and *TPA*; the independent variable *time* is eliminated (cf. Figs. 10.1, 10.2, and 10.6). Because the system representation con-

Table 10.2.

Dep. Var.	Indep. Var.	$\hat{b}_0$	$\hat{b}_1$	p-values	R^2
g_{11}	$\log(\alpha_1)$	−0.3726	−0.0815	<0.001	0.512
g_{12}	$\log(\alpha_1)$	−0.0940	−0.1333	<0.001	0.972
h_{21}	$\log(\beta_2)$	1.6436	−0.0499	0.005	0.357
h_{22}	$\log(\beta_2)$	0.1399	−0.1411	<0.001	0.974
$\log(\alpha_1)$	$\log(\beta_2)$	0.8910	0.1426	0.078	0.163
$h_{21} - g_{11}$	$\log(\alpha_1/\beta_2)$	2.1248	0.0383	0.062	0.181
$h_{22} - g_{12}$	$\log(\alpha_1/\beta_2)$	0.2148	0.1434	<0.001	0.967

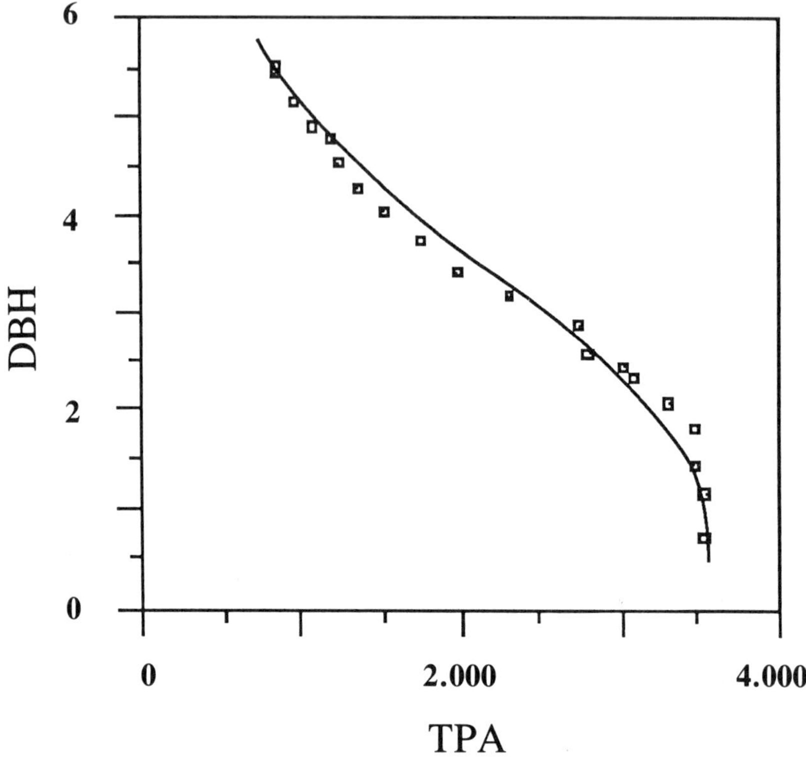

FIGURE 10.6. Observed and fitted trajectories of size vs. density.

sists of two autonomous differential equations with strictly positive solutions, the trajectory of *DBH* vs. *TPA* can be obtained directly from dividing the first by the second equation.

For simplicity, let's abbreviate $X_1(t)$ as $x(t)$ and $X_2(t)$ as $y(t)$. Then the trajectory of x vs. y is given as

$$\frac{dx}{dt}\bigg/\frac{dy}{dt} = \frac{dx}{dy} = \alpha_1 \beta_2^{-1}\, x^{g_{11} - h_{21}} y^{g_{12} - h_{22}}$$

This representation of the trajectory can be streamlined with the introduction of new parameters $a = g_{11} - h_{21}$, $b = g_{12} - h_{22}$, and $\gamma = \alpha_1 \beta_2^{-1}$, which yields

$$\frac{dx}{dy} = \gamma x^a y^b$$

This differential equation can be solved by separation of variables and straightforward integration. For $a = 1$, the solution is the logarithm of x. For $a \neq 1$ the solution is

$$\frac{1}{1 - a} x^{1 - a} = \gamma \frac{1}{b + 1} y^{b + 1} + C$$

from which we can solve for x in terms of y:

$$x = \left[\gamma \frac{1 - a}{b + 1} y^{b + 1} + C(1 - a) \right]^{1/1 - a}$$

This equation is of the form

$$x = [py^s + q]^r$$

where the parameters can be expressed in terms of the original parameters and the initial values $x(t_0)$ and $y(t_0)$ as

$$p = \frac{\alpha_1}{\beta_2} \left[\frac{1 - g_{11} + h_{21}}{1 + g_{12} - h_{22}} \right]$$

$$q = C[1 - g_{11} + h_{21}]$$

$$r = \frac{1}{1 - g_{11} + h_{21}}$$

$$s = 1 + g_{12} - h_{22}$$

where

$$C = \frac{x(t_0)^{1 - a}}{1 - a} - \frac{\gamma}{b + 1} y(t_0)^{b + 1}$$

Using the trajectorial solution, parameter estimation becomes nonlinear, but the relationship between *DBH* and *TPA* is expressed in the form of an algebraic equation instead of a differential equation. Furthermore, there is no need to measure or estimate slopes.

BMDP AR was used to estimate p, q, r, and s. Two options for making "educated guesses" for the initial estimates of p, q, r, and s are available. One can use results from the linear regression to compute initial estimates for p, r, and s. The linear regression does not help with the specification of an initial q, but one may use one observed data pair (x_0, y_0) to estimate C, and thus q. Alternatively, if one is confident in the observed data pair (x_0, y_0), one can express q in terms of p, r, s, x_0 and y_0:

$$q = x_0^{1/r} - py_0^s$$

Substitution of this expression in the trajectorial equation leads to the three-parameter model

$$x = \left[py^s + [x_0^{1/r} - py_0^s]\right]^r$$

For the nonlinear estimations, we selected the values x_0 and y_0 close to the center of the observed time period. The results are presented in Table 10.3. One trajectorial fit with the three-parameter model is shown in Fig. 10.6.

Interpretation of Results

Not counting initial values x_0 and y_0, the final Half-system model we discussed has three parameters. The residual errors in Table 10.3 indicate that the data fit

Table 10.3.

Tmt	Blk	$\log \hat{p}$	$\hat{r}$	$\hat{s}$	Res. SS
1	1	5.2035	0.2096	0.1109	1.1957
1	2	4.6720	0.2660	−0.2705	1.0944
1	3	0.7243	0.1163	2.1060	6.6608
1	4	6.4948	0.3314	−1.5082	1.3804
2	1	5.3514	0.3560	−0.9258	2.0332
2	2	9.0189	0.2147	−1.9159	0.5138
2	3	1.8438	0.2674	0.4255	2.5253
2	4	4.5021	0.2833	−0.3912	0.9299
3	1	6.3380	0.2415	−1.0077	2.4310
3	2	11.0961	0.2227	−2.6719	0.3895
3	3	3.9937	0.3250	−0.5363	0.4101
3	4	5.3038	0.3326	−0.9844	0.9106
4	1	8.7125	0.2324	−1.9080	0.0727
4	2	10.8158	0.2063	−2.4260	0.3002
4	3	8.3616	0.2548	−1.8901	0.5974
4	4	4.9629	0.3346	−0.9767	0.6751
5	1	11.1090	0.2174	−2.5533	0.1881
5	2	7.8675	0.2770	−1.7954	0.7078
5	3	4.7352	0.4162	−1.0167	0.5102
5	4	7.4616	0.3166	−1.7424	0.6938

is reasonably good, and numerical solutions based on the estimated parameters confirm this result.

The residual error decreases from treatment 1 (1,000 trees per acre) to treatments 4 and 5 (8,000 or 16,000 trees per acre). This indicates that the model is more appropriate for higher densities. While this may be surprising at first sight, one can give a possible explanation. In plots with low tree density, the trees grow faster and their growth curves start to approach the final phase of decreasing growth rate. By contrast, trees in dense plots grow slower and at the end of the observation period they still are in their initial phase of growth, which is more accurately modeled with the Half-system model.

The parameter r does not vary much throughout all treatments, and the variation between treatments is not higher than within treatments. Writing the trajectorial solution as

$$x^{1/r} = py^s + q$$

one sees that the parameter r is associated with tree size. At the beginning of the experiments, all trees have the same size, independent of the density. While the size difference between trees of different treatments increases with time, its effect on self-thinning is less significant than the decreasing density.

The parameters $\log p$ and s in Table 10.3 differ considerably within and between treatments. However, they vary in a systematic fashion, and a plot of s vs. $\log p$ shows a clear linear relationship ($p < 0.001$; $R^2 = 0.88$). Furthermore, along the regression line, the pairs $(s, \log p)$ are more or less ordered according to treatment, with decreasing s and increasing $\log p$ for increasing tree density. Treatments 4 and 5 do not differ statistically, which means that during the observation period a density of 8,000 trees per acre is as influential as a density of 16,000.

DISCUSSION

To the uninitiated, the estimation of S-system parameter values from the available *Loblolly* growth data may seem to be a routine problem. The data set is rather complete, the model structure is well defined, and several software packages claim to perform the estimation procedure. However, we have seen that the problem is far from trivial. Since only the first part of the growth dynamics is actually known, the S-system model is ill determined, and technical difficulties lurk in almost all parts of the analysis. Some of these difficulties can be circumvented by prudent choices of algorithms and technical parameters; other seem to be intrinsic.

Beyond experience with particular models and software packages, several strategies may have the potential of improving the estimation process:

1. The complexity of the estimation problem increases dramatically with the number of parameters to be estimated. Any constraints or bounds on parameters are extremely important.
2. It seems mandatory to determine good start values before parameter values in S-system differential equations can be estimated reliably. The experienced user may be able to find reasonable start values by trial and error and by running ESSYNS repeatedly.
3. A promising method for determining start values is the transformation of differential equations into algebraic equations. This requires determining slopes by hand or with nonparametric methods, but it drastically reduces computation times of the estimation process. Even if the results from this procedure are not optimal, they can be used as start values for the estimation with differential equations.
4. If in the relevant domain the S-system model is dominated by only one term in each equation (the Half-system form), the estimation process reduces to multiple linear regression, which obviously offers significant advantages. Again, the results of this regression may not be optimal but they are sufficient as start values in the original differential equation model. In autonomous models with only two variables, the Half-system form allows the analytical computation of trajectorial solutions that contain about half as many parameters as the original model.
5. The commercially available parameter estimation packages use fourth- and fifth-order Runge-Kutta-Fehlberg routines to solve the differential equations. These routines are significantly inferior to the solver in ESSYNS which typically solves S-systems one or two orders of magnitude faster than any available Runge-Kutta code (cf. Chapter 7 and [38, 46]). It is planned to integrate the efficient numerical algorithm of ESSYNS with the attractive features of BMDP and SCoP.

SUMMARY

S-system and Half-system models were developed to understand the growth and dynamics of even-aged *Loblolly* pine populations. The models capture observed phenomena, such as limited growth and synergistic interactions between individual trees. Of particular importance for this project are the mutual dependencies between size and density of *Loblolly* pine trees. Quantitative results were obtained from the estimation of parameter values for the S-system or Half-system models. Various estimation methods were evaluated and recommendations for future estimation procedures were made.

ACKNOWLEDGMENTS

This research was supported by the USDA Forest Service, Southeastern Forest Experiment Station, Structure, Growth, and Yield of Loblolly Pine Stands in the Southeastern United States Research Unit, Charleston, South Carolina. In particular, we thank Dr. William Harms of the Southeastern Forest Experiment Station for his continued help during the execution of this work.

11. Estimating Parameters of S-Systems

THOMAS JOHNSON

S-systems provide flexible forms for modeling nonlinear dynamic processes. The computational advantages of S-systems and the availability of the ESSYNS program (cf. Chapter 7 and [30, 38]) to facilitate computations on PC-class microcomputers are sure to increase the use of nonlinear differential equation models of dynamic processes. With this development, it will be important to be able to estimate the parameters of such models. Nonlinear regression techniques available in standard statistical packages will do the job for the special cases that can be solved in terms of known analytic functions. But important S-system models move beyond these special cases with so-called closed-form solutions.

Except for the rare cases where the mathematical model and measurements fit with no appreciable error, the form of the differences between the mathematical model and the measurements influences the properties of the estimates of parameters obtained by fitting the model to data. When the mathematical model is a system of differential equations, the statistical model is a system of stochastic differential equations.

A system of stochastic differential equations (SDEs) is usually written

$$dX = f(X, \theta)\, dt + B(X, t)\, dW \tag{11.1}$$

with X the $n \times 1$ vector of system variables, θ a $p \times 1$ vector of parameters, B an $n \times m$ matrix, and W an m-dimensional stochastic process that has the Martingale properties.

I have developed a procedure that combines results from several areas of statistical theory and numerical analysis to produce a new method for estimating the parameters of a system of nonlinear stochastic differential equations [31]. The resulting estimates are assured desirable asymptotic properties under assumptions that may be more generally applicable than those of existing methods. Basically, this method gives a way to move to the discrete times of the observations, whether or not observations are equally spaced, while retaining the identity and interpretation of the parameters from the original nonlinear system.

The advantage that this procedure has for modelers using the S-system is the generality that eliminates the necessity of restricting the model to a form that can be linearized over the entire range (e.g., see [156]) or with complicated (although analytic) solutions (e.g., see [24]). If the processes in the system are smooth enough for this procedure to work, many modeling compromises and considerable tedious algebra can be avoided. The advantage that the S-system has as a representation of the response function $f(X, \theta)$ is the ease of programming the model for the algorithm in addition to the interpretability of the model.

After summarizing the technique for estimating parameters of a system of nonlinear stochastic differential equations, in the remainder of this paper I concentrate on an example of modeling the crooked growth of pine trees. This example demonstrates a discovery of misspecification of a model in the forestry literature.

A SUMMARY OF THE ESTIMATION TECHNIQUE

The essential features of my technique for estimating the parameters of a system of nonlinear stochastic differential equations is to use quasi linearization to obtain one-step-ahead predictions of observations and error variances. This conditional least-squares technique obtains parameter estimates that minimize the sum of the errors in these one-step-ahead predictions.

Quasi linearization is a well-established approach to the treatment of nonstochastic differential equations [160–161], and without the explicit introduction of a stochastic process as in Eq. (11.1), the convergence properties for parameter estimation using quasi linearization have been established [154]. The quasi linearization procedure is to expand the function $f(X, \theta)$ in Eq. (11.1) in a first-order Taylor series about each observation. When there are no replications or cross-section components in the data, quasi linearization is expressed as

$$\begin{aligned} f(X, \theta) &= f_k + A_k(X - X_k) \\ &= a_k + A_k X \end{aligned} \tag{11.2}$$

where $f_k = f(X_k, \theta)$, $A_k = \{\partial f_i/\partial X_j/X_k\}$, and $a_k = f_k - A_k X_k$ within time interval k such that $t_k < t < t_{k+1}$.

With quasi linearization, the system (11.1) is approximated by

$$dX = (a_k + A_k X)\,dt + B_k\,dw \tag{11.3}$$

with $B_k = B(X_k, t_k)$ and w a stochastic process that approximates the process W in (11.1).

The crucial assumption for this method to produce recognizable results is to assume that *after* quasi linearization, B_k is a constant matrix in the interval following t_k and the stochastic process w is a Wiener process (i.e., a Gaussian white noise).

Since (11.2) is a nonlinear nonhomogeneous stochastic differential equation system, the solution X_{k+1} for $t_k < t_{k+1}$ with $k = t_{k+1} - t_k$ is

$$X_{k+1} = \exp\left(A_k(\delta_k)\right)X_k + \int_0^{\delta_k} \exp\left(A_k(\delta_k - s)\right)a_k\,ds + \int_0^{\delta_k} \exp\left(A_k(\delta_k - s)\right)B\,dw_s \tag{11.4}$$

[153, 155] where $\exp(A_k)$ is the matrix exponential function [157]. And, with the w a Wiener process, X_{k+1} has a normal distribution with

$$E(X_{k+1} \mid X_k) = \exp\left(A_k(\delta_k)\right)X_k + \int_0^{\delta_k} \exp\left(A_k(\delta_k - s)\right)a_k\,ds \tag{11.5}$$

$$\operatorname{cov}(X_{k+1}) = \int_0^{\delta_k} \exp(A_k s)\,U \exp(A_k' s)\,ds$$

$$U = BB' \tag{11.6}$$

where the prime ′ denotes transpose. If these matrix exponential functions can be computed reliably and economically, a nonlinear least-squares algorithm can be used to estimate the parameters using conditional least squares. The appendix gives some key results for computing the required matrix exponentials.

Conditional least squares is a procedure of obtaining the parameter estimates that minimizes the sum of squared residuals of the observed system variables X_k from their expected values conditional on all previous observations, that is, they are one-step-ahead predictions. The most straightforward way to apply conditional least squares is to discretize the model to get

$$X_{k+1} = X_k + f(X_k, \theta)(t_{k+1} - t_k) + \epsilon'_{k+1} \tag{11.7}$$

which is the first differencing model.

There are two frequently encountered reasons for unobservable system variables being included in a continuous time model. First, a transformation (such as recasting into an S-system) can introduce variables that cannot be observed directly (cf. Chapter 12). Reducing a higher order model and observing the integral of a flow variable are also important special cases of such transformations. Second, if the observed errors are serially correlated, the model must be modified to have the remaining errors (dW) follow a Wiener process. I have shown [159] how even a single unobservable variable can lead to autoregressive error terms.

Regardless of the source of important unobserved variables, the effects coming through these variables must be included if all the information in the observations is to be used to make the forecasts required by conditioned least squares. For a nonlinear model, this requires a procedure such as the extended Kalman Filter [158]. In my algorithms I have used the extended Kalman Filter suggested by Jazwinski since it too depends upon linearization at each observation.

AN EXAMPLE WITH CROOKED TREES

In his dissertation, Smith [162] needed to model the crooked growth of pine trees as a step in developing his procedure for obtaining maximum lumber yield. The measure of crookedness is the sweep, which is the distance of the pith from the straight line from the pith at the base to the top of the tree. Denoting the sweep by Z, the model that Smith drew from the literature is

$$d^2Z/dt^2 + \theta_1\, dz/dt + \theta_2 Z = \theta_3 \tag{11.8}$$

with initial conditions $Z(0)$, $\dot{Z}(0)$.

To get this model into a first-order system of the form (11.1), let

$$\begin{aligned} X_1 &= Z,\ X_2 = dZ/dt \\ dX_1 &= X_2\, dt \\ dX_2 &= (\theta_3 - \theta_1 X_1 - \theta_2 X_2)\, dt \end{aligned} \tag{11.9}$$

with initial conditions $X_1(0) = Z(0)$, $X_2(0) = \dot{Z}(0)$.

Since the sweep is observed at intervals by sawing cross sections from the tree, the rate of change of sweep, X_2, was not observed. Thus the transformed system presents a filtering problem.

Equation (11.9) can then be easily recast as an S-system in the manner of Savageau and Voit (Chapter 12 and [46]) by letting $Y_1 = X_1 = Z$, $Y_2 Y_3 = X_2 = \dot{Z}$. This gives the S-system

$$
\begin{aligned}
dY_1 &= Y_2 Y_3 \, dt \\
dY_2 &= (\theta_3 Y_3^{-1} - \theta_1 Y_1 Y_3^{-1}) \, dt \\
dY_3 &= -\theta_2 Y_3 \, dt
\end{aligned}
\tag{11.10}
$$

with $Y_1(0) = Z(0)$, $Y_2(0) = 1$, $Y_3(0) = \dot{Z}(0)$ where the choice of specification of initial conditions on Y_2 and Y_3 to give $(Y_2 \cdot Y_3)(0) = \dot{Z}(0)$ provides the dimensionality restriction. However, including a random error process leads to a little extra complication. If the process is additive in the original equation (11.8), the resulting first-order system (11.9) becomes

$$
\begin{aligned}
dX_1 &= X_2 \, dt \\
dX_2 &= (\theta_3 - \theta_1 X_1 - \theta_2 X_2) \, dt + \beta \, dw \\
X_1(0) &= Z(0), \; X_2(0) = \dot{Z}(0)
\end{aligned}
\tag{11.9$'$}
$$

The recasting transformation then leads to

$$
\begin{aligned}
dY_1 &= Y_2 Y_3 \, dt \\
dY_2 &= (\theta_3 Y_3^{-1} - \theta_1 Y_1 Y_3^{-1}) \, dt \\
dY_3 &= -\theta_2 Y_3 \, dt + Y_2^{-1} \beta \, dw \\
Y_1(0) &= Z(0), \; Y_2(0) = 1, \; Y_3(0) = \dot{Z}(0)
\end{aligned}
\tag{11.10$'$}
$$

where the random term is $Y_2^{-1} \beta \, dw$.

The model of equation (11.8) is simple enough to be solved in terms of known functions so that the parameters could be estimated with conventional nonlinear least squares. The tree that Smith had coded #67 produced intriguing results typical of several trees in his sample. The conventional nonlinear least-squares estimation procedure gave the results

$$\hat{\theta}_1 = .139 \qquad \hat{\theta}_2 = .110 \qquad \hat{\theta}_3 = .169$$

$$\text{SSE} = 3.22 \qquad \text{D. W.} = 3.25$$

The Durbin Watson statistic of 3.25 gives rejection of uncorrelated errors in favor of negative serial correlations at the $\alpha = 0.01$ level. Such significant serial correlation means that the model (11.8) is missing some important aspect of the changing sweep of tree #67. This misspecification, and especially the negative serial correlation of errors, leads to some interesting problems in using my conditional least-squares technique.

Since the system is linear in the variables in the form of equation (11.9′), I proceeded without recasting into the S-system of equation (11.10′). With conditional least squares and the Kalman filter for updating the unobserved $X_2 = dZ/dt$, the parameter estimates were

$$\hat{\theta}_1 = .622 \quad \hat{\theta}_2 = .991 \quad \hat{\theta}_3 = .793 \quad \text{SSE} = 5.54$$

Using the information from the most recent observation, instead of fitting one smooth line to all observations, caused the criterion SSE to *increase* from 3.22 to 5.54. This tells us that the model is misspecified because for a correctly specified model, the conditional least squares with Kalman filter make optimal use of the recent information.

The negative serial correlation of error terms implied by the Durbin-Watson statistic of 3.25 in the conventional nonlinear least-squares estimation led me to allow for serially correlated errors in the conditional least-squares method. Serially correlated errors can be modeled in a stochastic differential equation system by including a new vector of unobservable variables in the following manner:

$$\begin{bmatrix} dX \\ dv \end{bmatrix} = \begin{bmatrix} ao + AX + Cv \\ -Fv \end{bmatrix} dt + \begin{bmatrix} B & O \\ O & I \end{bmatrix} \tag{11.11}$$

In Eq. (11.11), v is a new vector of unobservable variables, and I is the identity matrix. In the present problem, only one v was used, so $I = 1$. However, this addition led to a reduction in the criterion sum of squared errors to only 5.25—still much larger than the 3.22 obtained with the conventional technique.

I next allowed for an autoregressive moving average component in the observable $X_1 = Z$. This is accomplished by introducing a new parameter in the equation for dX_1 to give the following model:

$$\begin{aligned} dX_1 &= (1 + \theta_4)X_2\, dt \\ dX_2 &= (\theta_3 - \theta_1 X_1 - \theta_2 X_2)\, dt + \beta\, dw \end{aligned} \tag{11.12}$$

This led to a statistically insignificant reduction in the SSE to 4.72. From these results, we must conclude that the model misspecification is not as simple as a conventional first-order serial correlation or an autoregressive moving average problem.

Finally, I recognized that the Kalman gain matrix is multiplied by one in the conditional least-squares approach, while it is multiplied by zero in the conventional unconditional least-squares technique. Therefore, I included a multiplier of the Kalman gain as a parameter to be estimated. With this parameter labeled as θ_5, the conditional least-squares technique gave

$$\hat{\theta}_5 = -1.77 \quad \text{SSE} - 1.001$$

The reduction in SSE from 3.22 to 1.001 gave an F-test statistic of 28.9 with 1 and 13 degrees of freedom. The corresponding $\alpha = 0.01$ critical value for this F is 9.65, so θ_5 is very significantly different from zero and it is in the opposite direction from the conditional least squares value of $+1$.

This negative multiplier on the Kalman gain matrix indicates that the most recent observation on the observable sweep is actually misleading in predicting the next observation by the received model. Therefore, I consider that for some class of trees, such as troublesome #67, the simple linear differential equation model of Eq. (11.9′) is badly misspecified. The problem cannot be glossed over by some time-honored statistical trick, such as a first-order autocorrelation on an autoregressive moving average specification of the stochastic process. A more complete mechanistic model of the reason for the crooked growth of pine trees is required to understand the shape of these trees. An S-system is surely the leading contender for such a model.

SUMMARY

The technique of conditional least squares with quasi linearization is reviewed briefly. This technique is applicable to well-behaved systems of nonlinear stochastic differential equations, including S-systems. An example is given for estimating the parameters in a model for the crooked growth of pine trees. In this case, recasting the original model as an S-system complicates the random error term. However, the results of the estimation show that the original equation is badly misspecified for the data analyzed. Remodeling the process with a more flexible S-system seems indicated.

APPENDIX

Computation of Integrals and Derivatives of Matrix Exponentials

Van Loan [163] shows that if

$$C = \begin{bmatrix} A_1 & B_1 & C_1 \\ 0 & A_2 & B_2 \\ 0 & 0 & A_3 \end{bmatrix} \begin{matrix} \} \, n_1 \\ \} \, n_2 \\ \} \, n_3 \end{matrix}$$

then for $t > 0$

$$\exp (Ct) = \begin{bmatrix} F_1(t) & G_1(t) & H_1(t) \\ 0 & F_2(t) & G_2(t) \\ 0 & 0 & F_3(t) \end{bmatrix}$$

where

$$F_j(t) = \exp (A_j t) \qquad j = 1, 2, 3$$

$$G_j = \int_0^t \exp \big(A_j(t - s)\big) B_j \exp (A_{j+1}) s \, ds \qquad j = 1, 2$$

$$H_1 = \int_0^t \exp \big(A_1(t - s)\big) C_1 \exp (A_3) s \, ds$$

$$+ \int_0^t \int_0^s \exp \big(A_1(t - s)\big) B_1 \exp \big(A_2(s - r)\big) B_2 \exp (A_3 r) \, dr \, ds$$

Van Loan carries these results out to a fourth block column headed by D_1 in the C matrix and K_1 in the e^{Ct} matrix. Thus, for the matrix $\exp (A_k d_k)$ and the integral required in (2.4) for the expectation of x_{k+1} given x_k, let

$$C_{1k} = \begin{bmatrix} A_k & I \\ 0 & 0 \end{bmatrix}$$

and compute $\exp (C_{1k} d_k)$ so that $F_{11k} = \exp (A_k d_k)$, and

$$G_{1k} = \int_0^{\delta_k} \exp \big(A_k(\delta_k - s)\big) \, ds$$

which gives the desired integral as

$$\int_0^{\delta_k} \exp\left(A_k(\delta_k - s\right)a_k \, ds = G_{1k}a_k$$

Similarly, the integral for the covariance is obtained by letting

$$C_2k = \begin{bmatrix} -A_k & U \\ 0 & A_k' \end{bmatrix}$$

so that

$$\int_0^{\delta_k} \exp\left(A_k s\right) U \exp\left(A_k' s\right) ds = F_{11k}G_{2k}$$

The Gauss-Newton algorithm for computing the least-squares estimates of parameters requires the first derivatives of the expected values of x_k with respect to the parameters. Usually it is preferable to compute derivatives from analytic expressions rather than to take numerical differences. Therefore, we want an expression for the derivatives $d(\exp(A\delta))/dq$ where A is a function of q.

For ease of notation, let A be a function of a scalar parameter q and find the total derivative $d \exp(A(q)\delta)/dq$. Define $D = A(q + h) - A(q)$ so that $A + D = A(q + h)$. From a result of Golub and Van Loan [163: p. 396],

$$\exp\left((A + D)\delta\right) - \exp\left(A\delta\right) = \int_0^{\delta} \exp\left(A(\delta - s)\right)D \exp\left((A + D)s\right) ds$$

Therefore,

$$\begin{aligned} d \exp\left(A\delta\right)/d\theta &= \lim_{h \to 0} \int_0^{\delta} \exp\left(A(\delta - s)\right)D \exp\left((A + D)s\right) ds/h \\ &= \int_0^{\delta} \exp\left(A(\delta - s)\right) \lim_{h \to 0} D \exp\left((A + D)s\right) ds/h \end{aligned}$$

Thus,

$$d \exp\left(A\delta\right)/d\theta = \int_0^{\delta} \exp\left(A(\delta - s)\right)(dA/d\theta) \exp\left(As\right) ds$$

since $\lim AB = (\lim A)(\lim B)$.

Note that this derivative takes the familiar form

$$d \exp (A\delta)/dq = dA\delta/dq \exp (A\delta)$$

if and only if A and dA/dq commute, since from the definition of $\exp (A\delta)$, we have $\exp(-As) \exp(As) = I$.

To compute the derivative with respect of θ_j, use the Van Loan result again letting

$$C_{3kj} = \begin{bmatrix} Ak & \partial A/\partial\theta_j \\ 0 & A_k \end{bmatrix}$$

with G_{3kj} from $\exp(C_{3kj}\delta)$, so that $\partial \exp(A\delta)/\partial\theta_j = G_{3kj}$.

Finally, computing the derivative of the expected value of x_k with respect to θ_j requires computing

$$\partial\left[\int_0^\delta \exp_j \left(A_k(\delta - s)\right) ds\right]\Big/\partial\theta_j = \int_0^\delta \left[\partial \exp\left(A_k(\delta - s)\right)/\partial\theta_j\right] ds$$

which, with the substitution $t = \delta - s$,

$$= \int_0^\delta \int_0^\tau \exp\left(A_k(\tau - r)\right)(\partial A_k/\partial\theta_j) \exp\left(A_k r\right) dr\, d\tau$$

If we let

$$C_{4kj} = \begin{bmatrix} 0 & I & 0 \\ 0 & A_k & \partial A_k/\partial\theta_j \\ 0 & 0 & A_k \end{bmatrix}$$

The Van Loan result gives the desired derivative from the H matrix from $\exp(C_{4kj}\delta)$, since this gives

$$H_{4kj} = \int_0^\delta \exp\left(0(\delta - s)\right)0 \exp\left(A_k s\right) d\tau +$$

$$\int_0^\delta \int_0^\tau \exp\left(0(\delta - \tau)\right)I \exp\left(A_k(\tau - r)\right)(\partial A_k/\partial\theta_j) \exp\left(A_k r\right) dr\, d\tau$$

These results give all of the elements to compute the analytic expressions in

$$\begin{aligned}
\partial \hat{x}_{k+1}/\partial \theta_j = {} & \left(\partial \exp\left(A_k/\partial \theta_j\right)\hat{x}_k + \exp\left(A_k \delta\right)\partial \hat{x}_k/\partial \theta_j\right. \\
& + \left\{\partial\left[\int_0^{\delta} \exp\left(A_k(\delta - s)\right) ds\right]\Big/\partial \theta_j\right\} a_k \\
& + \left\{\int_0^{\delta} \exp\left(A_k(\delta - s)\right) ds\right\} \partial a_k/\partial \theta_j
\end{aligned}$$

which is what is required in the Gauss-Newton algorithm.

Part III

Mathematical Properties of S-Systems

12. Canonical Nonlinear Forms: Recasting

EBERHARD O. VOIT

In the typical modeling situation, the essential components of a phenomenon are known and there is also *qualitative* information about the dynamic processes that govern the phenomenon. For instance, experiments may have shown that a component Y increases when another component X decreases or that a component Z is almost unaffected by fluctuations in X. In complex systems, the topology of interactions between the components and the qualitative responses of system variables to perturbations may be all that is known. In other systems, responses of system variables to different types of input may be measurable and provide additional *quantitative* information. Several of the preceding chapters showed that topological knowledge of the investigated network structure is sufficient to set up an S-system representation in symbolic form. Quantitative information can be translated into parameter values or, at least, used to establish constraints between the system parameters. The resulting S-system is mathematically guaranteed to be an exact description at an operating point of our choice and usually quite a good approximation in the vicinity of this point.

In this and the following chapters, we shall discuss a different aspect of S-systems and some other canonical nonlinear forms that has been called *recasting*. Recasting is the transformation of a set of differential equations or nonlinear functions into an S-system or a related canonical form that produces *exactly* the same solution as the original. Thus, in contrast to the modeling with S-systems, where the resulting set of differential equations is a power-law *approximation* of the original phenomenon, recasting generates a *canonical form realization* that is mathematically equivalent with respect to the solution, even though the appearance of the resulting representation may be very different from the original equation.

Recasting presents a fascinating branch of canonical nonlinear analysis with numerous theoretical and practical implications. The theoretically most intriguing fact is that virtually all nonlinearities that can be represented as a set of ordinary differential equations can also be represented as S-systems. Thus, the S-system structure is extraordinarily general and provides a novel tool for the classification and generalization of nonlinearities. Recasting enormously increases the importance of understanding in detail the mathematical properties of S-systems. It is also of practical interest since it has been shown that differential equations often can be analyzed very efficiently once recast as S-systems; for instance, the speed of numerical solution in some cases can be increased by one or two orders of magnitude (cf. Chapter 13 and [30, 38, 46]).

The recasting idea has accompanied the development of the Power-Law Formalism from the beginning (cf. Chapter 1). Phenomena like chemical kinetics, exponential growth, radioactive decay, and the original predator-prey models of Lotka [167] and Volterra [178] were immediately seen to be special cases of S-systems. The next significant step toward more general recasting was the insight that the introduction of new variables and differentiation in many cases transformed a nonlinearity into the S-system form. These techniques were used to recast growth functions [3, 19, 41] and central probability distributions [42]. They demonstrated that all these functions are special cases of an S-system with only one or two variables. This is quite surprising, since growth and probability functions in their original form have quite a different mathematical appearance. The S-system representation of this class of functions also contains numerous growth and probability functions without particular names and, in fact, it constitutes a continuous space of such functions in which the known special cases are embedded (cf. the Conclusion).

Many physical laws, for instance those describing electrical circuits, gravitation, unforced vibrations, cooling, and dilution processes, were recognized to be recastable as S-systems, as were some Bessel functions and elliptic integrals (cf. [24, 111]). The recastability of these special functions destroyed the hope that one might be able to find general closed-form solutions to S-systems in terms of elementary functions, since elliptical integrals are known to possess no such solutions. Even though some classes of S-systems were solved analytically [24], the lack of general solutions made it even more important to devise efficient numerical solutions, as they are now implemented in the S-system solver ESSYNS (cf. Chapter 7 and [7]).

The initial steps of the recasting process typically lead to differential equations whose right-hand sides contain several power-law terms. Thus, the last major step toward general S-system recastability was to devise a method to reduce the number of terms in a differential equation to just one difference of power-law terms, that is, to the S-system form. With this problem solved, it appears now that virtually any finite set of ordinary differential equations and virtually any other set of nonlinearities that can be represented in terms of ordinary differential equations can be recast as an S-system [46].

This chapter summarizes the concepts of recasting. It discusses the basic recasting rules and illustrate the recasting process with some examples. It then compares S-systems resulting from alternative recasting strategies and indicates how recasting can be employed to transform S-systems into alternative canonical nonlinear forms with the same scope and generality. Even though the alternative canonical forms generate the same solution as the original differential equation, they are not equivalent in every aspect. For instance, they often differ in the number of variables and equations or with respect to numerical efficiency. This makes it necessary to discuss criteria of optimal recasting and to compare the alternative canonical forms in terms of these criteria.

The subsequent chapters build on this chapter by introducing extensions and applications. Chapter 13 discusses the recasting of probability distribution functions and illustrates some of the exceptional properties of complicated problems once they are recast as S-systems. It also provides refined recasting techniques that can significantly improve the numerical properties of the S-system realization. Chapter 14 shows how recasting can be used as a modeling tool and thus it integrates recasting concepts and the analysis of organizationally complex systems discussed in earlier chapters. Chapter 15 discusses some algebraic properties of canonical nonlinear forms. It provides insight into the structure of these forms and complements this chapter by showing how the original equations can be retrieved from a recast system.

RECASTING TECHNIQUES

The key tool of recasting is the chain rule, since it decomposes nested functions into products of functions. Specifically, a new variable is defined for each nested function in the original differential equation and then it is expressed in the form of a differential equation. Because of the chain rule, its derivative is equal to a product of functions. For some elementary functions one additional equation is sufficient, while others, such as trigonometric functions, require two (cf. Chapter 14). Repeated application of the chain rule typically leads to sums and differences of products of power-law functions.

For instance, to recast the differential equation

$$\dot{Z} = Z^2 \exp(Z), \quad Z(t_0) = Z_0$$

one defines $X_1 = Z$, $X_2 = \exp(X_1)$, and according to the chain rule, the S-system realization is

$$\dot{X}_1 = X_1^2 X_2$$

$$\dot{X}_2 = X_2 \dot{X}_1 = X_1^2 X_2^2$$

The initial values for X_1 and X_2 are given as $X_1(t_0) = Z_0$ and $X_2(t_0) = \exp Z_0$.

The second recasting step is to assure that all variables are nonnegative, according to the definition of S-system variables (cf. Chapter 2). (This condition could be relaxed but it would prevent us from using any of the widely used analytical or numerical methods that are based on the logarithmic transformation of variables.) There are basically three strategies for enforcing the positivity of variables. If a variable X_i is strictly negative, a simple sign change is sufficient; the signs of all power-law terms in which X_i appears have to be adjusted correspondingly. If X_i changes sign within the relevant range of the independent variable, but it always is greater than a constant, $c < 0$, X_i is replaced with a new variable $X_i' = X_i + c'$, where $c' \geq |c|$. This linear shifting is intuitively straightforward but it sometimes leads to considerable complication of the system equations. A combination of sign change and shifting can be applied to variables with values in the range $(-\infty, +|c|)$. Finally, if a variable X_i assumes all values in the range $(-\infty, \infty)$, or if the range of its values is unknown, X_i can be replaced with the difference of two new positive variables, for example, $X_i = X_{n+2} - X_{n+1}$, where $X_{n+1} = (1 + X_i^2)^{1/2}$ and $X_{n+2} = X_i + X_{n+1}$. In particular cases, *ad hoc* transformations are possible and often preferable (cf. Chapter 13 and [48, 49]).

When all functions except for power-law functions are decomposed and all variables have been made positive, the resulting equations typically consist of the sums and the differences of several power-law terms. Systems of this type are known as *Generalized Mass Action* (GMA) systems or *Multinomial* systems (see the section *Canonical Forms*). The reduction of GMA equations to the S-system form is possible *via* the product rule of differentiation. Specifically, if the i-th equation is in GMA form, one replaces the variable X_i with the product of two new variables, for example, $X_i = X_{n+1}X_{n+2}$. Differentiation leads to the sum $\dot{X}_{n+1}X_{n+2} + X_{n+1}\dot{X}_{n+2}$, and one equates one positive and one negative power-law term of the i-th equation with $\dot{X}_{n+1}X_{n+2}$, which corresponds to the S-system form, and the remaining terms with $X_{n+1}\dot{X}_{n+2}$. Iteration of the procedure transforms all GMA equations into S-system equations.

If the i-th equation of a GMA-system contains many terms, one can either reduce it successively by replacing X_i with a product of two new variables whose differential equations have to be reduced further, or one can directly replace X_i with a product of several new variables. For instance, if the i-th equation contains four positive and three negative power-law terms, replacing X_i with a product of four new variables reduces the equation in one step. In the resulting system, three of the new variables contain an alpha term and a beta term, whereas the fourth equation contains only an alpha term. The missing beta term sometimes leads to undesired problems with numerical stability. This can be avoided when in the original GMA equation one of the negative power-law terms is simply split into a sum of two terms. Instead of introducing products of new variables, it is sometimes useful to replace a variable with the ratio of new variables or, generally, with a product of variables, raised to suitable powers.

No matter which of the recasting rules is applied, the initial conditions of the S-system realization are determined by the original initial values and the definition of the ancillary variables. This assignment of initial values is unique, except when a variable is replaced with the product of new variables. In the latter case, one initial value can be chosen freely but the product of initial values must be equal to the initial value of the replaced variable. When a variable is replaced with the product of m new variables, $m - 1$ degrees of freedom are generated.

Recasting is not only possible for sets of explicit first-order ordinary differential equations but also for higher-order systems and certain types of implicit, delay or integro-differential equations (cf. [46]). It is sometimes even useful to recast differentiable functions not originally expressed as differential equations because the differential form may provide more numeric advantages (cf. Chapter 13) or insight into the investigated phenomenon (e.g., [46: p. 95]). In this case, recasting is comparable with the realization of proper rational transfer functions in the form of linear state-variable descriptions [165]. Furthermore, the recasting of nonlinear functions can be employed as a tool for the modeling of complex phenomena (cf. Chapter 14).

Numerous examples for the recasting of differential equations can be found in the literature [41–50], and particularly in Appendix A of [46]. As one illustration, let's recast a set of equations that generates deterministic chaos [176]:

$$\dot{x} = -y - z$$

$$\dot{y} = x + ay$$

$$\dot{z} = bx - cz + xz$$

Since the first equation contains two negative terms and the second equation contains two positive terms, it is convenient to replace y with $-w$. All variables are known to be bounded but they can become negative, and so, in order to assure positivity, we shift them by large enough positive constants p, q, and r:

$$X_1 = x + p$$

$$X_2 = w + q$$

$$X_3 = z + r$$

Upon looking at the first differential equation of the system, it is immediately clear that $q = r$ is a good choice, since the shifting constants cancel out. Furthermore, the second differential equation becomes

$$\dot{X}_2 = -X_1 + p + a(X_2 - q)$$

which is especially simple if we make $p = aq$. This is always possible, since the only condition on the shifting constants is that they are large enough.

The third differential equation, expressed in terms of X_3 becomes

$$\dot{X}_3 = X_1 X_3 - k_1 X_1 + k_2 - k_3 X_3$$

where $k_1 = (q - b)$, $k_2 = (pq - pb + cq)$, and $k_3 = (c + p)$. This differential equation has the form of a GMA equation and it is reduced to S-system form by replacing X_3 with the product of two new variables: $X_3 = X_4 X_5$. X_4 and X_5 are defined as

$$\dot{X}_4 = X_1 X_4 - k_1 X_1 X_5^{-1}$$

$$\dot{X}_5 = k_2 X_4^{-1} - k_3 X_5$$

and one directly confirms that their product satisfies the differential equation for $\dot{X}_3$. Upon replacing X_3 with $X_4 X_5$ in the differential equations for X_1 and X_2 and renumbering, one obtains an S-system representation that produces exactly the same chaotic motions as the original system. Rössler's [176] numerical example, in which $a = 0.36$, $b = 0.4$, and $c = 4.5$, in S-system form reads

$$\dot{X}_1 = X_2 - X_3 X_4 \qquad X_1(0) = 18$$

$$\dot{X}_2 = 0.36 X_2 - X_1 \qquad X_2(0) = 47$$

$$\dot{X}_3 = 1117.8 X_4^{-1} - 49.6 X_1 X_4^{-1} \qquad X_3(0) = 1$$

$$\dot{X}_4 = X_1 X_4 - 22.5 X_4 \qquad X_4(0) = 50$$

where the shifting constants were chosen as $p = 18$ and $q = 50$. All variables are positive, but the original variables are easily retrieved as auxiliary variables $X_5 = x = X_1 - 18$, $X_6 = y = X_2 - 50$, $X_7 = z = X_3 X_4 - 50$. Numerical simulations confirm that this S-system produces the same trajectories as the original system in [176] (cf. Fig. 12.1).

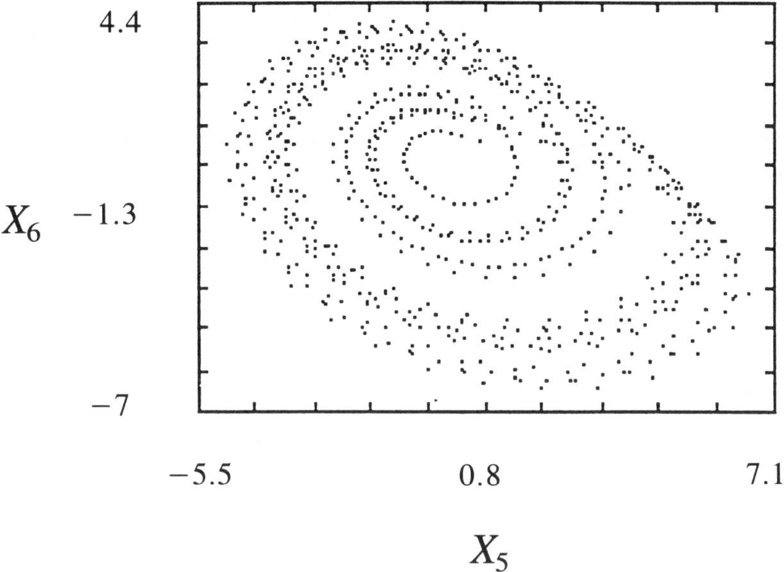

FIGURE 12.1. Trajectorial solution to Rössler's example, recast as an S-system.

ALTERNATIVE RECASTING STRATEGIES

The basic recasting techniques are sufficient to transform virtually any ordinary differential equation into an S-system. However, they do not prescribe a unique recasting strategy. Rather, they can be applied in different ways that lead to different S-system realizations. These realizations have in common the S-system structure and a solution that exactly corresponds to the original differential equation. Otherwise it is often difficult to detect similarities between S-systems that derive from the same original equation *via* alternative recasting procedures (however, see Chapter 15). The following examples illustrate the results of alternative recasting strategies.

Recasting of Polynomials

As an example, let's recast the one-variable polynomial equation

$$\dot{X} = \sum_{j=0}^{n} a_j X^j$$

in different ways.

The first method is based on the successive reduction of the sums of power-law terms. The first two new variables can be defined, for instance, as $X = X_2/X_1$. The differential equation for X_1 then is $\dot{X}_1 = -a_n X_1 X^{n-1}$, and upon replacing X with X_2/X_1, we obtain a pair of differential equations, the first one being an S-system equation and the second one containing a polynomial of degree $n - 1$:

$$\dot{X}_1 = -a_n X_1^{-n} X_2^{n-1}$$

$$\dot{X}_2 = X_1 \prod_{j=0}^{n-1} a_j X^j$$

This reduction of terms in the polynomial can be continued until the S-system form is obtained. If some of the coefficients a_j are negative, pairs of one positive and one negative term can be reduced in one step. For $n = 4$ and under the assumption that a_0 and a_2 are negative and the remaining coefficients positive, the resulting S-system is

$$\dot{X}_1 = -a_4 X_1^4 X_2^3 X_3^3$$

$$\dot{X}_2 = a_2 X_1 X_2^2 X_3 - a_3 X_1^2 X_2^3 X_3^2$$

$$\dot{X}_3 = a_0 X_1^{-1} X_2^{-1} - a_1 X_3$$

A second method of recasting is factorization. If all factors have the form $(X + q_i)$, one can define new variables as $X_i = X + q_i$. The resulting S-system then reads

$$\dot{X}_i = \dot{X} = \prod_{i=1}^{n} X_i$$

All equations of this S-system have the same form but they typically differ with respect to their initial conditions: $X_i(t_0) = X(t_0) + q_i$. If there are complex roots, they will occur in conjugate pairs. The factors $(X^2 + p_1 X + p_2)$ can be recast by defining the new variables as $X_j = X + p_1$ and $X_{j+1} = XX_j + p_2$. The differential equation for X_{j+1} then is

$$\dot{X}_{j+1} = \dot{X}X_j + X\dot{X}_j = (X + X_j) \prod_{i=1}^{n} X_i$$

which reduces to two S-system equations when $X_{j+2} = X + X_j$.

A third method is based on rewriting the original polynomial equation in the *Horner scheme*

$$\dot{X} = \left(\cdot \cdot \cdot \left((a_n X + a_{n-1})X + a_{n-2}\right)X + \cdot \cdot \cdot\right)X + a_0$$

and defining new variables as $X_n = a_n X + a_{n-1}$, $X_{n-1} = XX_n + a_{n-2}$, $X_{n-1} = XX_{n-i+1} + a_{n-i-1}$. The original equation then reduces to a set of differential equations that contain polynomials of at most degree $n - 1$, which can be reduced analogously.

A fourth method is based on repeated differentiation, which is similar to the realization of transfer functions as linear systems (cf. [165]). We define $X_1 = \dot{X}$ and X_i $(i = 2, \cdot \cdot \cdot, n)$ as the $(i - 1)$-th derivative of the polynomial with respect to X. If $n = 4$, this procedure yields the following four-variable S-system:

$$\begin{aligned}
\dot{X}_1 &= (4a_4X^3 + 3a_3X^2 + 2a_2X + a_1)\dot{X} &&= X_1X_2\\
\dot{X}_2 &= (12a_4X^2 + 6a_3X + 2a_2)\dot{X} &&= X_1X_3\\
\dot{X}_3 &= 24a_4X + 6a_3)\dot{X} &&= X_1X_4\\
\dot{X}_4 &= 24a_4\dot{X} &&= 24a_4X_1
\end{aligned}$$

These four different S-system representations are equivalent in the sense that they generate exactly the solution of the original polynomial equation, as long as the appropriate initial values are specified, which are determined through the definition of the new variables. If the initial conditions are ignored, each representation produces a different manifold that is suspended in a space of higher dimension. The intersection of all these manifolds contains the solution of the original polynomial equation and, in particular, the appropriate initial values.

While the four representations generate exactly the desired solution, they have different structural and numerical properties that can affect the efficiency of the solution. For instance, the last method leads to a representation consisting of four equations that each have only one term. Such a system has no steady state in which all variables are non-zero, and it is often prone to computational inaccuracy or instability. Comparisons between S-systems that result from alternative recasting strategies will be discussed in more detail in the *Optimal Recasting Strategies* section.

Recasting an S-System Equation

A second demonstration of the nonuniqueness of recastability is the application of different recasting strategies to S-system equations themselves. This process generates new S-systems with the same solution as the original equation but often with a different appearance and different structural and numerical properties. In the following paragraphs, such differences are illustrated with the recasting of a single S-system equation in three ways. The chosen recasting strategies lead through different types of S-systems (cf. the *Canonical Forms* section). First, they result in different *Half-systems* in which each equation contains only one power-law term. Then so-called *Binary systems* are obtained in which all exponents are equal to 0 or 1. Finally, further recasting leads to systems in *Lotka-Volterra* form.

The straightforward reduction method for sums of power-law terms transforms the one-variable S-system

$$\dot{X} = \alpha X^g - \beta X^h$$

into the S-system

$$\dot{Y} = \alpha Y^g Z^{1-g}$$
$$\dot{Z} = \beta Y^{h-1} Z^{2-h}$$

from which X is recouped as $X = YZ^{-1}$. The resulting system has only one term in each equation and it is called a *Half-system* (cf. the *Canonical Forms* section). At the cost of additional equations, the exponents in this system can be reduced to 0, 1, or 2, when one further defines $U = Y^{h-2}$, $V = Z^{1-h}$, $R = Y^{g-1}$, and $S = Z^{-g}$:

$$\dot{Y} = Y[\alpha RSZ]$$
$$\dot{Z} = Z[\beta UVY]$$
$$\dot{U} = U[\alpha(h-2)RSZ]$$
$$\dot{V} = V[\beta(1-h)UVY]$$
$$\dot{R} = R[\alpha(g-1)RSZ]$$
$$\dot{S} = S[-\beta g\, UVY]$$

As a matter of fact, when one defines the additional variables $W = V$ and $T = R$, all exponents become either zero or one, and the system assumes the form of a *Binary Half-system* (cf. the *Canonical Forms* section).

The previous system has a peculiar form in which each variable appears on the right-hand side of its differential equation, multiplied with a constant and with either of the triplets UVY or RSZ. When these products are replaced with new variables $P = UVY$ and $Q = RSZ$, the result is a two-variable S-system that, at the same time, is a *Lotka-Volterra system* (LV-system; cf. the *Canonical Forms* section):

$$\dot{P} = \beta(1 - h)P^2 - \alpha(1 - h)PQ$$
$$\dot{Q} = \beta(1 - g)PQ - \alpha(1 - g)Q^2$$

A second recasting strategy for the original S-system equation is factoring out the α-term and defining the remainder as a new variable $Y = (1 - \beta\alpha^{-1}X^{h-g})$:

$$\dot{X} = \alpha X^g Y$$
$$\dot{Y} = \beta(g - h)X^{h-1}Y$$

This system can also be transformed further with the additional definitions $U = X^{h-2}$, $V = X^{g-1}$. The result is a Half-system with four variables:

$$\dot{X} = \alpha VXY$$
$$\dot{Y} = \beta(g - h)UXY$$
$$\dot{U} = (h - 2)X^{h-3}\alpha X^g Y = \alpha(h - 2)UVY$$
$$\dot{V} = (g - 1)X^{g-2}\alpha X^g Y = \alpha(g - 1)V^2XY$$

Further transformation like in the previous case (i.e., $P = XU$, $Q = VY$) yields

$$\dot{P} = \alpha(h - 1)PQ$$
$$\dot{Q} = \beta(g - h)PQ - \alpha(1 - g)Q^2$$

which again is an LV-system and also an S-system as long as the terms in the second equation have opposite signs.

A third strategy is to define $P = X^{h-1}$ and $Q = X^{g-1}$ in the original S-system equation. Again, the equations simultaneously have S-system and LV-system form:

$$\dot{P} = \beta(1 - h)P^2 - \alpha(1 - h)PQ$$
$$\dot{Q} = \beta(1 - g)PQ - \alpha(1 - g)Q^2$$

It is interesting to note that the first and third LV-systems in P and Q are the same and are quite similar to the second one. This may indicate that not just the solution but also other properties are preserved in transitions between S-systems and LV-systems. Nevertheless, the fact that the intermediate systems also produce the same solution indicates that it is difficult to determine equivalence classes or normal forms of S-systems or LV-systems that correspond to the same original set of differential equations (however, see Chapter 15 for approaches to addressing such questions).

These examples may provide an intuitive feeling for the large number of possible recasting strategies for a given S-system. They may also illustrate some of the relationships between S-systems, Half-systems, Binary systems, and LV-systems that are discussed in the section *Canonical Nonlinear Forms*. In fact, transformations of this type have led to the observation that all these canonical forms are equivalent (cf. [45]).

Nontrivial Equivalences

The recasting strategies of the previous sections were different but they were all based on a more or less straightforward application of the recasting rules in different order. In specific cases, alternative recasting strategies can employ nontrivial types of coordinate transformations or definitions of new variables that are not so obvious. For instance, the pair of differential equations

$$\dot{Y} = Y(1 - Y^2 - Z^2) - Z$$
$$\dot{Z} = Z(1 - Y^2 - Z^2) + Y$$

could be recast with some of the rules discussed previously, but it is also possible to use the polar coordinate transformation $X_1^2 = Y^2 + Z^2$, $X_2 = \mathit{arctan}(Y/Z)$, which leads to the S-system [177: p. 214].

$$\dot{X}_1 = X_1 - X_1^3$$
$$\dot{X}_2 = -1$$

Other nontrivial recasting strategies have been involved in the S-system analysis of noncentral distributions. Procedures in that context involve the recasting of special functions, such as Bessel functions and the Gamma function, and the representation of infinite series as sets of differential equations. Some of these methods are discussed in Chapter 13 and [48–49].

CANONICAL NONLINEAR FORMS

The previous sections have demonstrated that the recasting process is not unique and can lead to numerous alternative representations. While all of the resulting sets are in S-system form, some classes of S-systems can be identified that have particular properties and, in some sense, a simpler or more stringent structure. For instance, we have seen that an S-system equation can be recast in such a way that the result is an S-system with only one term in each equation. Since these classes can be derived from S-system *via* recasting it is proven that they have the same scope as S-systems and include exactly the same nonlinearities as special cases. All classes of differential equations resulting from recasting can thus be considered *canonical nonlinear forms* or S-systems.

It is a matter of convention to call a particular structure a *canonical form*. We shall consider here some systems that in our *opinion* are "prominent," which is typically characterized by the fact that their general structure can be described in a few words. Admittedly, this "definition" of *prominent* is as vague as the notion of canonicity itself. It is open to arbitrariness and invites the construction of further canonical (?) forms. Only time can tell which of these forms will be really useful and thus deserve individual recognition.

It might be tempting to select the canonical form with the simplest structure and discard all other forms. However, simplification in structure is almost always accompanied by a larger number of variables. Furthermore, some of the simpler canonical forms are prone to numerical disadvantages, so that as of yet, no decision can be made about which of the alternative forms is best. In this section, we shall list some prominent canonical forms. The subsequent section will deal with a comparison of these forms with respect to different criteria of quality.

Generalized Mass Action Systems

It is evident that S-systems are immediate special cases of *Generalized Mass Action* (GMA) systems (or *Multinominal systems*; cf. [73]) in which the right-hand sides are composed of sums and differences of several power-law terms (cf. Table 12.1). Furthermore, recasting proves that each GMA-system can be transformed equivalently into an S-system through the introduction of products of new variables. Thus, GMA-systems include all S-systems as special cases, and S-systems include all GMA-systems as special cases. Consequently, both systems encompass as special cases exactly the same types of functions and differential equations. During the recasting of GMA-systems as S-systems the number of variables increases, but it seems noteworthy that the number of power-law terms remains the same while the distribution of terms is different. For instance, straightforward recasting of a GMA equation with one positive and three negative terms leads to an S-system with three equations. One equation has an alpha and a beta term and two equations have only one beta term in each (see also Chapter 15).

Like S-systems, GMA-systems can be derived as general approximations of dynamical systems (cf. Chapters 1–3, [27–29, 35–37]). In contrast to the approximation with S-systems, the change in each system variable X_i is described as a *sum* of all influences that increase or produce X_i, diminished by a *sum* of all influences that decrease or degrade X_i. Then, each term in either sum is approximated individually with a power-law function, and this procedure immediately leads to a GMA-system.

Half-Systems

The recasting rule for reducing GMA-systems to S-systems can be applied to S-systems as well and it leads to special S-systems whose right-hand sides contain only one power-law term that may be either positive or negative. This type of system is called a *Half-system* (cf. Table 12.1). Similar to the comparison between GMA-systems and S-systems, the comparison between S-systems and Half-systems shows that Half-systems usually consist of more equations with a simpler structure, and the total number of power-law terms is the same. Advantages and disadvantages of the Half-system form are discussed in the *Optimal Recasting Strategies* section.

Table 12.1. Canonical Nonlinear Forms for Differential Equations

Generalized Mass Action System:

$$\dot{X}_i = \sum_{k=1}^{m} \alpha_{ik} \prod_{j=1}^{n} X_j^{g_{ijk}} - \sum_{k=1}^{m} \beta_{ik} \prod_{j=1}^{n} X_j^{h_{ijk}}$$

S-system:

$$\dot{X}_i = \alpha_i \prod_{j=1}^{n} X_j^{g_{ij}} - \beta_j \prod_{j=1}^{n} X_j^{h_{ij}}$$

Half-system:

$$\dot{X}_i = \gamma_i \prod_{j=1}^{n} X_j^{g_{ij}}$$

Binary System:

$$\dot{X}_i = \alpha_i \prod_{j \in \mathfrak{J}} X_j - \beta_i \prod_{j \in \mathfrak{J}'} X_j$$

Binary Half-system:

$$\dot{X}_i = \gamma_i \prod_{j \in \mathfrak{J}} X_j$$

Lotka-Volterra System:

$$\dot{X}_i = X_i\left(a_i + \sum_{j=1}^{n} b_{ij} X_j\right)$$

α_{ik}, α_i, β_{ik}, β_i non-negative real; g_{ijk}, h_{ijk}, g_{ij}, h_{ij}, γ_i, a_i, b_{ij} real; $\mathfrak{J}, \mathfrak{J}' \subseteq \{1, \ldots, n\}$

Like S-systems and GMA-systems, Half-systems can be interpreted as approximations of dynamical systems (cf. Chapters 1, 2, and 10). In this case, the change in each system variable X_i is described as one aggregated function and this function is approximated with a power-law function. The result is a set of equations with one power-law term in each equation and, thus, in Half-system form.

Binary Systems

S-systems may seem to draw their generality from the exponents g_{ij} and h_{ij}, which by definition can be any real number that remains essentially unaltered when GMA-systems are reduced to S-systems or S-systems are reduced to Half-systems. However, one can reduce an S-system exactly to a special S-system in which all exponents are zero or unity; such a system has been called a *Binary System* (cf. Table 12.1). The binary form is easy to obtain: for each g_{ij} and h_{ij} that is not equal to zero or one, one simply defines a new variable that replaces X_i, raised to the particular g_{ij} or h_{ij}. The number of variables in most cases increases drastically, but the structure of the resulting system is significantly simpler than that of an S-system (cf. Appendix B of [46]).

Binary Half-systems

It is obvious that recasting as a Half-system and the subsequent introduction of new variables according to the previous paragraph lead to a so-called *Binary Half-system* in which each equation has only one power-law term and all exponents are equal to zero or one (cf. Table 12.1). The resulting structure is completely characterized by one matrix consisting of zeroes and ones, representing the exponents, a vector of multipliers (alphas and betas), and a vector of initial values. Of course, the multipliers could be replaced with additional variables whose derivatives are equal to zero and whose initial values are equal to the multiplier (see also Chapter 16 for the elimination of multipliers). To recoup all features of the original system, one also needs to record which variables in the Binary system and the original system correspond. From a theoretical standpoint, Binary Half-systems may be most appealing. However, the simple structure is often paid for by many variables and numeric disadvantages (cf. the *Optimal Recasting Strategies* section).

Lotka-Volterra Systems

During the 1920s and 1930s, Lotka [167] and Volterra [178] studied a pair of differential equations and its use for investigating the dynamics in predator-prey systems. In its original form as well as in its generalization to n interacting populations

$$\dot{X}_i = X_i\left(a_i + \sum_{j=1}^{n} b_{ij}X_j\right) \qquad i = 1, \ldots, n$$

the change in each population size X_i is given as a linear production term plus a sum of bilinear terms reflecting encounters, associations, cooperation, or competition, depending on the subject under consideration. The need for other types of interactions in chemical kinetics led to a temporary neglection of Lotka-Volterra (LV) equations, but recently these equations have been revived in analyses of ecological networks and other cooperative, competitive, and chaotic dynamic systems (e.g., see [164, 166–174]). As a matter of fact, the majority of models for predator-prey interactions and population dynamics, and their applications, for instance, to fishery biology and agriculture are still based on LV-systems (e.g., see [175]).

Similar to the twofold usage of S-systems for modeling and recasting, LV-systems also provide a canonical form for functions and differential equations that have the range of generality that S-systems have. This property was elaborated by Peschel and Mende [73, 169–172] and called the *Structure Design Principle* (SDP). The primary goal of their approach was to develop a highly structured representation for nonlinear aspects of technical systems because it was realized that a unified representation for nonlinear systems was superior to *ad hoc* formulations in terms of technical design and computational effort. Even though technical systems were of primary interest, it was observed from the beginning that the same principles would also address natural networks.

Like S-system recasting, SDP employs the introduction of new variables and the chain rule. The major difference is that SDP is based on *logarithmic* differentiation. For instance, to apply SDP to the harmonic oscillator

$$X_1 = \sin(t)$$

one computes the logarithmic derivative

$$\frac{d \log X_1(t)}{dt} = \frac{\cos(t)}{\sin(t)}$$

The right-hand side is called X_2 and, again, it is logarithmically differentiated:

$$\frac{d \log X_2(t)}{dt} = -\frac{\sin(t)}{\cos(t)} - X_2$$

Finally, the term sin $(t)/\cos(t)$ is replaced with X_3, and logarithmic differentiation yields

$$\dot{X}_1 = X_1 X_2$$

$$\dot{X}_2 = -X_2 X_3 - X_2^2$$

$$\dot{X}_3 = X_2 X_3 + X_3^2$$

Many examples and extensions of SDP can be found in Peschel and Mende [73].

LV-systems stress the additive nature of the underlying processes whereas S-systems and Half-systems emphasize their aggregate, and, hence, multiplicative nature. Both systems show a rigid formal structure, have special cases in common, and are themselves distinct special cases of GMA-systems. Otherwise, no relationships or similarities are evident. Nonetheless, LV-systems can be recast as S-systems, and the application of SDP to S-systems yields LV-systems. Therefore, the two systems are equivalent in the sense that they both encompass exactly the same spectrum of nonlinearities (cf. [45, 73, 171]). The full consequences of this equivalence are not yet understood.

OPTIMAL RECASTING STRATEGIES

Since the recasting process is not unique, the question of optimal recasting is at hand. As simple and obvious as the question is, it is by no means easy to answer. The first requirement is to have a set of criteria of optimality which *a priori* is not available and furthermore can depend on the particular application. Without claiming completeness in any respect, in the following paragraphs, we discuss the alternative canonical forms in light of some of the more obvious criteria.

Number of Equations

A criterion of optimality that transcends the disciplines of science is simplicity. As early as the fourteenth century, Ockham proposed that one representation or theory should be considered better than another if it was simpler though it produced the same results (cf. [179]). Applying this dogma to recasting, one could argue that among equivalent S-system analogues the best system is the one with the fewest equations. In fact, this criterion often is a good guide. The fewer equations there are the less analytic and numeric effort typically is necessary and, as a consequence, computational speed is held to a minimum (cf. [49]). One could surmise that a smaller system also contains fewer parameter values and initial conditions, and thus all estimation procedures would be simpler. This conclusion, however, is erroneous because all S-systems that are equivalent to the same original set of equations are uniquely determined by this original, and even though parameters may be aggregated or appear in several places of the resulting S-system, the number of degrees of freedom is the same (at least if one does not count choices in initial values when a variable is replaced with the product of new variables (cf. the *Recasting Techniques* section).

The number of equations is not always the best criterion of optimality. For instance, many growth functions in current use can be written as S-systems with either one or two equations [19, 41]. The smallest S-system that includes these functions as special cases has two equations. However, in each case, at least one alpha term or beta term is missing, so the system essentially is characterized by a total of only one, two, or three power-law terms. Nevertheless, since different terms of the S-system vanish for different special cases, one has to perform classifications and comparisons with respect to all four terms of the two-variable S-system and a correspondingly high number of parameters. As an alternative, one can recast the growth functions as three-variable Half-systems that have a higher number of equations but contain a total of only three power-law terms, thus allowing for a comparison in a lower dimensional parameter space (cf. the *Conclusion* and [39]).

As of yet, no theorems are available about the minimum number of equations needed to recast a differential equation into one of the alternative canonical forms.

Simplicity of Structure

Simplicity is not only manifest in the number of equations but also in their structure. It was shown that every S-system can be recast as a Half-system, a Binary system, or a Binary Half-system. These transformations usually require an increase in the number of equations, but the resulting equations are less complex. In the Half-system form, each equation contains only one power-law term, and a simple logarithmic transformation reduces the right-hand sides to sums in the logarithms of the system variables. As promising as this may sound, however, no significant, generally applicable advantages of this transformation have been detected so far. The exponents in a Binary System are all zeroes or ones, which can be regarded as a different type of reduction in complexity since exponentiation reduces to multiplication. It is an appealing thought that virtually all ordinary differential equations can be represented and analyzed in this simple form. But again, this promising simplification has not yet born significant fruit, for instance, in terms of solution speed.

The obvious disadvantage of Half-systems and Binary systems is that the number of equations often increases considerably. A second drawback is that Half-systems can only exhibit steady states in which at least one variable is equal to zero. These steady states and their stability cannot be evaluated with the standard methods of S-system analysis [2, 9] (for exceptions, cf. Chapter 15). On the other hand, trajectories of Half-systems are again in Half-system form, and if a Half-system derives from the recasting of an S-system, the trajectories are related to the steady states of that S-system (cf. Chapter 1).

Speed of Solution

Computational speed in solving equivalent S-systems is not just determined by the number of equations. As a matter of fact, those differences are often minimal in comparison with other factors that affect speed. When the efficient solver ESSYNS (Chapter 7, [7, 38]) is used to solve S-systems, the primary factors are stiffness and closeness of variables to zero. Stiffness is most often due to α and β values that differ by several orders of magnitude. Values of variables close to zero cause problems since the numerical algorithm makes use of logarithmic transformations that are not well behaved in the vicinity of zero. Consequently, solution time often is drastically increased. Recasting a system with such a problem into a system whose variables are sufficiently different from zero can speed up numerical solutions by some orders of magnitude. When solution speed is of critical importance, strategies of avoiding zero are crucial (cf. Chapter 13).

Recasting to some degree allows shifting parameter values into initial values and *vice versa*. For instance, the simple Half-system

$$\dot{X} = \alpha X^g \qquad X(0) = X_0$$

can be transformed with the definition $Y = cX$ into

$$Y = c^{1-g}\alpha Y^g \qquad Y(0) = cX_0$$

Thus, with the choice $c = \alpha^{1/(g-1)}$ one obtains

$$\dot{Y} = Y^g \qquad Y(0) = \alpha^{1/(g-1)}X_0$$

The parameter α has disappeared from the differential equation itself and has become part of the initial condition. Although such a transformation may appear trivial, it can be used either to move initial conditions away from zero or to alleviate stiffness.

Qualitative Behavior

Even though S-systems, Half-systems, Binary systems, and LV-systems are equivalent with respect to solution, they exhibit distinct numerical and qualitative features. For instance, steady states and stability usually can be evaluated for S-systems with analytical methods that fail in the case of GMA-systems or Half-systems. Half-systems that are obtained from S-systems through recasting do possess the same steady states, but their determination requires the incorporation of all algebraic constraints that were introduced in the form of ancillary variables (cf. Chapter 15). Without these constraints, the set of steady-state values is quite different from that of the original S-system.

Even more pronounced are differences between S-systems and LV-systems. For instance, direct application of the Structure Design Principle can lead to "parasitic" poles that are generated by the transformation and, of course, are numerically undesired; the introduction of additional variables helps to avoid such poles [73: p. 106]. We have not yet encountered the introduction of poles as the result of S-system recasting; however, we insist on the positivity of variables, which also may require additional variables.

Outside their mathematical and numerical properties, S-systems and LV-systems differ with respect to meaning and interpretability. LV-systems emphasize the additive nature of complex systems whereas S-systems emphasize their synergistic, multiplicative nature. As a consequence, LV-systems are perceived as the more natural representations of systems that are composed of subsystems that respond more or less independently of each other, as is often the case in ecological systems. In contrast, associative systems, in which many components are interacting to produce a response, are more appropriately represented with S-systems; examples are plentiful in biochemistry and molecular biology.

While recasting often drastically alters the appearance of the original nonlinearity, crucial properties of the solution are preserved. For instance, one can show that some types of parameter groups of the original differential equation are invariant. This is an interesting observation because some groups can be evaluated with straightforward elementary means for the GMA- and S-system forms although they are quite difficult to address in general (cf. Chapter 15). Another qualitative property is the bifurcation structure of a differential equation. The development of simple criteria for codimension-1 Hopf bifurcations in S-systems (cf. Chapter 16) is an important step toward the fuller mathematical characterization of S-system properties. It remains to be explored to what extent these bifurcation criteria, combined with recasting in one of the alternative canonical forms, are useful for general bifurcation analysis.

The Meaning of Parameters

When an S-system is developed as an approximative power-law model, the parameters have a clearly defined meaning: The exponent g_{ij} describes the influence of component X_j on the production of X_i, h_{ij} describes the effect on degradation and the multipliers α_i and β_i are rate constants reflecting the speed of the production and degradation of X_i (cf. Chapter 2). The parameters in

GMA- and Half-system models can be interpreted analogously. The parameters in Lotka-Volterra models also have a distinct meaning as rates of proliferation (a_i) or of interactions between species (b_{ij}). Recasting usually confuses the original meaning of variables and parameters. Ancillary variables are introduced that only have mathematical significance, and the parameters of the resulting system typically are combinations of parameters of the original system that are difficult to interpret. Because language and terminology influence thinking, this disappearance of the original parameters can be a negative side effect. However, such rearrangement can also have positive consequences. It may yield new insight and, in fact, lead to parameters that can be measured experimentally or have a meaning that is easily understood, whereas the parameters of the original system were aggregate quantities that were difficult to measure.

These arguments show that it is useful to maintain the various canonical forms as separate entities. Nevertheless, the equivalence between these systems is of great importance. It allows us to study one system with methods of analysis developed for the other. The equivalence also unites the ranges of application and classes of special cases of each formalism, thus increasing the power and generality of both. This should render it possible to translate the mathematical properties of one system and all the theorems related to it into the other formalism. Mathematical analysis of a particular problem then can utilize results and methods from each formalism. For certain steps, one representation might be used because of its advantages. One then transforms the problem into the alternative formalism whenever it is advantageous to do so.

CONCLUSION

Recasting is a new tool for the analysis of differential equations. The general idea and the mechanics of recasting are easily learned, but its implications and applications are far from understood. Currently, four aspects of recasting are being investigated; they deal with questions of classification, mathematical analysis, numerical analysis, and modeling.

Classification

The recasting process itself provides an operational proof that sets of ordinary differential equations consisting of nested elementary functions can be recast in S-system form or in any of the other alternative canonical forms. This generality seems to render it possible to recast virtually every type of differentiable nonlinearity as an S-system and classify it in this form. However, we have seen that recasting is not unique and that the result is highly dependent on the recasting strategy chosen. Consequently, the same original nonlinearity is likely to appear in many S-system classes, which makes it very difficult to define mutually exclusive equivalence classes.

In special cases, the classification in terms of a canonical form has proven useful. For instance, Savageau [19, 41] has shown that all prominent growth functions are special cases of a two-variable S-system. They are even easier to classify as special cases of the two-variable Half-system

$$\dot{X}_1 = \alpha_1 X_1^{g_{11}} X_2^{g_{12}}$$
$$\dot{X}_2 = -X_2^{h_{22}}$$

in which X_1 represents the growing entity and X_2 is an ancillary variable [cf. 39]. In this Half-system form, the growth functions can be embedded in a three-dimensional parameter space of exponents (Fig. 12.2) that illustrates their re-

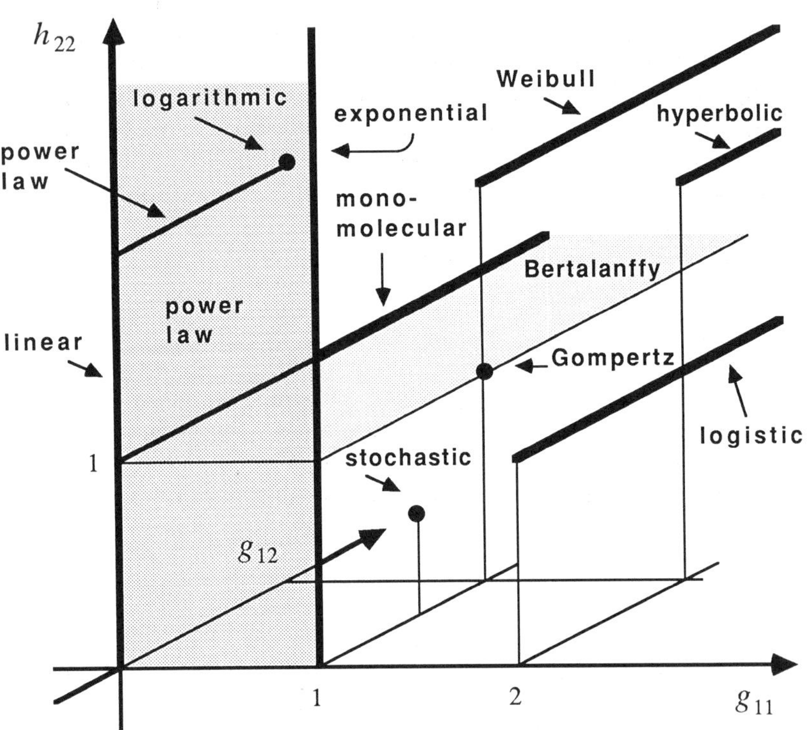

FIGURE 12.2. Growth functions, recast in Half-system form, can be embedded in a three-dimensional parameter space that shows their close similarities. Some growth functions in this representation correspond to single points (e.g., Gompertz growth), others to lines (e.g., logistic growth), still others to infinite strips (e.g., Bertalanffy growth). Besides the prominent special cases, the Half-system contains a continuum of other growth functions without particular names.

lationships to one another and shows a multitude of further growth functions without a particular name. It is interesting to note that those special cases of the two-variable S-system that have a simple analytical solution have actually been published by various authors during the past hundred years, while all other cases lead to quite complex solutions [24].

A specific problem of general classification is the thorough comparative evaluation of alternative canonical forms. One can expect that not a single form is optimal for all purposes, and one has to develop guidelines for *a priori* decisions about which form should be used under given circumstances.

Mathematical Analysis

Several of the methods for analyzing S-systems fail if some of the equations have only one power-law term. For instance, questions of stability cannot be answered if some of the steady-state values are equal to zero. Thus, alternative methods must be developed that can be employed to systems whose properties do not satisfy the regularity conditions of full S-systems.

An intriguing area for future research is the qualitative analysis of differential equations recast as S-systems. Preliminary work has shown that one-parameter stretching groups and translation groups of S-systems can be detected with elementary matrix operations (cf. Chapter 15). These group properties are invariant under recasting operations and, for instance, are the same for S-systems and equivalent Half-systems. This invariance is to be expected since recasting preserves the solution of the original problem. It remains to be seen whether recasting will reduce the complexity in detecting group properties of nonlinear systems in general.

While the basic steady-state and stability properties of S-systems have long been known (e.g., see [9, 14]), other qualitative features such as bifurcations have only been investigated empirically (e.g., see [2]). Chapter 16 is the first systematic step into the largely unexplored area of the qualitative analysis of S-systems. The beautifully simple conditions for codimension-1 Hopf bifurcations of S-systems raise our hope that the rigid S-system structure will also allow us to develop simple conditions for other types of generic dynamic responses.

Numerical Analysis

Recasting in combination with the efficient S-system solver in ESSYNS can be employed to analyze nonlinear problems that require speed and high accuracy (cf. [38, 46]). It is evident that for such problems the choice of the best recasting techniques is crucial. The primary criteria of optimality in this case are given, and it is a matter of refining the basic methods in order to achieve optimal performance. The immediate goal of this research is to extend the current repertoire of recasting techniques and to develop guidelines for optimal employment of these methods in a given application. These guidelines would suggest how to avoid initial conditions close to zero, how to develop optimal strategies for assuring the positivity of variables, how to set up measures preventing numerical inaccuracy and instability, and how to hold the number of variables to a minimum.

A particularly fruitful area of application that is just beginning to emerge is computational statistics. Chapter 13 and [48–49] discuss some of these applications and stress techniques for improving efficiency. In [49], the noncentral t distribution is recast in two drastically different ways and, thus, it allows for a direct evaluation of alternative recasting strategies. While the accuracy of the resulting systems is indistinguishable, significant differences exist with respect to computational speed.

Modeling

The power-law formalism provides a variety of methods for the design and analysis of S-system representations for complex phenomena. These representations can be evaluated symbolically with paper and pencil or with a computer algebra program (Chapter 6) and numerically with the program package ESSYNS (Chapter 7). Formulating a model in S-system form thus offers considerable advantages.

Sometimes one intends to analyze a complex phenomenon with an S-system model but wants to include exact mathematical representations for subsystems in non-S-system form. For instance, an ecological model may significantly depend on the circadian change in solar radiation. The required inclusion of an appropriate oscillation would destroy the S-system form and make analytical or numerical S-system methods of analysis inapplicable. In such a situation it is helpful to recast the oscillation and thus regain the S-system form.

A specific application of recasting as a modeling tool is the improvement of accuracy. For instance, if a biochemical network is represented with a power-law approximation in S-system form, the range of valid representation usually is rather wide. Nevertheless, the approximation may not be accurate enough over the total range of interest, and one might want to replace the power-law representation with another rate law, for example, a Michaelis-Menten rate law. Inclusion of this rate law in its usual form would destroy the S-system form, but upon recasting the S-system form is regained throughout (cf. Chapter 14).

Although the use of recasting as a modeling tool is rather obvious in concept, many details need to be worked out and experience must be gained for optimally combining strategies of modeling and recasting. A few examples for the application of this technique are discussed in Chapter 14 and [39].

SUMMARY

Virtually any nonlinearity that can be expressed as a set of ordinary differential equations can be recast as a canonical nonlinear form that exhibits exactly the same solution as the original. Upon recasting, differential equations and differentiable nonlinear functions can be compared, classified, and generalized. Often they can be evaluated more efficiently in canonical form than in their original form. The chapter reviews the basic recasting methods and compares the results of alternative recasting strategies. The application of recasting methods to S-systems transforms these into different S-systems or into alternative canonical forms for nonlinearities. The evaluation of advantages and disadvantages of some prominent canonical forms helps in the development of optimal recasting strategies.

ACKNOWLEDGMENTS

The author is grateful to Louis Cyr and Volker Weinberger for their constructive criticism of the manuscript. This work was supported in part by U.S. Public Health Service Grant GM 30054-05 from the National Institutes of Health.

13. Applications of Recasting in Statistical Distributions

PHILIP F. RUST

The relevance of differential equations to statistical distributions has long been known. The continuous distributions of the Pearson family solve second-order ordinary differential equations; moreover, several classes of densities were defined by Pearson in terms of the numeric values of certain coefficients. Subsequent investigators have proposed other distribution families, and some of these have also been defined in terms of differential equations. However, such equations have generally experienced limited use, and their value has been perceived to be definitional rather than analytic.

The pioneering breakthrough was made by Savageau [42] who pointed out that a great variety of statistical distributions could be recast as S-systems. Specifically, they could be redefined in terms of a set of simultaneous nonlinear first-order ordinary differential equations in the form

$$\dot{X}_i = \alpha_i \prod_{j=1}^{n} X_j^{g_{ij}} - \beta_i \prod_{j=1}^{n} X_j^{h_{ij}}, \qquad i \in \{1, \ldots, n\} \tag{13.1}$$

Here the X_i are positive real variables, α_i and β_i are non-negative real coefficients, and the exponents are real. If t denotes the independent variable then $\dot{X}_i = dX_i/dt$, and n is the number of equations. Savageau was able to show that the entire Pearson family is a subset of the densities that may be modeled with S-systems. Several other families are likewise contained within the S-system structure. Many of the well-known distributions can in fact be expressed in terms of only two S-system equations, implying that a subsequent analysis of these will not be difficult. Finally, Savageau outlined how new densities could be constructed within the S-system context, whether or not the corresponding closed form is obtainable.

One of the simplest of these densities is that of the exponential distribution, whose density may be written $X_1(t) = be^{-bt}$. Since $dX_1/dt = -b^2e^{-bt}$, the corresponding equation in S-system form is $\dot{X}_1 = -bX_1$.

THE NORMAL DISTRIBUTION

The normal distribution is probably the most important of all statistical distributions. Unlike the exponential, however, a single S-system equation is insufficient to model the $N(\mu, \sigma^2)$ distribution. Two equations provide a straightforward solution: if one defines $X_1(t) = (2\pi\sigma^2)^{-1/2} \exp[-(t - \mu)^2/2\sigma^2]$ and $X_2(t) = (t - \mu)/\sigma$, then the corresponding S-system is

$$\dot{X}_1 = -\frac{1}{\sigma} X_1 X_2$$

$$\dot{X}_2 = \frac{1}{\sigma}$$

This system is completely determined except for the choice of initial conditions. If one initializes at the mode, i.e., $t_0 = \mu$, then $(X_1)_0 = (2\pi\sigma^2)^{-1/2}$, $(X_2)_0 = 0$ would follow. To solve an S-system with the efficient computer program ESSYNS (Chapter 7), none of the dependent variables may vanish. Thus, we rather initialize at a larger value, say $t_0 = \mu + \sigma$. This time $X_2(t_0) = 1$, while $(X_1)_0 = (2\pi\sigma^2)^{-1/2} \exp(-1/2) = 0.241970724519/\sigma$. From here t can be incremented upwards towards ∞, or downwards to $\mu + \epsilon$.

Usually one will be interested in computing the distribution also below the mode. In the above S-system, this would cause a problem because X_2 would be negative below the mode. Fortunately, the recasting procedure is not unique and, thus, a search for an alternative is appropriate. A simple modification is to add a constant $c > 0$ to X_2. This modification does not affect the equation of $\dot{X}_2$ but it does affect its initial value and the equations for $\dot{X}_1$. The modified S-system is

$$\dot{X}_1 = \frac{c}{\sigma} X_1 - \frac{1}{\sigma} X_1 X_2$$

$$\dot{X}_2 = \frac{1}{\sigma} \tag{13.2}$$

and if $t_0 = \mu$, then $(X_1)_0 = (2\pi\sigma^2)^{-1/2}$ and $(X_2)_0 = c$. In solving for increasing t, one may arbitrarily set $c = 1$ since this value will not matter. If t is to be run in the negative direction, $X_2(t = \mu - c\sigma) = 0$ is a barrier and a sufficiently large value of c needs to be selected.

Another modification can serve to eliminate even this problem. Let the right and left halves of the density be modeled with separate variables, as follows

$$X_1 = f^+(t) \qquad \dot{X}_1 = \frac{c}{\sigma} X_1 - \frac{1}{\sigma} X_1 X_3$$

$$X_2 = f^-(2t_o - t) \qquad \dot{X}_2 = \frac{1}{\sigma}\left[c + \frac{2}{\sigma}(t_o - \mu)\right] X_2 - \frac{1}{\sigma} X_2 X_3$$

$$X_3 = c + \frac{t - \mu}{\sigma} \qquad \dot{X}_3 = \frac{1}{\sigma}$$

$$X_4 = \sigma X_3 + (\mu - c\sigma) = t$$

$$X_5 = -\sigma X_3 + (2t_o - \mu + c\sigma) = 2t_o - t$$

Here X_4 and X_5 are auxiliary algebraic variables which ESSYNS is able to compute without directly associated differential equations. A simultaneous plot of X_1 on X_4, and X_2 on X_5, will display the entire normal curve. It is not necessary to assign a large value for c. Note that the variables are renumbered and that the derivative of $f^-(2\mu - t)$, the left branch, becomes essentially equivalent to $\dot{X}_1$ due to symmetry, when $t_o = \mu$.

In addition to modeling the $N(\mu, \sigma^2)$ density, we can easily obtain the cumulative distribution function, since the density is the derivative of the c.d.f. Returning to (13.2) for illustrative purposes, we can formulate density and cumulative as

$$X_1 = f(t) \qquad \dot{X}_1 = \frac{c}{\sigma} X_1 - \frac{1}{\sigma} X_1 X_2$$

$$X_2 = \frac{t - \mu}{\sigma} + c \qquad \dot{X}_2 = \frac{1}{\sigma}$$

$$X_3 = F(t) \qquad \dot{X}_3 = X_1$$

The initial value of X_3 if $t_0 = \mu$ is $(X_3)_0 = 0.5$. Other recasting strategies for the normal distribution are doubtless possible.

THE CENTRAL CHI-SQUARE DENSITY

The central chi-square density is well known to be $f(v;\ t) = \exp(-t/2)(t/2)^{v/2}/[t\Gamma(v/2)]$. Savageau [42] has formulated it as the S-system:

$$X_1 = t \qquad \dot{X}_1 = 1$$

$$X_2 = f(v;\, t) \qquad \dot{X}_2 = \tfrac{1}{2}(v - 2)X_1^{-1}X_2 - \tfrac{1}{2}X_2 \tag{13.3}$$

This representation is valid except for $v < 2$, because then α_2 is no longer positive. A different recasting strategy yields the generally valid S-system

$$X_1 = t \qquad \dot{X}_1 = 1$$

$$X_2 = tf(v;\, t) \qquad \dot{X}_2 = \frac{v}{2}X_1^{-1}X_2 - \frac{1}{2}X_2 \tag{13.4}$$

The density itself is recouped as $X_1^{-1}X_2$. When v is small, it is convenient to initialize with $t_0 = 2$, $(X_1)_0 = 2$, $(X_2)_0 = [e\Gamma(v/2)]^{-1}$. When v is large, the density is close to zero at $t = 2$, and a more appropriate value would be $t_0 = v$.

Since the chi-square distribution is bounded at the origin, one can run the system downward by a finite amount, whereas in the positive direction t becomes indefinitely large. When the extreme values are of particular interest, two separate runs can be employed to obtain the complete curve: when v is small, it is probably easiest to run (13.4) first from $t_0 = 2$ to ϵ, and then extend to a large positive value that may be of interest. For larger values of v, the distribution resembles the normal, and a strategy similar to that given for the normal leads to the S-system:

$$X_1 = t \qquad \dot{X}_1 = 1$$

$$X_2 = tf^+(v;\, t) \qquad \dot{X}_2 = \frac{v}{2}X_1^{-1}X_2 - \frac{1}{2}X_2$$

$$X_3 = 2t_0 - t \qquad \dot{X}_3 = -1$$

$$X_4 = tf^-(v;\, t) \qquad \dot{X}_4 = \frac{1}{2}X_4 - \frac{v}{2}X_3^{-1}X_4$$

$$X_5 = X_1^{-1}X_2$$

$$X_6 = X_3^{-1}X_4 \tag{13.5}$$

Both branches will be generated regardless of the direction in which the system is run; however, in running it to the negative direction one needs to avoid zero values for X_1 and X_3.

Systems (13.4) can be expanded to include the cumulative by adding the equation $\dot{X}_3 = X_1^{-1}X_2$. An initial condition for the central chi-square cumulative is generally obtainable from commercial packages or the computer program

RTAIL [182]. In the context of (13.5), the left tail c.d.f. would have the derivative $-X_3^{-1}X_4$.

THE NONCENTRAL CHI-SQUARE DISTRIBUTION

First Solutions

The noncentral chi-square distribution is substantially more complicated than its central counterpart, and it affords an opportunity to demonstrate the ability of S-systems to handle some of the more difficult problems. Several representations for the noncentral chi-square density are possible. One of these, which expresses this density as a product, is

$$f(v, \lambda; x) = \frac{1}{2}\left(\frac{x}{\lambda}\right)^{n/2} I_n(\sqrt{(\lambda x)})e^{-(\lambda + x)/2}, \qquad n = (v - 2)/2$$

where $I_n(\cdot)$ is the modified Bessel function which, represented as $y(t)$, satisfies the second-order differential equation $t^2y'' + ty' - (t^2 + n^2)y = 0$.

Applying the transformation $t = \sqrt{(\lambda x)}$, the density becomes

$$f(t^2/\lambda) = \frac{1}{2}\left(\frac{t}{\lambda}\right)^n I_n(t) \exp\left[-\frac{\lambda^2 + t^2}{2\lambda}\right]$$

The recasting strategy for this equation is based on expressing the density as a product of three factors, each of which is not difficult to differentiate. The full S-system is:

$$X_1 = t \qquad \dot{X}_1 = 1$$

$$X_2 = \frac{1}{2}(t/\lambda)^n \qquad \dot{X}_2 = nX_1^{-1}X_2$$

$$X_3 = \exp\left[-\frac{\lambda^2 + t^2}{2\lambda}\right] \qquad \dot{X}_3 = \frac{1}{\lambda}X_1X_3$$

$$X_4 = I_n(t) \qquad \dot{X}_4 = X_5$$

$$X_5 = I_n't \qquad \dot{X}_5 = X_1^{-2}X_4X_6 - X_1^{-1}X_5$$

$$X_6 = t^2 + n^2 \qquad \dot{X}_6 = 2X_1$$

$$X_7 = F(v, \lambda; t) \qquad \dot{X}_7 = \frac{2}{\lambda}X_1X_2X_3X_4$$

$$X_8 = f(v, \lambda; t) = \frac{2}{\lambda}X_1X_2X_3X_4 \tag{13.6}$$

Here X_7 and X_8 are the noncentral cumulative and density respectively. They include not only the three factors but also the derivative of the argument t^2/λ. Initial conditions for the modified Bessel functions $I_n(t)$ and $I_n'(t)$ are tabulated for certain arguments [180] or are obtainable from the recursion relationship

$$I_n'(t) = I_{n-1}(t) - nt^{-1}I_n(t)$$

or from the direct summation of the infinite series. When v is large we select $t_0 \approx v/3$ as an initial value since in its vicinity the values of $|\log I_n(t)|$ are small, even though they are large elsewhere.

If the behavior of $I_n(t)$ should pose a problem, one may note that

$$t^2x'' + (t + 2t^2)x' + (t - n^2)x = 0$$

is an altered differential equation that is solved by $e^{-t}I_n(t)$, which will here be denoted as $W_n(t)$. Based on $W_n(t)$, the resulting S-system is somewhat different:

$$X_1 = t \qquad \dot{X}_1 = 1$$

$$X_2 = \frac{1}{2}(t/\lambda)^n \qquad \dot{X}_2 = nX_1^{-1}X_2$$

$$X_3 = \exp\left[-\frac{(t-\lambda)^2}{2\lambda}\right] \qquad \dot{X}_3 = X_3 - \frac{1}{\lambda}X_1X_3$$

$$X_4 = W_n(t) \qquad \dot{X}_4 = X_5X_6$$

$$X_5 = W_n'(t)X_6^{-1} \qquad \dot{X}_5 = 2X_5 - X_1^{-1}X_4X_6^{-1}$$

$$X_6 = W_n'(t)X_5^{-1} \qquad \dot{X}_6 = n^2X_1^{-2}X_4X_5^{-1} - X_1^{-1}X_6$$

$$X_7 = F(v, \lambda; t) \qquad \dot{X}_7 = \frac{2}{\lambda}X_1X_2X_3X_4$$

$$X_8 = f(v, \lambda; t) = \frac{2}{\lambda}X_1X_2X_3X_4$$

If we, without loss of generality, select $(X_6)_0 = X_6(t_0) = 1$, then the initial conditions for X_3, X_4, and X_5 will require modification as compared with (13.6), while the other starting values will be the same.

Initialization of variable X_7 may be difficult computationally. However, since X_7 does not appear on the right-hand side of any equation, any error in $(X_7)_0$ merely translates thc solution by the same amount. Therefore, we could guess $(X_7)_0 = 1$ and observe the limit of $X_7(t)$ either as $t \to \infty$ or $t \to 0$. Since the true limits are known to be 1 and 0 respectively, the proper value of $(X_7)_0$ is obtained immediately.

One drawback of the representation in terms of $I_n(t)$ is that with λ appearing in the denominator, the behavior of the noncentral distribution as $\lambda \to 0$ may lead to numerical problems.

Second Solutions

Another approach permits one to link the central and noncentral chi-square distributions more closely. It is based on an alternative form of the density:

$$f(v, \lambda; t) = \frac{e^{-(\lambda+t)/2}\,(t/2)^{v/2}}{t\Gamma(v/2)}\,\zeta(t)$$

where

$$\zeta(t) = \sum_{j=0}^{\infty} \frac{\Gamma\left(\frac{v}{2}\right)\lambda^j t^j}{\Gamma\left(\frac{v}{2}+j\right)2^{2j}j!}$$

satisfies the equation

$$t\zeta''(t) + \frac{v}{2}\zeta'(t) - \frac{\lambda}{4}\zeta(t) = 0$$

A second-order differential equation corresponds to two first-order equations that then are recast, and the resulting S-system, including two auxiliary equations for the densities, is

$$\begin{aligned}
X_1 &= t & \dot{X}_1 &= 1\\
X_2 &= e^{-t/2}(t/2)^{v/2}/\Gamma(v/2) & \dot{X}_2 &= \frac{v}{2}X_1^{-1}X_2 - \frac{1}{2}X_2\\
X_3 &= F(v; t) & \dot{X}_3 &= X_1^{-1}X_2\\
X_4 &= \zeta(t) & \dot{X}_4 &= X_5\\
X_5 &= \zeta'(t) & \dot{X}_5 &= \frac{\lambda}{4}X_1^{-1}X_4 - \frac{v}{2}X_1^{-1}X_5\\
X_6 &= F(v, \lambda; t) & \dot{X}_6 &= e^{-\lambda/2}X_1^{-1}X_2X_4\\
X_7 &= f(v; t) = X_1^{-1}X_2 & &\\
X_8 &= f(v, \lambda; t) = e^{-\lambda/2}X_1^{-1}X_2X_4 & &
\end{aligned} \tag{13.7}$$

With multiple S-system formulations for the same problem, it is natural to question, from a specific or a general viewpoint, which representation is best (see also Chapter 12). Not only may the number of dependent and auxiliary variables differ, but the expressions on the right-hand sides of the S-system equations can affect the speed and the accuracy of the solution. A criterion more scientific than technical in nature is the amenability to further analyses. In this case, the fact that the central and noncentral chi-square c.d.f.'s have been simultaneously recast in one system offers additional possibilities, as we shall see below.

INVERSION OF S-SYSTEMS

S-system inversion pertains to change-of-variable transformations that are of value in a wide range of applications. Within the present context, suppose there exists at least one solution point $(t^*, \boldsymbol{X}(t^*))$ which is of particular interest. If t^* is known then one can merely run the independent variable up to that point, but if it is unknown such evaluation may prove awkward. However, when $X_k(t^*)$ is known for some $k \in \{1, \ldots, n\}$, this point will be easily located if one can run the system with X_k as the independent variable. This can be accomplished through application of the chain rule to the S-system, if either $\alpha_k = 0$ or $\beta_k = 0$.

For example, suppose we want to solve (13.7) in terms of central chi-square quantiles. Since $\dot{X}_3 = X_1^{-1}X_2$ (and thus $\beta_3 = 0$), we can define

$$\frac{dX_i}{dt}\bigg/\frac{dX_3}{dt} = \frac{dX_i}{dX_3} = \mathring{X}_i$$

where $\mathring{X}_i$ denotes a derivative taken with respect to a new variable. The inversion of (13.7) then becomes

$$\begin{aligned}
\mathring{X}_1 &= X_1X_2^{-1} \\
\mathring{X}_2 &= \frac{v}{2} - \frac{1}{2}X_1 \\
\mathring{X}_3 &= 1 \\
\mathring{X}_4 &= X_1X_2^{-1}X_5 \\
\mathring{X}_5 &= \frac{\lambda}{4}X_2^{-1}X_4 - \frac{v}{2}X_2^{-1}X_5 \\
\mathring{X}_6 &= e^{-\lambda/2}X_4
\end{aligned} \qquad (13.8)$$

It may be noted that the third equation in (13.8) is now superfluous and can be deleted. Although variables are automatically renumbered in ESSYNS upon deletion of an equation, we prefer to preserve clarity here by retaining the original subscripts.

In the event that α_k and β_k are both non-zero, set $\dot{X}_k$ equal to a product of $n(k)$ new variables, where $n(k)$ will depend on the complexity of the expression being replaced.

The initial conditions of all the dependent variables remain the same as they were prior to the inversion, except that the new independent variable is initialized at $X_k(t_0)$. In the present example, the solution generates quantiles of the chi-square distribution as $X_1(X_3)$; this is possible since $X_1 = t$ in (13.7). For systems where t is not represented by a variable, one can retain the kth variable in the inverted system and initialize it at t_0.

Since $X_6 = F(v, \lambda; t)$, it is possible to express the noncentral quantiles in terms of the central quantiles. In other words, we can obtain the power of the chi-square test with specified v and λ. More specifically, the power equals $1 - X_6(X_3 = 1 - \alpha)$, since such an hypothesis will be rejected for large values of the chi-square statistic.

In similar fashion, one can apply the chain rule to express the central quantiles in terms of noncentral quantiles; dividing (13.7) by $\dot{X}_6$ yields the following inverted system:

$$
\begin{aligned}
X_1 &= t & \mathring{X}_1 &= X_1 X_2^{-1} X_4 \\
X_2 &= e^{-(t+\lambda)/2}\left(\frac{t}{2}\right)^{v/2} \Big/ \Gamma\left(\frac{v}{2}\right) & \mathring{X}_2 &= \frac{v}{2} X_4^{-1} - \frac{1}{2} X_1 X_4^{-1} \\
X_3 &= F(v; t) & \mathring{X}_3 &= e^{\lambda/2} X_4^{-1} \\
X_4 &= \zeta(t) & \mathring{X}_4 &= X_1 X_2^{-1} X_4^{-1} X_5 \\
X_5 &= \zeta'(t) & \mathring{X}_5 &= \frac{\lambda}{4} X_2^{-1} - \frac{v}{2} X_2^{-1} X_4^{-1} X_5
\end{aligned}
\qquad (13.9)
$$

Note that the definition of X_2 has been changed slightly to absorb the factor $e^{-\lambda/2}$. This was done to avoid the ubiquitous appearance of $e^{-\lambda/2}$ in the inverted system upon division by $\dot{X}_6$. Although X_7 is not shown in (13.9) due to its status as an auxiliary variable, the coefficient $e^{\lambda/2}$ is needed there for the central density. Furthermore, the initial value of X_2 must be correspondingly modified.

Quantiles of the noncentral chi-square distribution are evaluated in $X_1(X_6)$. In addition, we can obtain the inverse of the power function by computing $1 - X_3(X_6)$; that is, with the power of the chi-square test specified, $1 - X_6$, for a given v and λ, we can determine what α-level this test is using. These accomplishments of S-system methodology have no counterpart in classical statistical theory.

SIMPLE TRANSFORMATIONS OF DENSITIES

Some densities are related to one another through simple functional relationships. Manipulations on algebraic expressions for densities have long been available, and they are preserved in the S-system context. Perhaps the best known example is that the square of the standard normal distribution equals the chi-square distribution with $v = 1$. To illustrate this relationship, the first step in the transformation process is to append new variables (say, X_3 and X_4) to the original system (say, with variables X_1 and X_2), whose form is known:

$$\begin{aligned} X_1 &= t & \dot{X}_1 &= 1 \\ X_2 &= z(t) & \dot{X}_2 &= -X_1 X_2 \\ X_3 &= t^2 & \dot{X}_3 &= 2X_1 \\ X_4 &= f(t^2) & \dot{X}_4 &= -X_2 - X_1^{-2} X_2 \end{aligned} \qquad (13.10)$$

(For this purpose, the equations for initial variables need not address issues of initial conditions). In the second step, one formulates the derivatives of X_3 and X_4. After differentiating the transformation (t^2) to be performed on the density, we do the same for the relationship $f(t^2)\,\dot{X}_3 = cz(t)$. Here $c = 2$ since both positive and negative tails of the standard normal are mapped onto the single tail of the resultant density. Thus, $X_4 = 2X_2(\dot{X}_3)^{-1} = X_1^{-1}X_2$, and its derivative reduces as in the last equation of (13.10).

The third step makes a change in the independent variable. Division by $\dot{X}_3$ means that $dX_4/dX_3 = -0.5X_1^{-1}X_2 - 0.5X_1^{-3}X_2$. The fourth step rewrites the right-hand side in terms of X_3 and X_4 rather than X_1 and X_2. Using $X_1^2 = X_3$ and $X_2 = X_1X_4$, we arrive at $dX_4/X_3 = -0.5X_4 - 0.5X_3^{-1}X_4$. Recalling that $v = 1$, we observe that (13.3) has been recovered. It may be converted to S-system form as shown in (13.4). Initial conditions are transformed using relationships employed in the second step. Cumulatives are best handled after the S-system for densities has been optimized.

CALCULATION OF MOMENTS

Moments of distributions, especially the mean and variance, are a fundamental part of statistical analysis. For a simple exposition on the calculation of moments, reconsider the central chi-square system (13.4). The m-th moment of a distribution generally may be expressed $\int_{-\infty}^{\infty} t^m f(t)dt$. Expressions of this form are easily written as differential equations. For instance, the first and second moments can be added directly to (13.4), with the result:

$$
\begin{aligned}
X_1 &= t & \dot{X}_1 &= 1 \\
X_2 &= tf(v;\, t) & \dot{X}_2 &= \frac{v}{2} X_1^{-1} X_2 - \frac{1}{2} X_2 \\
X_3 &= \int_0^t sf(v;\, s)\, ds & \dot{X}_3 &= X_2 \\
X_4 &= \int_0^t s^2 f(v;\, s)\, ds & \dot{X}_4 &= X_1 X_2 \\
X_5 &= X_4 - X_3^2 & & \qquad (13.11)
\end{aligned}
$$

Here the first moment (the mean) is $X_3(t = \infty)$, and the second moment is $X_4(t = \infty)$. The second *central* moment (the variance) is expressed as $X_4(\infty) - [X_3(\infty)]^2$. Clearly, other moments can be constructed by inserting different values for the exponent. It is noted that the exponent is not constrained to positive integers and that extensions to other weighting functions are possible. As yet, statistical science has not assigned a particular value to fractional and generalized moments.

To illustrate the use of system (13.11), consider the case $v = 15$, with the argument initialized at $t_0 = 0.02$. The associated initial conditions are:

$$
\begin{aligned}
X_1(0.02) &= 0.02 \\
X_2(0.02) &= 5.2908353005877 \times 10^{-19} \\
X_3(0.02) &= 1.2449 \times 10^{-21} \\
X_4(0.02) &= 2.621 \times 10^{-24}
\end{aligned}
$$

Since X_2 appears on the right-hand side of most of the differential equations in (13.11), it requires greater precision than X_3 and X_4. Upon running the system from 0.02 towards ∞, we obtain the asymptotic results $X_3(\infty) = 15$ and $X_5(\infty) = 30$ exact to fourteen decimal places.

Clearly, for such a simple example one would doubtlessly prefer existing methods of solution. The S-system approach can, however, be recommended for complex problems involving the evaluation of moments that tax the traditional methodologies (cf. [49]).

THE GAMMA FUNCTION

A great array of mathematical functions are related to differential equations and can be recast as S-systems. The gamma function arises frequently in pure and applied mathematics, and it plays a distinctive role in the field of statistical distributions. There is no known differential equation with the gamma function $\Gamma(p)$ as its solution [181]. Nevertheless, the incomplete gamma function

$$\Gamma(p, t) = \int_0^t s^{p-1} e^{-s}\, ds = e^{-t} \sum_{k=1}^{\infty} t^{p+k} \Big/ \prod_{j=0}^{k} (p+j)$$

solves the very simple equation $d\Gamma(p, t)/dt = t^{p-1}e^{-t}$. The corresponding S-system follows immediately:

$$\begin{aligned} X_1 &= t & \dot{X}_1 &= 1 \\ X_2 &= e^{-t} & \dot{X}_2 &= -X_2 \\ X_3 &= \Gamma(p, t) & \dot{X}_3 &= X_1^{p-1} X_2 \end{aligned} \qquad (13.12)$$

For example, if $p = 7$ with $t_0 = 0.01$, the initial conditions are $(X_1)_0 = 0.01$, $(X_2)_0 = .990049833749168$, and $(X_3)_0 = 1.41637994 \times 10^{-15}$. Upon running ESSYNS towards ∞, or until the asymptote is attained ($t = 70$ in this case), we obtain $X_3(\infty) = \Gamma(7) = 719.999999999993$. Alternately, if one starts at $t_0 = 1$, we initialize with $(X_1)_0 = 1$, $(X_2)_0 = e^{-1}$, and

$$(X_3)_0 = e^{-1} \sum_{j=0}^{\infty} \left[\prod_{j=0}^{k} (p + j) \right]^{-1} = 0.0599336274873766$$

The infinite series for $(X_3)_0$ does not converge quite as rapidly if $t_0 = 1$, but this is more than compensated for by the more accurate final result $\Gamma(7) = 720.000000000000$.

A more complicated and very different S-system yielding the gamma function may be derived through the transformation $s = \tan\theta$.

THE BETA FUNCTION

The beta function is closely related to the gamma through the formula

$$B(m, n) = \frac{\Gamma(m)\Gamma(n)}{\Gamma(m+n)} = \int_0^1 t^{m-1}(1-t)^{n-1}\, dt$$

and it is useful in solving F distributions. It is also of interest for analyses of the t-distribution where it appears in the expression

$$\frac{\Gamma\left(\frac{v+1}{2}\right)}{\Gamma\left(\frac{v}{2}\right)} = \frac{\sqrt{\pi}}{B\left(\frac{v}{2}, \frac{1}{2}\right)}$$

This formula is useful when v is large enough to produce computer overflows on gamma function evaluations.

In general, however, simultaneous modeling of three gamma functions permits the construction of a beta function. Yet other approaches are available. Since the integral vanishes at both endpoints, we split $B(m, n)$ into two integrals:

$$B_1 = \int_{1/2}^{1} t^{m-1}(1-t)^{n-1}\, dt \quad \text{and} \quad B_2 = \int_{1/2}^{1} (1-t)^{m-1} t^{n-1}\, dt$$

that are represented by the following S-system:

$$\begin{aligned}
X_1 &= t & \dot{X}_1 &= 1 \\
X_2 &= 1 - t & \dot{X}_2 &= -1 \\
X_3 &= B_1 + 1 & \dot{X}_3 &= X_1^{m-1} X_2^{n-1} \\
X_4 &= B_2 + 1 & \dot{X}_4 &= X_1^{n-1} X_2^{m-1} \\
X_5 &= X_3 + X_4 - 2 & &
\end{aligned} \tag{13.13}$$

With $t_0 = 0.5$, the other initial conditions are $(X_1)_0 = (X_2)_0 = 0.5$, $(X_3)_0 = (X_4)_0 = 1$. As t approaches 1, X_5 approaches $B(m, n)$. Because of the singularity at 1, ESSYNS does not yield results at $t = 1$, but it is possible to run the system close enough to obtain the result asymptotically.

If $B(v/2, 1/2)$ is desired, the convergence of (13.13) is less good. The beta integrals, when transformed by $t = \cos^2 \theta$, become

$$B(m, n) = 2 \int_0^{\pi/2} \cos^{2m-1} \theta \sin^{2n-1} \theta \, d\theta$$

and

$$B\left(\frac{v}{2}, \frac{1}{2}\right) = 2 \int_0^{\pi/2} \cos^{v-1} \theta \, d\theta$$

The latter may be recast in S-system form as

$$\begin{aligned} X_1 &= \cos \theta & \dot{X}_1 &= 1 - X_2 \\ X_2 &= \sin \theta + 1 & \dot{X}_2 &= X_1 \\ X_3 &= B\left(\frac{v}{2}, \frac{1}{2}\right) + 1 & \dot{X}_3 &= 2X_1^{v-1} \end{aligned}$$

where $X_1(0) = X_2(0) = X_3(0) = 1$. Although the specified value of X_3 is attained theoretically at $\pi/2$, running the system up to $\theta = 1.55$ yields fourteen place accuracy unless v is quite small.

INDEPENDENT VARIABLE TRANSFORMATIONS FOR TAILS

In the following, we shall consider several transformations on independent S-system variables with the goal of facilitating computations along horizontal asymptotes.

As above, we specify $X_1 = t + c$, $\dot{X}_1 = 1$, whereas $X_2, \ldots, X_n$ take on more general forms within the S-system structure (13.1). These variables are sometimes employed in recasting heavy-tailed distributions for which the cumulative distribution function does not closely approximate its limit except for very large values of the argument. This involves some risk of round-off error because many steps are required to follow the independent variable out to very large values. Also, repeated system runs may be necessary because it often is uncertain how large the argument must become. Therefore, it becomes reasonable to consider which transformations of the independent variable would be helpful in such situations.

A fairly simple approach would be to define a new variable $X_{n+1} = (X_1 - 1)/X_1$, which would approach unity as t become indefinitely large, although initially $X_1 \leq 1$ would not be permitted. More precisely, X_1: $[c, \infty)$ maps onto X_{n+1}: $[1 - 1/c, 1)$. Clearly, $\dot{X}_{n+1} = X_1^{-2}$. The next step is to invert the system via the chain rule, i.e., let $\mathring{X}_i = (dX_i/dt)/(dX_{n+1}/dt)$ so that X_{n+1} becomes the new independent variable. Thus, $\mathring{X}_1 = X_1^2$ and

$$\mathring{X}_i = \alpha_i X_1^{g_{i1}+2} \prod_{j=2}^{n} X_j^{g_{ij}} - \beta_i X_1^{h_{i1}+2} \prod_{j=2}^{n} X_j^{h_{ij}}, \quad i \in \{2, \ldots, n\} \qquad (13.14)$$

This transformation is therefore made without increasing the total number of equations; the variable t is no longer in the picture.

Another possibility is to use the transformation that maps $[c, \infty)$ onto $[\tan^{-1} c, \pi/2)$. An advantage of this approach is that the admissible domain is somewhat expanded; namely, we assume $X_1 > 0$ rather than $X_1 > 1$. A disadvantage is that the system is expanded slightly, since now

$$\begin{aligned} X_{n+1} &= 1 + X_1^2 & \dot{X}_{n+1} &= 2X_1 \\ X_{n+2} &= \tan^{-1} X_1 & \dot{X}_{n+2} &= X_{n+1}^{-1} \end{aligned} \tag{13.15}$$

A similar application of the chain rule produces the transformed system

$$\begin{aligned} \mathring{X}_1 &= X_{n+1} \\ \mathring{X}_i &= \alpha_i X_{n+1} \prod_{j=1}^{n} X_j^{g_{ij}} - \beta_i X_{n+1} \prod_{j=1}^{n} X_j^{h_{ij}} \qquad i \in \{2, \ldots, n\} \\ \mathring{X}_{n+1} &= 2X_1 X_{n+1} \end{aligned} \tag{13.16}$$

Here X_{n+2} is the new independent variable.

Systems (13.14) and (13.16) have comparable rates of asymptotic convergence. Different rates become possible when the latter is modified with $X_{n+1} = 1 + X_1^{2p}$, $X_{n+2} = \tan^{-1}(X_1^p)$, where $p > 0$, and different constants p permit various rates of convergence, depending on the problem. The corresponding differential equation becomes $\dot{X}_{n+2} = pX_1^{p-1}X_{n+1}^{-1}$, with the result that (13.16) may be generalized as

$$\begin{aligned} \mathring{X}_1 &= p^{-1} X_1^{1-p} X_{n+1} \\ \mathring{X}_i &= \frac{\alpha_i}{p} X_1^{g_{i1}+1-p} X_{n+1} \prod_{j=2}^{n} X_j^{g_{ij}} - \frac{\beta_i}{p} X_1^{h_{i1}+1-p} X_{n+1} \prod_{j=2}^{n} X_j^{h_{ij}} \\ \mathring{X}_{n+1} &= 2X_1^p X_{n+1} \end{aligned} \tag{13.17}$$

The above transformed systems (13.14), (13.16), and (13.17) involve running the independent variable close up to, but not including, a limiting constant (i.e., 1 or $\pi/2$). How close may be a matter of trial and error, particularly for $\pi/2$ where logarithmic closeness is not merely a function of the number of decimal places. There is also some risk of an inadvertent entry of too large a final value.

The transformation X_1: $[c, \infty)$ onto X_{n+1}: $[\log c, \infty)$ sidesteps these issues.

If one is prepared to accept the restriction $X_1 > 1$, then appending $X_{n+1} = \log X_1$, $\dot{X}_{n+1} = X_1^{-1}$ leads to a transformed S-system with $\overset{\circ}{X}_1 = X_1$ and

$$\overset{\circ}{X}_i = \alpha_i X_1^{g_{i1}+1} \prod_{j=2}^{n} X_j^{g_{ij}} - \beta_i X_1^{h_{i1}+1} \prod_{j=2}^{n} X_j^{h_{ij}}, \qquad i \in \{2, \ldots, n\} \quad (13.18)$$

In this case no finite value of the new independent variable actually attains the asymptote, but the approach can be considerably faster than it is for an untransformed system.

The increased speed of this convergence is retained by $X_{n+1} = p \log X_1$. However, this generalization entails a disadvantage that is shared by (13.17), namely that the power parameter p influences a substantial number of S-system coefficients and exponents. A need to adjust p implies many corresponding redefinitions. This problem may be avoided by an alternate modification of the logarithmic transformation:

$$\begin{aligned} X_{n+1} &= k + k^{-1} \log X_1 & \dot{X}_{n+1} &= k^{-1} X^{-1} \\ X_{n+2} &= k^{-1} X_{n+1}^{-1} \log X_1 & \dot{X}_{n+2} &= X_1^{-1} X_{n+1}^{-2} \end{aligned} \quad (13.19)$$

Here k is a positive constant, and since this corresponds to the mapping of X_1: $[c, \infty)$ onto X_{n+2}: $[\log c/\{k^2 + \log c\}, 1)$, we assume $X_1 \geq 1$. When the chain rule is applied to make X_{n+2} the new independent variable, the resulting S-system is

$$\begin{aligned} \overset{\circ}{X}_1 &= X_1 X_{n+1}^2 \\ \overset{\circ}{X}_i &= \alpha_i X_1^{g_{i1}+1} X_{n+1}^2 \prod_{j=2}^{n} X_j^{g_{ij}} - \beta_i X_1^{h_{i1}+1} X_{n+1}^2 \prod_{j=2}^{n} X_j^{h_{ij}}, \\ &\qquad i \in \{2, \ldots, n\} \\ \overset{\circ}{X}_{n+1} &= k^{-1} X_{n+1}^2 \end{aligned} \quad (13.20)$$

To run (13.20) for various values of k, one needs only to alter α_{n+1} and the initial values of X_{n+1} and (often) X_{n+2}. Faster convergence rates are sometimes obtained by selecting smaller values of k. One may run X_{n+2} arbitrarily close to unity.

Still another modification may be termed the pth root transformation. More particularly, we introduce the variable $X_{n+1} = pX_1^{1/p}$, $p \neq 0$; usually, one will select $p > 1$. Upon division by $\dot{X}_{n+1} = X_1^{(1-p)/p}$, an S-system takes on the form

$$\mathring{X}_1 = X_1^{(p-1)/p}$$

$$\mathring{X}_i = \alpha_i X_1^{g_{i1}+(p-1)/p} \prod_{j=2}^{n} X_j^{g_{ij}} - \beta_i X_1^{h_{i1}+(p-1)/p} \prod_{j=2}^{n} X_j^{h_{ij}} \quad (13.21)$$

Large values of p tend to give results similar to those of the logarithmic transformation. Only exponents of X_1 are affected by this method and no equations are added. One merely needs to evaluate endpoints for the run parameter.

Example: Noncentral Chi-Square Distribution

Some of the above-mentioned techniques may be illustrated with a special case, namely the chi-square distribution with $v = 6$ and $\lambda = 7.12277324467228$. With these parameters the chi-square test has power 0.25 if $\alpha = 0.01$; its power curve appears in Fig. 13.1.

If system (13.8) is run out to $t = 48.282$, beginning from $t_0 = 2$, $F(v, \delta; t) = .99990002$ is a result. To achieve this involves 7.52 secs and nineteen steps. Six-decimal accuracy is obtained at $F(v, \delta; 63.05) = .99999900002$ in twenty-two steps (8.35 secs), and eight decimals are found at $F(v, \delta; 76.897) = .9999999900005$ in twenty-four steps (8.95 secs). The reported times refer to a 6 MHz IBM AT.

If the transformation corresponding to (13.14) is used, then using sixteen steps and 8.24 secs give us four-decimal accuracy. Here $(X_1 - 1)/X_1$ is run

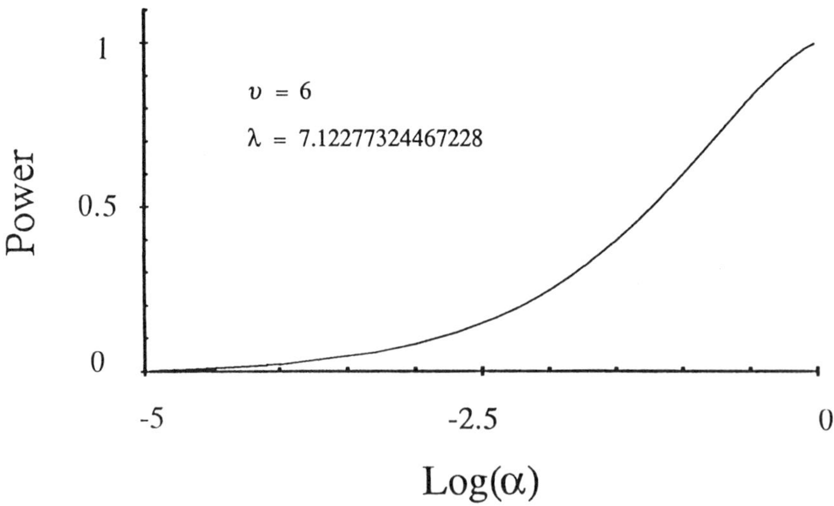

FIGURE 13.1. Power of the χ^2 Test

from 0.5 to 0.979288348. The six-decimal result requires seventeen steps in 8.78 secs, and eight decimals, eighteen steps in 9.28 secs. The final independent variable values are .984139572 and .986995592. In this example, some h_{i1} exponents are transformed from 0 to 2 and thus they slightly lengthen the execution time. Local error tolerance remains set at 10^{-16}.

The arctangent transformation, as given in (13.15) and (13.16), does not differ greatly from the previous one. From its initial value of 1.107148718, the new independent variable rises to 1.550087635, 1.554937228, and 1.557792651 respectively. The three precisions involve fifteen steps in 8.84 secs, seventeen steps in 10.10 secs, and eighteen steps in 10.71 secs, respectively. This method involves the addition of a seventh system equation.

If the arctangent method is modified as given in (13.17), the runs behave somewhat differently. For instance, if $p = 0.5$ is selected, the value of t_0 becomes 0.955316618. To obtain four significant digits, the run up to 1.427862300 uses twenty-one steps and 7.63 secs. The results for six digits are 1.445517681 with twenty-three steps and 8.29 secs, while to attain eight-figure accuracy we need 1.457249959 with twenty-four steps and 8.67 secs. In this instance, this modification does not radically change behavior. A choice of $p = 2$ is inferior with this example.

The log transformation involves relatively few parameter changes and an initial run parameter of ln 2, and final values ln 48.282, ln 63.050, and ln 76.897, respectively. The number of steps taken to solve are 28, 32, and 34, while the elapsed times are 4.73, 5.33, and 5.61 seconds, for the three levels of precision. This noncentral chi-square is solved considerably faster with the log transformation, but more steps are required. Supported by experience with other problems, this method is often recommendable.

A generalization of the log transform appearing in (13.19) and (13.20) may be attempted with $k = 4$, so the run parameter is initialized at .041522858. To obtain four, six, and eight digits of accuracy, we require twenty-two, twenty-six, and twenty-eight steps taking 6.92, 7.97, and 8.46 secs. Other results are obtained with other choices of k. In any case, a seventh system equation must be introduced.

Finally, setting $p = 2$ in (13.21) yields the square root transformation, with the new independent variable starting at $2\sqrt{2}$. Accuracy of four, six, and eight digits is obtained with nineteen, twenty-two, and twenty-four steps with times of 6.04, 6.81, and 7.25 seconds.

Example: Cauchy Distribution

The Cauchy distribution, otherwise known as the t-distribution with $v = 1$, is noted for having particularly heavy tails. Its S-system formulation may therefore be given as:

$$
\begin{array}{lll}
X_1 = t + c & \dot{X}_1 = 1 & (X_1)_0 = 2 \\
X_2 = 1 + t^2 & \dot{X}_2 = 2X_1 - 2c & (X_2)_0 = 2 \\
X_3 = f(t) & \dot{X}_3 = 2cX_2^{-1}X_3 - 2X_1X_2^{-1}X_3 & (X_3)_0 = 1/(2\pi) \\
X_4 = F(t) & \dot{X}_4 = X_3 & (X_4)_0 = 3/4
\end{array}
\qquad (13.22)
$$

We have taken $c = 1$ rather than $c = 0$ to avoid zero initial conditions for some transformations, and because it conforms to common practice. Inversion of (13.22) shows that the 99.99%, 99.9999%, and 99.999999% percentile points are 3183.098757, 318309.8863, and 31830990.87, respectively. If $(X_4)_0$ were unknown, its precise evaluation using the strategy of computing the limiting values of X_4 would be problematic. However, the use of any of the transformations described earlier would in fact be of assistance, although some would be of more value than others. A system with fewer equations than (13.8) may often lend itself well to such strategies since run-time behavior tends to be determined by the least well-conditioned variable.

In Table 13.1, the number of steps rather than CPU time are tabulated since the latter is influenced by the number of reporting points. However, an execution time table would follow a similar pattern, except that the log and modified log transformations are roughly equally fast. Both are three to five times faster than the untransformed S-system runs.

SUMMARY

Analysis of statistical distributions covers a diversity of characterizations, including densities, cumulatives, quantiles, moments, and relationships to other distributions. The S-system approach to these problems is demonstrated with a

Table 13.1. Number of ESSYNS Steps to Solve (13.22)

	Number of Places Accuracy for X_4		
Transformation	4	6	8
None	47	73	99
Ratio	32	52	72
Arctangent	33	52	72
Logarithm	20	23	25
Modified log ($k = 1$)	11	13	15
Square root	31	43	56

few examples. The noncentral chi-square is included as a representative of rather complicated distributions that are not easily analyzed with traditional methodology. In complicated cases, special functions are necessary as input for the set of differential equations. These special functions can also be evaluated in S-system form with high precision. Several transformations of S-systems are described that help to optimize representations of distributions with respect to numerical efficiency.

14. Recasting as a Modeling Tool

MITCHELLE S. MORRISON

Given the topological structure of a network, an S-system description usually can be derived in a straightforward fashion. Rather strict rules dictate which internal or external variables should appear in each power-law term of the system, and quantitative and qualitative knowledge about the processes that govern the network can be used to specify or at least constrain some of the parameter values of the S-system model (cf. Chapter 2). Once a phenomenon is represented in S-system form, it can be analyzed with paper and pencil or with symbolic and numeric computer programs (cf. Chapters 6 and 7). While there are definite advantages to using the direct modeling approach to obtain an S-system representation, there are some disadvantages.

Although there are many examples in biology where the resulting S-system representation has proven appropriate, one sometimes encounters situations in which the direct modeling strategy is not sufficient. Two situations are of primary importance: In one case, the accuracy of the representation is insufficient. In the other case, the network contains components whose input-output relationships are known but whose internal states are not well defined; such a network component is sometimes called a *black box*.

As was discussed in Chapters 2 and 3, S-system representations of biological systems frequently are very accurate over wide ranges of variation in the system variables. Usually, these ranges by far exceed ranges observed *in situ* (cf. [35:189]). However, it is obvious that there may be situations where the underlying power-law approximations become inaccurate if one deviates too far from the operating point. If accuracy throughout an extraordinarily wide range is required, one can subdivide the range into smaller intervals and develop S-system representations within each interval. However, this strategy has the obvious drawback that the resulting system description will consist of a series of models with different parameter values, instead of one model. Furthermore, in some cases not even a series of approximations is satisfactory. For instance, if the dynamic system under investigation is very sensitive, even minute deviations may cause the qualitative behavior to change. Later in this chapter we will encounter an example where an S-system exhibits spiralling trajectories instead of the limit cycles of the approximated system. These examples may not be relevant in the daily routine of modeling but they require our attention nevertheless.

A second problem with the direct modeling strategy occurs when one encounters ill-defined system components. For instance, such components can be encountered when the model receives time-varying input that is outside the focus of the model. As an illustration, suppose one is interested in a metabolic network in which one of the components fluctuates in a circadian fashion; the daily fluctuations are quantifiable even though it is difficult to account for them with a biochemical mechanism that can be included in the metabolic model. In such a case it is desirable to employ an S-system module whose structure is irrelevant but whose response appropriately represents the desired input. This module can be incorporated into the master model, governing its input.

This chapter describes how methods of recasting (cf. Chapter 12) can be employed to solve modeling problems of the above types. Since the recast system has the exact same solution as the original system, it exhibits all qualitative and quantitative properties of the original system. Thus, recasting is an important tool to use in the case where the underlying power-law approximation becomes inaccurate over the operating range of interest. Recasting is also a useful tool for including an ill-defined component in a model. Recasting used in conjunction with the direct modeling strategy allows the modeler to take full advantage of the various methods of S-system analysis.

RECASTING AS A TOOL FOR IMPROVING ACCURACY

Modeling with S-systems has proven to be appropriate for many problems in biology. Nevertheless, there are situations where the underlying power-law approximation becomes inaccurate if one deviates too far from the operating point. The following examples illustrate how recasting can be used to overcome the problem of insufficient accuracy. The first example is a case where the underlying power-law representation overshoots the approximated rate law. The second example is a structurally sensitive oscillator.

Example 1: A Chain of Michaelis-Menten Reactions

Consider a linear chain of essentially irreversible enzyme-catalyzed reactions *in vitro* of the form

$$S_1 \to S_2 \to S_3 \to P$$

and assume that it is appropriately representable by Michaelis-Menten rate laws. A dynamical system description is then

$$\dot{S}_1 = \qquad\qquad - \frac{V_{m1}S_1}{S_1 + K_{m1}} \tag{14.1}$$

$$\dot{S}_2 = \frac{V_{m1}S_1}{X_1 + K_{m1}} - \frac{V_{m2}S_2}{S_2 + K_{m2}} \tag{14.2}$$

$$\dot{S}_3 = \frac{V_{m2}S_2}{S_2 + K_{m2}} - \frac{V_{m3}S_3}{S_3 + K_{m3}} \tag{14.3}$$

$$\dot{P} = \frac{V_{m3}S_3}{S_3 + K_{m3}} \tag{14.4}$$

where (for $i = 1, 2, 3$) V_{mi} is the maximal velocity and K_{mi} is the substrate concentration that produces half-maximal binding. As discussed in Chapter 2, an S-system representation of the pathway can be obtained by deriving power-law approximations of the rate laws. For example, the rate of synthesis of the end product

$$v_p = \frac{V_{m3}S_3}{S_3 + K_{m3}} \tag{14.5}$$

in S-system representation is given as $\alpha_4 S_3^{g_{43}}$ where the apparent kinetic order g_{43} and the rate constant α_4 are defined as

$$g_{43} = \frac{K_{m3}}{S_3(Op) + K_{m3}}$$

and

$$\alpha_4 = v_p(Op)S_3(Op)^{-g_{43}} \tag{14.6}$$

Since many enzymes under physiological conditions operate around their Michaelis constant, $Op = K_{m3}$ often is a logical choice for the operating point, and this implies $g_{43} = 0.5$. Close to the operating point the power-law approximation is very accurate. However, if any of the substrates deviate far from the operating point, the power-law approximation considerably overshoots the Michaelis-Menten rate law.

If responses over wide ranges of variation in concentrations are of interest, the modeler may use recasting instead of power-law approximation. The result is an exact analogue to the Michaelis-Menten representation in S-system form. This is very convenient but it must be paid for with auxiliary variables and a system matrix that is singular.

To recast (14.1)–(14.4), one first renames the variables as $X_1 = S_1$, $X_2 = S_2$, $X_3 = S_3$, and $X_4 = P$. Then one defines ancillary variables $X_5 = X_1 + K_{m1}$, $X_6 = X_2 + K_{m2}$, and $X_7 = X_3 + K_{m3}$. It is obvious that the derivatives of X_5, X_6, and X_7 are the same as the derivatives of X_1, X_2, and X_3, respectively. The only difference lies in the initial values that are $X_i(0) = S_i(0)$ (for $i = 1, 2, 3$), $X_4(0) = P(0)$, and $X_j(0) = S_j(0) + K_{mj}$ (for $j = 5, 6, 7$). The result of recasting the kinetic equations (14.1)–(14.4) is

$$\dot{X}_1 = \qquad - V_{m1}X_1X_5^{-1} \tag{14.7}$$

$$\dot{X}_2 = V_{m1}X_1X_5^{-1} - V_{m2}X_2X_6^{-1} \tag{14.8}$$

$$\dot{X}_3 = V_{m2}X_2X_6^{-1} - V_{m3}X_3X_7^{-1} \tag{14.9}$$

$$\dot{X}_4 = V_{m3}X_3X_7^{-1} \tag{14.10}$$

$$\dot{X}_5 = \qquad - V_{m1}X_1X_5^{-1} \tag{14.11}$$

$$\dot{X}_6 = V_{m1}X_1X_5^{-1} - V_{m2}X_2X_6^{-1} \tag{14.12}$$

$$\dot{X}_7 = V_{m2}X_2X_6^{-1} - V_{m3}X_3X_7^{-1} \tag{14.13}$$

Because this S-system by definition is an exact representation of the system of Michaelis-Menten rate laws, the problem of insufficient accuracy is overcome: the S-system produces exactly the same responses over the entire range of substrate concentrations.

It is obvious that recast representations of rate laws can be mixed with power-law approximations. For instance, suppose that S_1 is the only substrate that cannot be restricted to a fairly small range around the operating point. The degradation of S_1 then can be represented with an exact S-system analogue of the Michaelis-Menten rate law while all the other processes are represented with power-law approximations.

Example 2: The Brusselator

Chemical systems are capable of exhibiting complex nonlinear dynamics. The best example presumably is the rather complicated Belousov-Zhabotinsky reaction. But even much simpler chemical systems can possess dynamical features that are mathematically very interesting. For instance, let's consider the hypothetical chemical pathway proposed by Prigogine and co-workers in Brussels [185, 187] which commonly is known as the "Brusselator" [195]:

$$A \underset{k_{-1}}{\overset{k_1}{\rightleftharpoons}} X \tag{14.14}$$

$$B + X \underset{k_{-2}}{\overset{k_2}{\rightleftharpoons}} Y + D \tag{14.15}$$

$$2X + Y \underset{k_{-3}}{\overset{k_3}{\rightleftharpoons}} 3X \tag{14.16}$$

$$X \underset{k_{-4}}{\overset{k_4}{\rightleftharpoons}} E \tag{14.17}$$

Four reversible reactions govern the dynamics of the two components of interest, X and Y. For most analyses of this scheme of reactions, one assumes an experimental design where the end products D and E are removed as soon as they are produced. The corresponding reactions then are essentially irreversible, and k_{-4} and k_{-2} can be set equal to zero. Furthermore, it is assumed for mathematical convenience that $k_{-1} \approx 0$ and $k_{-3} \approx 0$. The remaining parameters are not always numerically characterized, but a particular case that has been studied in the literature is $k_3 = B$, $k_1 = A$ and $k_2 = k_4 = 1$ [186, 188]. With these kinetic specifications, one obtains the kinetic equations (14.18) and (14.19).

$$\dot{X} = A + X^2Y - (B + 1)X \tag{14.18}$$

$$\dot{Y} = BX - X^2Y \tag{14.19}$$

Direct calculation confirms that the Brusselator has the single steady-state solution

$$X_0 = A, \; Y_0 = \frac{B}{A} \tag{14.20}$$

The dynamic behavior is more difficult to explore, but one can show that a critical condition between the external constituents A and B separates qualitatively different nonlinear responses. For $B < A^2 + 1$, the steady-state solution is stable and for any choice of initial conditions other than the steady state, solutions are damped oscillations. $B = A^2 + 1$ is the Hopf bifurcation point at which a stable limit cycle is generated. For $B > A^2 + 1$ and any choice of initial conditions, the Brusselator approaches a stable limit cycle [190].

Different strategies can be pursued to represent the chemical reactions (14.14)–(14.17) as an S-system. The straightforward approach is to use the modeling strategies outlined in Chapter 2 and to translate the reactions (14.14)–(14.17) directly into S-system equations of the form

$$\dot{X}_1 = \alpha_1 A^{g_{10}} X_1^{g_{11}} X_2^{g_{12}} - \beta_1 B^{h_{10}} X_1^{h_{11}} \tag{14.21}$$

$$\dot{X}_2 = \alpha_2 B^{g_{20}} X_1^{g_{21}} - \beta_2 X_1^{h_{21}} X_2^{h_{22}} \tag{14.22}$$

This S-system representation with the two variables $X_1 = X$ and $X_2 = Y$ is not the most general model possible, but it already reflects that X_2 does not directly affect its own synthesis, which implies that $g_{22} = 0$ and the degradation of X_1 is not directly influenced by X_2 and thus $h_{12} = 0$. Since A and B represent external variables that do not change during an experiment, they can be considered constants and absorbed into the α and β rate constants. The streamlined S-system representation is thus

$$\dot{X}_1 = \alpha_1 X_1^{g_{11}} X_2^{g_{12}} - \beta_1 X_1^{h_{11}} \tag{14.23}$$

$$\dot{X}_2 = \alpha_2 X_1^{g_{21}} - \beta_2 X_1^{h_{21}} X_2^{h_{22}} \tag{14.24}$$

where the rate constants α_i and β_i $(i = 1, 2)$ have been redefined to include the external variables A and B.

The kinetic orders of the chemical system immediately dictate the exponents of the second equation. As in the Brusselator equations (14.18) and (14.19), they are given as $g_{21} = 1$, $h_{21} = 2$ and $h_{22} = 1$. For the particular case in which $k_3 = B$ and $k_2 = 1$ (see above), the rate constants of the second equation are $\alpha_2 = B$ and $\beta_2 = 1$. The degradation term in the equation of X_1 is also the same as in the Brusselator equation (14.18), i.e., $\beta_1 = B + 1$ and $h_{11} = 1$.

For the synthesis of X_1, the situation is more complicated, and different options are available. One option is to use the power-law approximation of the expression $A + X^2 Y$ at an operating point of our choice (cf. Chapter 2). Because of its premier importance, we select as the operating point the steady state, which yields the parameter specifications

$$g_{11} = \frac{2B}{B + 1}, \; g_{12} = \frac{B}{B + 1}, \; \alpha_1 = \frac{(B + 1)A^{1/(B+1)}}{B^{B/(B+1)}} \tag{14.25}$$

Incorporation of these specifications yields the S-system representation

$$\dot{X}_1 = \alpha_1 X_1^{g_{11}} X_2^{g_{12}} - (B + 1)X_1 \tag{14.26}$$

$$\dot{X}_2 = BX_1 - X_1^2 X_2 \tag{14.27}$$

where α_1, g_{11}, and g_{12} are defined by (14.25).

The S-system (14.26)–(14.27), derived from the original chemical scheme through power-law approximation, is now completely characterized and can be analyzed analytically and numerically. Since the steady state was chosen as the operating point, the S-system and the Brusselator have the same steady state. However, as a local approximation, the power-law representation in this case is not sufficiently accurate, even close to the operating point. Numerical solution demonstrates that the temporal responses of X_1 and X_2 deviate enough from those of X and Y of the Brusselator to change the qualitative behavior: The Brusselator exhibits a limit cycle ($B > A^2 + 1$), whereas the S-system model (14.26)–(14.27) either produces nested centers ($B = A^2 + 1$), inward spirals ($B < A^2 + 1$), or outward spirals ($B > A^2 + 1$). It is not an issue here whether the Brusselator or the S-system is a more relevant model of the hypothetical chemical system (14.14)–(14.17), but the example illustrates a limitation of the power-law approximation as a method of estimating parameter values from the original dynamical system.

As an alternative to power-law approximation, one can insist that the S-system exhibits a limit cycle and, under this constraint, one can adjust the parameter values as well as possible. The starting point for this approach is the criterion for limit cycles of S-systems which is derived and discussed in Chapter 16. It turns out that this criterion can be satisfied even if we retain the equation for X_2 and also a first-order kinetic for the degradation of X_1. In other words, the parameters of the synthesis term of X_1 and the rate constant β_1 provide enough freedom to enforce responses in the form of limit cycles.

Beginning with the equations

$$\dot{X}_1 = \alpha_1 X_1^{g_{11}} X_2^{g_{12}} - \beta_1 X_1 \tag{14.28}$$

$$\dot{X}_2 = B X_1 - X_1^2 X_2 \tag{14.29}$$

we choose parameters g_{11}, g_{12}, α_1, and β_1 so that the system has the same steady state as the Brusselator and also qualitatively exhibits the same dynamic behavior. The steady-state requirement corresponds to the constraints

$$\beta_1 = \alpha_1 A^{g_{11} - g_{12} - 1} B^{g_{12}} \tag{14.30}$$

$$g_{11} - g_{12} - 1 \neq 0 \tag{14.31}$$

In order to obtain the desired dynamic behavior, one scales the system with respect to the steady state. Then choosing g_{11} as the bifurcation parameter (i.e., replacing g_{11} with $g_{11}(\mu)$ such that bifurcation occurs at $\mu = \mu_0$), the conditions

$$g_{11} = \frac{\beta_1 + A^2}{\beta_1} \tag{14.32}$$

and

$$g_{12} - g_{11} + 1 > 0 \tag{14.33}$$

must be satisfied. Substituting all parameters and constraints into the equation for the critical parameter α (cf. Chapter 16), one obtains for $\mu = \mu_0$ the equation

$$a = \frac{\beta_1 A^2 g_{12}^2}{8} [b_1 - A^2] \tag{14.34}$$

If $a < 0$ there exists a family of stable limit cycles for $\mu > \mu_0$. Since by definition β_1 is non-negative, the condition reduces to

$$\beta_1 < A^2 \tag{14.35}$$

Within the limits of (14.30)–(14.35), the unspecified parameter in (14.28)–(14.29) can be assigned arbitrarily. For instance, one may obtain a limit cycle with $A = 1$, $B = 2.5$, $g_{11} = 7/3$, $g_{12} = 2$, $\alpha_1 = 12$, and $\beta_1 = .75$. This system does not exactly represent the Brusselator, and the shapes of the two limit cycles are quite different. Nevertheless, the two models do have many properties in common; for instance, the functional dependence of X_2 on X_1 is the same, and both have the same steady state and exhibit a limit cycle.

A third alternative for representing the Brusselator as an S-system is recasting. By definition, the resulting S-system is an exact model and, in particular, it exhibits exactly the same limit cycle as the Brusselator. There is a price for these advantages; the recast S-system consists of three instead of two variables, and X_1 no longer appears explicitly but it must be retrieved from an auxiliary algebraic equation.

To recast the Brusselator equations (14.18) and (14.19) as an S-system, one begins by renaming the variables, e.g., $X = X_1$ and $Y = X_4$. Equation (14.19) already has the desired form. Equation (14.18) can be recast into an S-system by using the rule for reducing Generalized Mass Action equations (cf. Chapter 12): With $X_1 = X_2X_3$, one obtains the equation

$$\begin{aligned} \dot{X}_1 &= \dot{X}_2 X_3 + X_2 \dot{X}_3 \\ &= A - X_1 + X_1^2 X_4 - BX_1 \end{aligned} \tag{14.36}$$

Now one has some freedom in assigning terms to the new variables. For instance, one can set

$$\dot{X}_2 = AX_3^{-1} - X_1X_3^{-1} \tag{14.37}$$

and substituting $X_1 = X_2X_3$ leads to the equation

$$\dot{X}_2 = AX_3^{-1} - X_2 \tag{14.38}$$

Likewise the equation for $\dot{X}_3$ is obtained as

$$\dot{X}_3 = X_2 X_3^2 X_4 - BX_3 \tag{14.39}$$

The substitution $X_1 = X_2 X_3$ in $\dot{X}_4$ produces equation (14.40)

$$\dot{X}_4 = BX_2 X_3 - X_2^2 X_3^2 X_4 \tag{14.40}$$

and renumbering $X_2 \to X_1$, $X_3 \to X_2$, and $X_4 \to X_3$ yields the recast Brusselator in S-system form as

$$\dot{X}_1 = AX_2^{-1} - X_1 \tag{14.41}$$

$$\dot{X}_2 = X_1 X_2^2 X_3 - BX_2 \tag{14.42}$$

$$\dot{X}_3 = BX_1 X_2 - X_1^2 X_2^2 X_3 \tag{14.43}$$

with the initial conditions $X_1(0) = X(0)$, $X_2(0) = 1$, and $X_3(0) = Y(0)$. The original variables are regained as $X = X_1 X_2$ and $Y = X_3$. Simulations confirm that this S-system exhibits the same dynamical responses as the Brusselator (14.18)–(14.19) (cf. figs. 14.1 and 2).

RECASTING AS A TOOL FOR CREATING S-SYSTEM MODULES

An ill-defined system component is one in which the internal structure is not known or is irrelevant. Such a component is often characterized by its input-output relationships. Recasting is a useful tool for developing an S-system module to represent an ill-defined component. Once a module has been created, the modeler needs to know how to modify an S-system. For instance, a modeler may desire to change the amplitude or frequency of an oscillator or to shift the output horizontally or vertically. As examples of S-system modules, representations of a threshold function, a simple harmonic oscillator, and a van der Pol oscillator will be examined.

Example 1: A Simple Threshold Module

Thresholds are often encountered in biology; for instance, they play a significant role during the development and differentiation of an organism. Their general purpose is to trigger the onset of new system responses. The biochemical nature of a threshold effect or a trigger function is usually complex and often

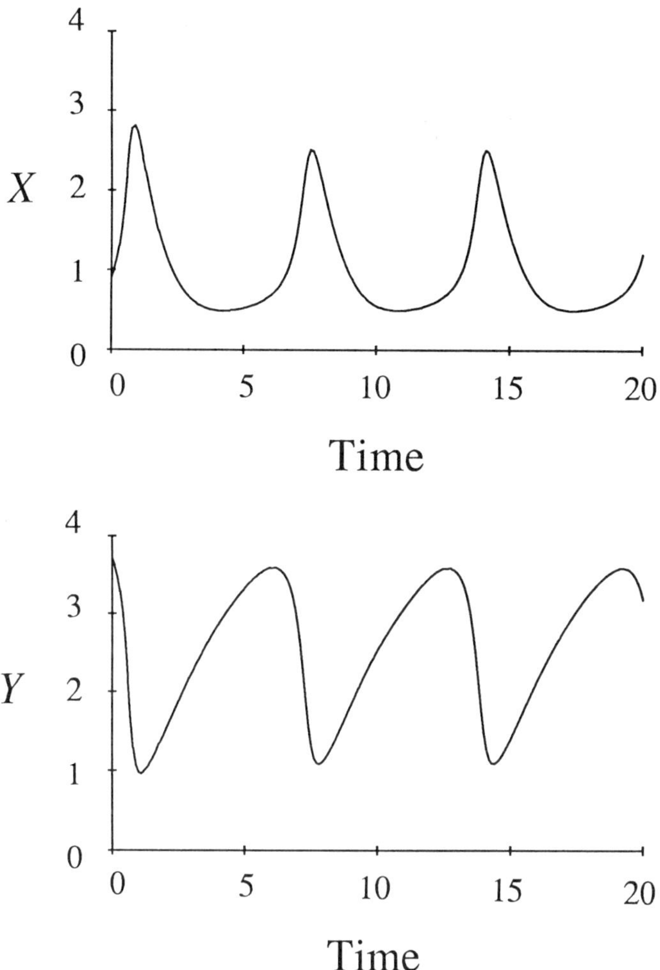

FIGURES 14.1 and 14.2. Simulations of the recast Brusselator exhibit the same dynamical responses as the Brusselator given by Eqs. (14.18)–(14.19). Figures 14.1 and 14.2 demonstrate the oscillations of the concentrations of X and Y in the time domain.

irrelevant, and in many model analyses it can be considered a black box. A trigger can readily be represented with a one-dimensional S-system of the form

$$\dot{X} = \alpha X^g - \beta X^h \tag{14.44}$$

It is convenient if $g < h$, since then the nonzero steady-state solution is stable. With an initial condition between 0 and 1, the solution of this equation exhibits an increasing sigmoid curve whose shape can be customized to fit a desired output. The nonzero steady state is the maximum threshold response. When $\alpha = \beta$, the maximum response is normalized to unity. α determines the rate of response, g determines the curvature of the rise from 0, and h determines the curvature of the approach to 1. Finally, the initial condition, $X(t_0)$ determines when the response "steps" from 0 to 1. A steep steplike function is characterized by a large exponent h (e.g., $\alpha = 0.1$, $g = 1$ and $h = 6$), and an s-shaped threshold function with a slower rise can be realized with $\alpha = 0.1$, $g = 1$ and $h = 2$ (cf. fig. 14.3).

Example 2: Harmonic Oscillator

An S-system module can be developed by recasting the mathematical representation of a phenomenon with the correct response characteristics. As an example, let's study the harmonic oscillator. Its analytical representation is given by the function in Eq. (14.45), where a determines the amplitude, b the frequency, c the phase shift, and d the vertical shift:

$$f(t) = a \sin b(t - c) + d \tag{14.45}$$

Equation (14.45) is recast as an S-system by differentiating with respect to t. The result is the S-system

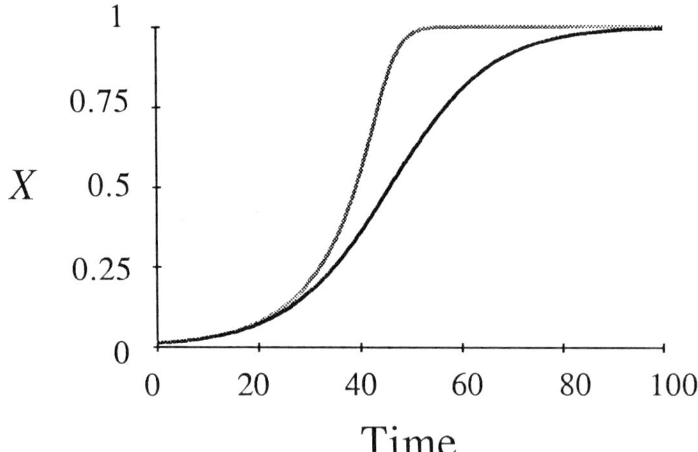

FIGURE 14.3. An example of a steep steplike threshold function is given by $\alpha = 0.1$, $g = 1$ and $h = 6$. An example of an s-shaped threshold function is obtained by $\alpha = 0.1$, $g = 1$, and $h = 2$.

$$\dot{X}_4 = \alpha_4 X_5 - \beta_4 \tag{14.46}$$

$$\dot{X}_5 = \beta_4 - \alpha_4 X_4 \tag{14.47}$$

where

$$X_4(t) = a \sin b(t - c) + d \tag{14.48}$$

$$X_5(t) = A \cos b(t - c) + d \tag{14.49}$$

The S-system parameters relate directly to the properties of the harmonic oscillator: α_4 determines the frequency, β_4/α_4 determines the vertical shift, and phase shift and amplitude are determined by the initial conditions. Thus, once recast, the harmonic oscillator can be customized for a particular application. An example will be given in a later section.

Example 3: van der Pol Oscillator

In 1927 van der Pol [196] showed that periodic oscillations in the current of vacuum tubes could be predicted from differential equations of the circuit if certain nonlinear terms were retained. In contrast to the harmonic oscillator, the resulting *relaxation oscillator* is a stable oscillator; the harmonic oscillator changes its amplitude upon perturbation whereas the relaxation oscillator returns to its original amplitude. Because of its mathematical simplicity, the relaxation oscillator became an appealing tool for modeling. For instance, van der Pol [197] himself applied it to cardiac rhythms, Wever and Lund [198] described circadian rhythms with it, and modifications of the van der Pol oscillator have been used in a great number of applications. In addition to its stability, the van der Pol equation has other interesting dynamical properties and it is a standard part of most texts on nonlinear dynamics.

Because the van der Pol oscillator has received much attention from mathematicians, engineers, and biologists, and often is considered the paradigm of a stable oscillator, it is used here to show how a mathematically defined phenomenon can be recast and used as an S-system module. The original second-order van der Pol equation

$$\ddot{Z} + \frac{g_2}{C}\left(3Z^2 - \frac{g_1}{g_2}\right)\dot{Z} + \frac{1}{CL}Z = 0 \tag{14.50}$$

contains four parameters whose names and meaning stem from the original application to electrical circuits: C denotes conductance; L denotes inductance; and g_1 and g_2 denote ion conductance. The second-order equation can be transformed by Lienard's transformation [191] into a first-order system of two variables:

$$\dot{Z} = \frac{Y}{C} + \frac{g_1 Z}{C} - \frac{g_2 Z^3}{C} \tag{14.51}$$

$$\dot{Y} = \frac{-1}{L} Z \tag{14.52}$$

This pair of equations can be recast as an S-system with three variables

$$\dot{X}_1 = \frac{1}{C} X_2 X_3 - g_2 X_1^3 X_2^{-2} \tag{14.53}$$

$$\dot{X}_2 = \frac{\mu}{C} X_2 \quad - \frac{3 g_2 a_1}{C} X_1 \tag{14.54}$$

$$\dot{X}_3 = \frac{a_1}{L} \quad - \frac{1}{L} X_1 X_2^{-1} \tag{14.55}$$

in which Y is represented by $X_3 - a_2$ and Z is retrieved from X_1 and X_2 as

$$Z = X_1 X_2^{-1} - a_1 \tag{14.56}$$

The "shifting parameters" a_1 and $a_2 = g_2 a_1^3 - g_1 a_1$ assure that the S-system variables are positive. The details of the recasting process are of secondary importance here, and are therefore presented in the Appendix. A graph of the (recast) van der Pol oscillator is given in Fig. 14.4.

Upon recasting, the modeler can use the van der Pol oscillator within the S-system environment. It is clear that the specifications of a particular oscillator, defined by the original parameters C, L, g_1 and g_2, translate directly into S-system parameters whereas the new parameters a_1 and a_2 are of a more technical nature.

In many circumstances it is desirable to modify the original oscillator in order to obtain stable oscillations in which the qualitative properties are preserved whereas the quantitative features are customized for a particular application. For instance, an application could require a stable relaxation oscillator that is asymmetric with respect to peaks and valleys, or it could be necessary to flatten the valleys, change the upward or downward slope, or alter the shape of the plateau. Such modifications can be achieved by defining additional variables that result from transformations of some of the S-system variables. Without exploring the full potential of this type of modification, the following examples illustrate two simple tools for generating customized relaxation oscillators. In the subsequent example the results will be applied.

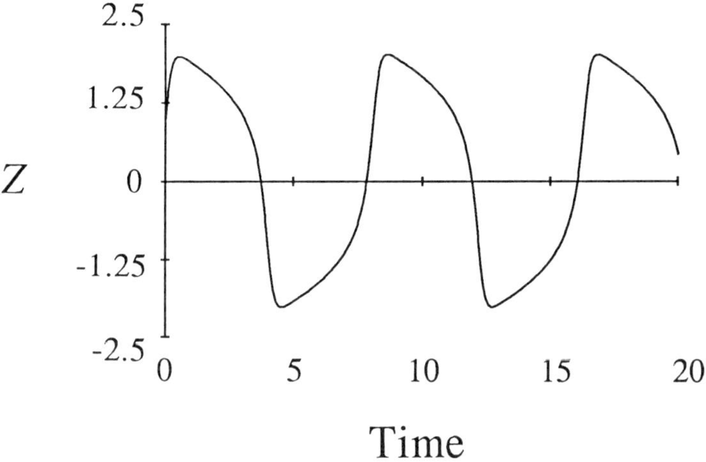

FIGURE 14.4. The recast van der Pol equation exhibits characteristic oscillations. The parameter values in this example are $\alpha_1 = 2.5$, $\alpha_2 = 13.125$, $\alpha_3 = 1$, $\beta_1 = 5/6$, $\beta_1 = 6.25$, and $\beta_1 = 0.4$.

One may begin with vertical shifting. The variable

$$Z_2 = X_1 X_2^{-1} \tag{14.57}$$

oscillates around a_1 exactly the same way the van der Pol oscillator oscillates around 0. Another simple transformation is (vertical) scaling; Z_2 may be scaled by a constant c:

$$Z_3 = cX_1 X_2^{-1} \tag{14.58}$$

In comparison with Z and Z_2, Z_3 has the same frequency but a modified amplitude.

Now suppose an application requires a modeler to produce an asymmetry between the peak and valley of the original oscillator. A possible strategy is to scale Z_2 in such a way that it oscillates between 0 and 1 and then to raise it to a real power p that vertically skews the variable. Thus, as the first step, employ Eq. (14.58) and select $c = (\max\{Z_2\})^{-1}$. In the second step, define a new variable

$$Z_4 = Z_3^p = c^p X_1^p X_2^{-p} \tag{14.59}$$

or redefine $Z_{3\,\text{new}} = Z_{3\,\text{old}}^p$. If the exponent p is greater than 1, the valley will become flatter; as p increases, Z_4 becomes more asymmetric: the peaks become steeper and the valleys become flatter. Fig. 14.5 shows an example of a modified oscillator.

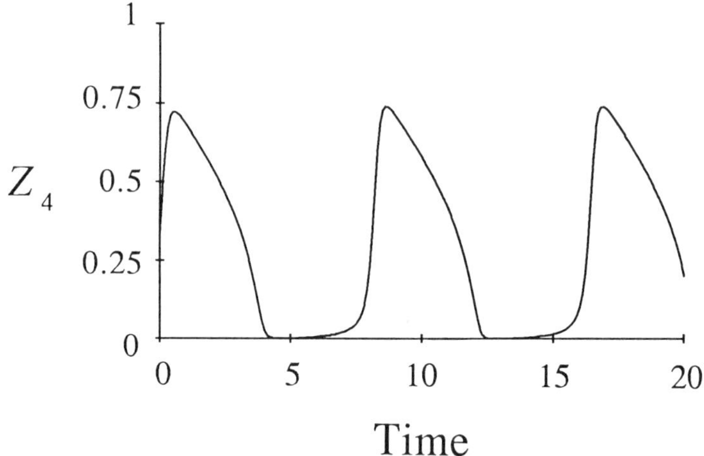

FIGURE 14.5. Modified van der Pol oscillator with $c = 0.2$ and $p = 3$ (cf. Eq. [14.59]).

Of course, it is possible to define many other modulations. It should be noted that one cannot simply scale X_1 or X_2 because a change in one of the system variables without a concomitant change in the other variables and/or parameter values usually destroys the stable oscillations exhibited by the van der Pol oscillator. The result may be an unstable oscillator or even a function with completely different dynamic characteristics.

CONNECTING MODULES

As previously mentioned, recasting can be used in conjunction with other methods of S-system modeling and systems analysis. Therefore, it is important to integrate recast submodels into an S-system master model. First the modules are customized, as illustrated, and then they are linked together. The output of the first module serves as the input for the second module. These principles will be illustrated with two examples. The harmonic oscillator will be triggered by a threshold module, and, as a more complex example, it will be shown how the van der Pol module can be triggered by an S-system model of a metabolic system.

Example 1: Triggering the Harmonic Oscillator

Suppose the onset of a harmonic oscillation is triggered by an outside force that is the output of another system. As shown above, the normalized trigger can be represented by a threshold module of the form

$$\dot{X}_6 = \alpha X_6 - \alpha X_6^6 \tag{14.60}$$

and the harmonic oscillator can be written with two S-system equations (cf. Eqs. (14.46) and (14.47)). To couple the trigger to the oscillator, one can simply include the trigger variable X_6, with a convenient power, in all four terms of the oscillator:

$$\dot{X}_4 = \alpha_4 X_5 X_6^2 - \beta_4 X_6^2 \tag{14.61}$$

$$\dot{X}_5 = \beta_4 X_6^2 - \alpha_4 X_4 X_6^2 \tag{14.62}$$

Fig. 14.6 illustrates the result: initially, the trigger function is close to zero, and the amplitude of the oscillator is essentially zero; as soon as the trigger reaches an appreciable size, the oscillations become apparent. The oscillation can be turned off with an additional module or with a function X_6 that returns to zero.

Example 2: Gastric Emptying

When food reaches the stomach it evokes a number of biochemical and neurophysiological responses. The food is mixed with gastric juices and this mixture, called *chyme*, affects the gastric smooth muscle to become active. The muscle activity further mixes the chyme and transports it toward the intestine. Once the food is sufficiently digested it is released through the pyloric sphincter into the duodenum.

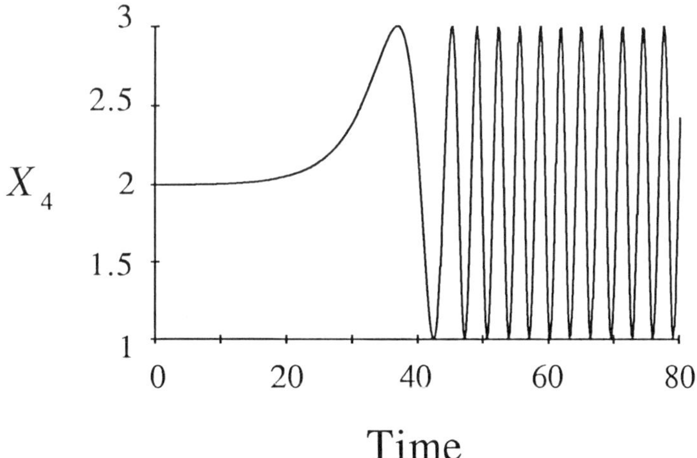

FIGURE 14.6. Triggering of a harmonic oscillator ($\alpha = 0.1$; cf. Eqs. [14.60]–[14.62]).

Without going into any detail, one can identify several control mechanisms that regulate the complicated processes in the stomach. The presence of food triggers the release of several neurohormonal agents that include neurotransmitters and endocrines. A well-balanced release and interaction of these agents regulates the acidity of the chyme. Acidity is instrumental in the modification of the electrical activity of the gastrointestinal smooth muscle. During fasting, this activity consists of "slow-wave" oscillations that are governed by the muscle itself. When triggered, however, the membrane potential of the smooth muscle starts to change rapidly, exhibiting fast spikes or "bursts," and the muscle contracts. The muscle activity in turn determines gastric flow and eventually depletes the stomach of the chyme [183, 193].

To model the process, three major components are necessary: a biochemical system, representing the interactions between food and hormonal agents; a system representing the slow wave during fasting; and a mechanism that, triggered by the biochemical system, produces bursts and eventually leads to gastric emptying and, thus, to the removal of chyme.

Of course, any representation of the biochemical part of the system will be fairly complicated, but modeling strategies in the first part of the book can be expected, at least in principle, to suffice for the formulation of an appropriate model. Thus, for the duration of this example, it is assumed that such a model is available and the focus is on the slow wave and its connection with the phenomenon of bursting.

Historically, the slow wave has been represented with the van der Pol oscillator [184, 192]. While several explanations have been given for the choice of this particular model, the most convincing one seems to be that the oscillator is stable and, at the same time, mathematically fairly simple. The shape of a van der Pol oscillation does not quite agree with the observed asymmetric oscillations, but as one of the above examples has shown, it is possible to modify the properties of the van der Pol oscillator within the framework of S-systems. Applying such modifications one can customize the shape of the oscillator to match experimental observations, at least qualitatively. Alternatively, one could design a stable oscillator from scratch, as was demonstrated in the Brusselator example. No matter which of these strategies one pursues, the second component of the control system of gastric emptying is available as an S-system module.

Experimental data demonstrate that muscle activity correlates to "bursting," which is reflected in fast spikelike oscillations that appear on top of the peaks of the slow wave, but not in its valleys. The actual shape of these spikes appears to be of secondary importance, and for simplicity they can be modeled as high-frequency harmonic oscillations. Again, this is not the only possible choice. Some authors in the past have suggested oscillations of the Hodgin-Huxley type [189], and if this more complicated oscillator were really appropriate, one could use it, recast as an S-system.

The missing link now is the connection of the modules and, in particular, the triggering of bursts on the peaks of the slow wave by the neurohumoral system. Suppose, X_6 is the output variable of the neurohumoral system, and for convenience, it is assumed to be scaled between zero and one. X_6 then can function as the switch to turn on and off the harmonic oscillator, as previously demonstrated. Eqs. (14.63) and (14.64) represent the harmonic S-system oscillator, coupled with the switching mechanism:

$$\dot{X}_4 = \alpha_4 X_5 X_6^k - \beta_4 X_6^k \tag{14.63}$$

$$\dot{X}_5 = \alpha_5 X_6^k - \alpha_4 X_4 X_6^k \tag{14.64}$$

where the exponent k can be used to customize the onset of oscillations.

This system can now be coupled with the slow wave, which may be represented as a modified van der Pol oscillator (Eq. [14.59]). By using X_4 or X_5 as a multiplier of the variable Z_4 that oscillates between zero and one, one can enforce that bursts are apparent only on top of the peaks of the slow wave but not in the valleys. Fig. 14.7 shows an example where the harmonic oscillator (X_4) has been switched on and then coupled with the slow wave by dividing Z_4 by X_4. The resulting variable $Z_5 = Z_4 X_4^{-1}$ exhibits responses that qualitatively coincide with clinical and experimental observations (cf. [194]). Z_5, expressed as $cX_1^p X_2^{-p} X_4^{-1}$, can be used as an input for other modules, for instance, those representing muscle activity.

Obviously, it is not the intention of this example to explain gastric emptying.

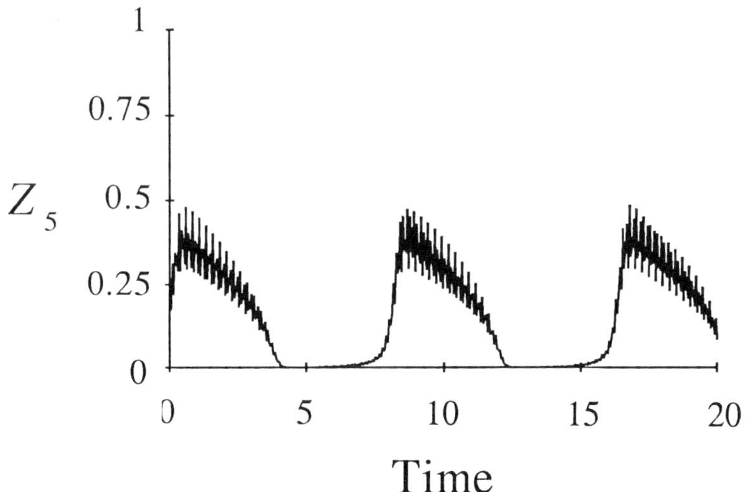

FIGURE 14.7. Example of a modified van der Pol oscillator (as in Fig. 14.5), coupled with a harmonic oscillator.

Rather, this complex phenomenon was chosen to demonstrate how modules at different levels of organization can be combined. The van der Pol and harmonic oscillators are not structured models of the true phenomena, but it has become clear that through recasting they can be incorporated into a structured biochemical model as black-box modules.

SUMMARY

The examples in this chapter demonstrate how recasting expands the tools available to the S-systems modeler. Recasting is an option when a power-law approximation is not sufficiently accurate over the range of interest. As with the Brusselator and van der Pol oscillators, recasting may introduce auxiliary variables, but the recast system has the advantage of being qualitatively as well as quantitatively accurate over any range of interest.

There are other times when it may not be desirable to use a direct modeling approach. This may be due to insufficient information, or it may be that the complexity thus introduced to the model is not warranted at that level of modeling. In such cases, recasting may be used to design a black-box module. Three modules are used as examples in this chapter: a threshold module, the harmonic oscillator, and the van der Pol oscillator. These modules can be cutomized and linked to form a network. This is demonstrated with an example from gastric electrophysiology.

APPENDIX

Recasting of the van der Pol Oscillator

The recasting process basically follows the steps discussed in Chapter 12. One begins by setting $X_1 = Z + a_1$, $X_2 = Y + a_2$, $X_{10} = Z_0 + a_1$, $X_{20} = Y_0 + a_2$ where a_1 and a_2 are chosen to assure the non-negativity of X_1 and X_2. Substitution yields the following system:

$$\dot{X}_1 = \frac{g_1 - 3g_2a_1^2}{C} X_1 + \frac{1}{C} X_2 + \frac{3g_2a_1}{C} X_1^2 - \frac{g_2}{C} X_1^3 + \frac{g_2a_1^3 - a_2 - g_1a_1}{C} \qquad (14.65)$$

$$\dot{X}_2 = \frac{a_1}{L} - \frac{1}{L} X_1 \qquad (14.66)$$

The first equation contains five terms with differing powers of X_1 and X_2, and if one blindly applied the rule for reducing Generalized Mass Action equations, the resulting S-system would have three or four equations representing X_1. However, one has some freedom in choosing a_1 and a_2, as long as they are large enough to assure the positivity of X_1 and X_2. Thus, with a prudent choice of a_1 and a_2, the number of equations can be kept to a minimum. Specifically, one sees easily that the constant term becomes zero when a_1 and a_2 are selected such that $(g_2a_1^3 - a_2 - g_1a_1)C^{-1} = 0$. The constant term associated with X_1 can be positive or negative, depending on the actual parameter values. It is advantageous to select a_1 large enough to assure that this term is negative, since then two terms are positive and two are negative, and the reduction to S-system form will result in only two equations. Thus, one chooses a_1 such that $g_1 - 3g_2a_1^2 = m < 0$, and Eq. (14.65) reduces to

$$\dot{X}_1 = \frac{1}{C} X_2 - \frac{\mu}{C} X_1 + \frac{3g_2a_1}{C} X_1^2 - \frac{g_2}{C} X_1^3 \tag{14.67}$$

Now this equation may be reduced by letting $X_1 = X_3^{-1}X_4$, which corresponds to the differential equation

$$\begin{aligned} \dot{X}_4 - X_1\dot{X}_3 &= X_3\dot{X}_1 \\ &= \frac{1}{C} X_2X_3 - \frac{\mu}{C} X_1X_3 + \frac{3g_2a_1}{C} X_1^2X_3 - \frac{g_2}{C} X_1^3X_3 \\ &= \left(\frac{1}{C} X_2X_3 - \frac{g_2}{C} X_1^3X_3\right) - \left(\frac{\mu}{C} X_1X_3 - \frac{3g_2a_1}{C} X_1^2X_3\right) \end{aligned} \tag{14.68}$$

X_3 and X_4 are not yet explicitly defined but they have to satisfy Eq. (14.68). Straightforward explicit definitions of $\dot{X}_3$ and $\dot{X}_4$ are given as

$$\dot{X}_3 = \frac{\mu}{C} X_3 - \frac{3g_2a_1}{C} X_1X_3 = \frac{\mu}{C} X_3 - \frac{3g_2a_1}{C} X_4 \tag{14.69}$$

and

$$\begin{aligned} \dot{X}_4 &= \frac{1}{C} X_2X_3 - \frac{g_2}{C} X_1^3X_3 \\ &= \frac{1}{C} X_2X_3 - \frac{g_2}{C} X_3^{-2}X_4^3 \end{aligned} \tag{14.70}$$

and their initial conditions can be selected as $X_3(0) = 1$ and $X_4(0) = X_1(0) + a_1$. The equation for X_2 already has the correct form, and renumbering $X_3 \rightarrow X_1$, $X_4 \rightarrow X_2$, and $X_2 \rightarrow X_3$ produces the final S-system (14.53)–(14.55).

15. Algebraic Properties of Canonical Forms

EBERHARD O. VOIT

When we discussed S-systems as representations for organizationally complex systems (cf. Part I), we alluded to the variety of analytical tools that help characterize their mathematical properties: Steady states, sensitivities, and local stability can be evaluated with paper and pencil, and because of the structural homogeneity of S-systems, their analysis also can be implemented in symbolic algebra programs (cf. Chapter 6). In addition, steady-state features and dynamic responses can be computed numerically with the software program ESSYNS (Chapter 7). When we look at this arsenal of tools we might be tempted to say that we are able to obtain most relevant information about a given S-system.

In Chapter 12, we discussed the equivalence between alternative canonical forms, and maybe at that point we first wondered whether we really understand these forms. For instance, we were able to prove that an S-system can be recast as a Half-system that represents exactly the same solution. However, the Half-system no longer allowed us to compute steady states with the symbolic means that we developed for S-systems.

In this chapter we eludicate some of the fundamental properties of canonical forms. These properties are related to *transformations of coordinates* (variables). The general idea is to switch to a coordinate system that is mathematically more convenient. This system may have a lower dimension or it may make properties of the S-system visible that previously were hidden. The coordinate transformations of interest form groups that are called *symmetry groups* or *Lie groups*, in honor of the Norwegian mathematician Sophus Lie who, about 100 years ago, introduced the extremely fruitful concept of transformation groups in the search for unifying methods of solving differential equations (e.g., see [201]).

The characterization of symmetry groups yields deep insight into the investigated differential equation, and some authors go so far as to claim that a "process which is modeled by a differential equation is understood only if its symmetries are completely known" [204: p. 458]. In fact, symmetries are directly related to *conservation laws*, that is, to quantities that are unchanged under the action of the differential equation (cf. [199, 202–203]). For instance, a valid model description of a closed system will possess a symmetry that reflects the conservation of energy and mass.

It is not the purpose of this chapter to review Lie groups and their applications to differential equations. Rather, I have selected a few properties from the theory of Lie groups that are most relevant and relatively easy to grasp, and I employ them exclusively in the algebraic characterization of Generalized Mass Action (GMA) systems that contain all previously discussed canonical forms as special cases.

TRANSFORMATIONS OF CANONICAL FORMS

When a GMA-system derives as a model of a real-world phenomenon, each variable has a well-defined meaning. In a biochemical context we may think of t as time and $X_i (i = 1, \ldots, n)$ as concentrations of enzymes, metabolites, or inhibitors. For a biochemical analysis we like to retain all variables in their original form because any change in the definition of a variable would require additional bookkeeping and would preclude an immediate interpretation. In contrast, for a mathematical analysis of the system, the given definition of variables may not be optimal and, in fact, a simple transformation of coordinates may enable us to exhibit features that otherwise would be hidden. For example, Chapter 16 discusses how easily some types of bifurcations are characterized once an S-system is represented in a new, and for this purpose, more appropriate coordinate system. In this chapter we discuss transformations of GMA-systems that simplify the representation of their flow. In particular, some GMA-systems admit transformations that allow the decoupling of simultaneous equations; in some cases we may even be able to solve the system analytically.

Chapter 12 discussed how recasting techniques can be employed to transform virtually any ordinary differential equation into an S-system or one of the alternative canonical forms. In most cases, this homogeneity in structure was paid for with an increase in the number of system variables and equations. The reduction in the number of equations that is discussed in this chapter in a sense is the reversion of recasting: We begin with an S-system or one of the other canonical forms and project it into a coordinate system of lower dimension.

While decoupling the set of simultaneous equations, we usually lose the structural homogeneity of the equations.

The strategy for decoupling sets of equations is as follows: First, we introduce *Lie groups of scaling transformations* that leave GMA-systems unaltered. From these transformations we derive functions that are *invariant* under the action of the GMA-system. The invariants can be used as new coordinates, and it will be shown that some of the new variables can be decoupled from the rest of the system. Thus, from an analytical standpoint, we are one step closer to solutions of canonical forms.

The transition from one coordinate system to another often requires a change in the independent variable as well. In the preceding chapters this was not an issue, and the notation $\dot{X}$ for dX/dt was unambiguous. In this chapter, however, we may become confused with different times scales, and thus we will identify the independent variable with the subscript 0, that is, we will use independent variables like X_0, Z_0, or ξ_0. Furthermore, we allow the independent variable to appear on the right-hand side of the differential equations.

The first goal is the derivation of Lie groups of scaling transformations in GMA-systems. To this end, a *symmetry matrix* is constructed in which each row corresponds to one product of power-law functions in the GMA-system. If the i-th equation of the GMA-system contains m_i terms (i.e., products of power-law functions), then the symmetry matrix consists of Σm_i rows. The elements in each row are defined in terms of the exponents associated with the variables $X_0, \ldots, X_n$, and thus the symmetry matrix has $n + 1$ columns. If the rank of the symmetry matrix is less than $n + 1$, the GMA-system admits a symmetry group.

To formalize this concept, we use the following notation. Let

$$\frac{dX_i}{dX_0} = \sum_{k=1}^{m_i} \gamma_{ik} \prod_{j=0}^{n} X_j^{g_{ikj}} \tag{15.1}$$

with $i \in \{1, \ldots, n\}$, $\gamma_{ik} \in R - \{0\}$, and $g_{ikj} \in R$ be a potentially non-autonomous GMA-system with explicit incorporation of the independent variable X_0. Each product of power-law functions

$$\gamma_{ik} \prod_{j=0}^{n} X_j^{g_{ikj}}$$

is called a *term* of the GMA-system. Let $M = \Sigma m_i$ be the total number of terms of the GMA-system. For convenience, we rename the terms as t_h $(h = 1, \ldots, M)$ and introduce the abbreviated notation

$$t_h = \gamma_h \prod_{j=0}^{n} X_j^{g_{hj}}$$

For each term, we define a row vector $\boldsymbol{c}_h$ with $n + 1$ components:

$$c_{hj} = \begin{cases} g_{hj} + 1 & \text{if } j = 0 \\ g_{hj} & \text{if } j = 1, \ldots, n; j \neq i \\ g_{hj} - 1 & \text{if } j = i \end{cases} \tag{15.2}$$

The *symmetry matrix* (C_{ij}) of the GMA-system is defined as the $(M \times (n + 1))$ matrix whose h-th row is $\boldsymbol{c}_h$:

$$(C_{ij}) = \begin{pmatrix} \boldsymbol{c}_1 \\ \boldsymbol{c}_2 \\ \vdots \\ \boldsymbol{c}_M \end{pmatrix}$$

Let v be a column vector with $n + 1$ real components $v_0, v_1, \ldots, v_n$ and $\boldsymbol{o}$ the zero vector of dimension M. Let p be a positive real number.

Theorem 1:

1. If the linear system

$$(C_{ij})v = \boldsymbol{o} \tag{15.3}$$

has a nontrivial solution $v = (v_0, v_1, \ldots v_n)^T$, then the transformations

$$Y_j = p^{v_j} X_j \tag{15.4}$$

$(j = 0, \ldots, n; p \in R)$ form a one-parameter Lie group of coordinate transformations that leave the GMA-system (15.1) unaltered. That is, the GMA-system is invariant under the action of the group.

2. Let $\rho = \text{rank}\,(C_{ij})$ and let $\{v_h\}$ be a set of $n + 1 - \rho$ linearly independent solutions to (15.3). Let v_h have the components v_{hj}. Then (15.1) admits the $n + 1 - \rho$ dimensional Lie group of transformations

$$Y_j = \prod_{k=1}^{r} p_k^{v_{kj}} X_j \qquad (15.5)$$

with $r = n + 1 - \rho$ parameters $p_1, p_2, \ldots, p_r \in R$.

Remarks:

1. The groups in 1 and 2 are called *stretching groups* or *scaling groups*, and the GMA-system (15.1) is called *quasi-homogeneous* when it admits such groups (cf. [199]).
2. For $r > 1$, a one-parameter group of transformations of type (15.4) can be generated from the r transformations (15.5) by setting $p = p_1 = p_2 = \ldots = p_r$.

The following theorems show that recasting operations that transform GMA-systems into S-systems, Half-systems, Binary systems, or Binary Half-systems do not destroy symmetries.

Theorem 2:

1. Addition of a new variable, defined as a power-law function of an existing GMA-system variable does not destroy the existence of scaling symmetries.
2. Substitution of a power-law function of an existing GMA-system variable with a new variable does not destroy the existence of scaling symmetries.

Theorem 3:

Substitution of a GMA-system variable with a product of new variables does not destroy the existence of scaling symmetries.

Corollary 1:

Substitution of a GMA-system variable with a product of new variables, raised to powers, does not destroy the existence of scaling symmetries.

Corollary 2:

Theorem 2 and Corollary 1 also hold for products of more than two variables.

Proofs are provided in the Appendix.

INVARIANTS OF GMA-SYSTEMS

Given a coordinate system and a group of transformations, *invariants* are defined as functions that remain unaltered under the action of the group. As an example, let's consider a closed chemical system in which mass is preserved. The invariant in this case is the sum of the mass of all constituents. No matter which reactions take place within the system, the sum remains the same and, thus, it is invariant. A simple mathematical example of an invariant is the function $x^2 + y^2$ which remains unaltered under the action of the group of rotations of the form

$$(x, y) \to (x_1, y_1) = (x \cos(p) - y \sin(p), x \sin(p) + y \cos(p))$$

where p is the group parameter that determines the angle of rotation. The invariance of the function $x^2 + y^2$ can easily be seen from straightforward substitution:

$$\begin{aligned} x_1^2 + y_1^2 &= x^2 \cos^2(p) - 2xy \cos(p) \sin(p) + y^2 \sin^2(p) \\ &\quad + x^2 \sin^2(p) + 2xy \cos(p) \sin(p) + y^2 \cos^2(p) \\ &= x^2 + y^2 \end{aligned}$$

In Theorem 1 we have developed criteria for the existence of scaling transformations of the type $Y_j = p^{v_j} X_j$ in GMA-systems. For nontrivial transformations, at least some v_k are different from zero and they satisfy the equation

$$\sum_{k=0}^{n} v_k s_{ik} = 0 \tag{15.6}$$

where the s_{ik} are real numbers that are related to the parameters of the GMA-system (cf. Eq. (15A.1) of the Appendix). Invariants ξ_i with respect to one-parameter scaling transformations have to satisfy the condition

$$\xi_i(Y_1, \ldots, Y_n) = \xi_i(p^{v_1} X_1, \ldots, p^{v_n} X_n) = \xi_i(X_1, \ldots, X_n) \tag{15.7}$$

The most obvious functions satisfying the second equality are products of power-law functions in which the exponents are solutions to (15.6):

$$\xi_i = \prod_{k=0}^{n} X_k^{s_{ik}} \tag{15.8}$$

Straightforward substitution confirms that these power-law functions indeed are valid invariants:

$$\xi_i(Y_1, \ldots, Y_n) = \prod_{k=0}^{n} Y_k^{s_{ik}} = \prod_{k=0}^{n} (p^{v_k}X_k)^{s_{ik}} = p^{\Sigma v_k s_{ik}} \prod_{k=0}^{n} X_k^{s_{ik}}$$
$$= \xi_i(X_1, \ldots, X_n) \tag{15.9}$$

While power-law functions are the most obvious invariants, other types of invariants are possible. In particular, one can show that any differentiable function of an invariant is an invariant. The only functions we will employ later are power functions of invariants, and these are easily confirmed as invariants. Suppose ξ_i $(i = 1, \ldots, n)$ are invariants of a GMA-system that admits the transformation $Y_j = p^{v_j}X_j$. Let q be a real number. Then

$$\xi_i^q(Y_1, \ldots, Y_n) = \prod_{k=0}^{n} Y_k^{qs_{ik}} = p^{q\Sigma v_k s_{ik}} \prod_{k=0}^{n} X_k^{qs_{ik}}$$
$$= \xi_i^q(X_1, \ldots, X_n) \tag{15.10}$$

which holds because

$$\sum_{k=0}^{n} q v_k s_{ik} = 0 \tag{15.11}$$

and because of Eq. (15.6). At first sight, it might be surprising that all differentiable functions of invariants also are invariants. On the other hand, recalling the example of a closed, mass-conserving chemical system, it is intuitively clear that any well-behaved function of the (constant) sum of mass also is constant and, thus, an invariant.

The most convenient functionally independent set of group invariants presumably is

$$\xi_i = X_i X_n^{-v_i/v_n} \tag{15.12a}$$

$(i = 0, \ldots, n - 1)$ with the additional definition

$$\xi_n = X_n \tag{15.12b}$$

These invariants offer the advantage that the original variables X_i are easily retrieved from the reduced system in ξ_i coordinates, because we have

$$X_i = \xi_i \xi_n^{v_i/v_n}$$

$$X_n = \xi_n \tag{15.13}$$

In definition (15.12), the variables X_n and ξ_n play a special role. We shall see later that ξ_n is eliminated when we decouple simultaneous equations. Recalling again the example of a closed chemical system, the choice of X_n could be compared to solving the conservation law in terms of X_n and substituting $X_n = Total - (X_1 + \ldots + X_{n-1})$.

As an alternative to (15.12), or in the case of $v_n = 0$, one can select another index j with $v_j \neq 0$ and employ the invariants

$$\xi_i = X_i X_j^{-v_i/v_j}$$

$$\xi_j = X_j \tag{15.14}$$

($i = 0, \ldots, n; i \neq j$) or one can renumber the equations such that $v_n \neq 0$.

DECOUPLING OF SIMULTANEOUS GMA-SYSTEM EQUATIONS

A prominent reason for finding invariants is that the choice of convenient coordinates transforms the system in such a way that one or more of the original, simultaneous equations can be decoupled from the rest of the system. Representing the original GMA-system (15.1) in terms of the new variables ξ_i, as defined in (15.12), and using Eq. (15A.1) of the Appendix, we obtain for $i = 0, \ldots, n-1$ and $v_n \neq 0$

$$\xi_n^{v_0/v_n} \frac{d\xi_i}{dX_0} = \sum_{k=1}^{m_i} \gamma_{ik} \prod_{j=0}^{n-1} \xi_j^{g_{ikj}} - \frac{v_i}{v_n} \xi_i \sum_{k=1}^{m_i} \gamma_{nk} \prod_{j=0}^{n-1} \xi_j^{g_{nkj}} \tag{15.15}$$

If $v_0 = 0$, then $\xi_0 = X_0$, and Eq. (15.15) reduces to

$$\frac{d\xi_i}{d\xi_0} = \sum_{k=1}^{m_i} \gamma_{ik} \prod_{j=0}^{n-1} \xi_j^{g_{ikj}} - \frac{v_i}{v_n} \xi_i \sum_{k=1}^{m_i} \gamma_{nk} \prod_{j=0}^{n-1} \xi_j^{g_{nkj}} \tag{15.16}$$

To decouple one equation in the case of $v_0 \neq 0$, one divides Eq. (15.15) for $i = 1, \ldots, n-1$ by the equation $d\xi_0/dX_0$, which eliminates ξ_n.

In either case, ξ_n does not appear on the right-hand side of the resulting equation. Thus, the system can be solved without the equation for ξ_n, while ξ_n itself is determined in a second step by quadrature (cf. [203: p. 158f]). Specifically, ξ_n is obtained from the GMA-system equation of $X_n (= \xi_n)$ in (15.1) upon substitution of the new variables ξ_i for the old variables X_i. For $v_0 = 0$, the result is the separable equation

$$\frac{d\xi_n}{d\xi_0} = \xi_n \sum_{k=1}^{m_i} \gamma_{nk} \prod_{j=0}^{n-1} \xi_j^{g_{nkj}} \tag{15.17}$$

For $v_0 \neq 0$, $d\xi_n/d\xi_0$ also is separable, but the right-hand side of (15.17) is divided by the right-hand side of (15.15) for $i = 0$.

If any of the variables ξ_i does not appear on the right-hand side, the system can be decoupled further, since the remaining equations can be solved without the equation for ξ_i. Furthermore, if ξ_0 only appears in the equation $d\xi_0/dX_0$, the system can be reduced by either eliminating the equation $d\xi_0/X_0$ or by selecting a new invariant ξ_j as differentiation variable and by expressing the system in the form $d\xi_i/d\xi_j$.

Example 1: The Complete Solution of an S-System

The S-system

$$\frac{dX_1}{dX_0} = X_1^2 X_2^{-1} - \beta X_1$$

$$\frac{dX_2}{dX_0} = \alpha X_1$$

represents a generalized Gompertz distribution (cf. [42]). Its symmetry matrix is

$$(C_{ij}) = \begin{pmatrix} 1 & 1 & -1 \\ 1 & 1 & -1 \\ 1 & 0 & 0 \end{pmatrix}$$

which admits the solution $v_0 = 0$, $v_1 = v_2$. According to (15.12), convenient invariants are

$$\xi_0 = X_0$$

$$\xi_1 = X_1 X_2^{-1}$$

$$\xi_2 = X_2$$

and since $v_0 = 0$, we obtain immediately

$$\frac{d\xi_1}{d\xi_0} = \xi_1^2 - \beta\xi_1 - \alpha\xi_1^2$$

(cf. Eq. (15.16)). For $\alpha = 1$, the solution is

$$\xi_1 = \exp(-\beta\xi_0 + c)$$

and, according to (15.17)

$$\frac{d\xi_2}{d\xi_0} = \alpha\xi_1\xi_2 = \exp(-\beta\xi_0 + c)\xi_2$$

which is separable and yields

$$\xi_2 = \exp\left(-\beta^{-1}\exp(-\beta\xi_0 + c) + c'\right)$$

The original variables X_0, X_1 and X_2 are computed from these invariants as

$$\begin{aligned} X_0 &= \xi_0 \\ X_1 &= \xi_1\xi_2 = \exp(-\beta X_0 + c)\cdot\exp\left(-\beta^{-1}\exp(-\beta X_0 + c) + c'\right) \\ X_2 &= \xi_2 = \exp\left(-\beta^{-1}\exp(-\beta X_0 + c) + c'\right) \end{aligned}$$

For $\alpha \neq 1$, the solution is

$$\xi_1 = \beta\left[1 - \alpha - \exp(\beta\xi_0 + c'')\right]^{-1}$$

and the original variables are retrieved in the same way as before.

Example 2: Steady States in Half-Systems

Given that its system matrix is regular, the steady state of an S-system can be calculated through matrix inversion (cf. Chapter 2). However, when we recast some of its equations as Half-system equations, the system matrix becomes singular, and the inversion method fails. In this section we demonstrate with an example how the transformation to a coordinate system of invariants can be employed to retrieve the original steady state.

Let's look at the S-system

$$\begin{aligned} \frac{dZ_1}{dZ_0} &= 2Z_1^{0.5}Z_2^{0.4} - 3Z_1^2Z_2 \\ \frac{dZ_2}{dZ_0} &= 5Z_1Z_2 - 3Z_1^{-2}Z_2^{1.4} \end{aligned} \qquad (15.18)$$

with the nonzero steady state $Z_{1S} = 0.82260$, $Z_{2S} = 0.82897$. Let's assume we have reason to recast the second equation by replacing $Z_2 = Z_3^{2.5} Z_4^2$. Formulating the differential equations for Z_3 and Z_4 as discussed in Chapter 12 and renaming $X_0 = Z_0$, $X_1 = Z_1$, $X_2 = Z_3$, and $X_3 = Z_4$, we obtain

$$\frac{dX_1}{dX_0} = 2X_1^{0.5}X_2X_3^{0.8} - 3X_1^2X_2^{2.5}X_3^2$$

$$\frac{dX_2}{dX_0} = 2X_1X_2$$

$$\frac{dX_3}{dX_0} = -1.5X_1^{-2}X_2X_3^{1.8}$$

Obviously, the system matrix is singular and it cannot be inverted to yield the steady-state solution. Furthermore, we see by inspection that X_2 grows without bound while X_3 vanishes when X_0 approaches infinity. Numerical solution confirms this but it also shows that X_1 approaches a finite steady-state value different from zero. Transformation to a convenient system of invariants makes it possible to calculate this value with paper and pencil.

The system admits a scaling transformation since the symmetry matrix

$$(C_{ij}) = \begin{pmatrix} 1 & -0.5 & 1 & 0.8 \\ 1 & 1 & 0 & 0 \\ 1 & -2 & 1 & 0.8 \\ 1 & 1 & 2.5 & 2 \end{pmatrix}$$

has rank 3 and admits the solution $v_0 = 0$, $v_1 = 0$, $v_2 = -0.8v_3$. Typical invariants are thus, according to (15.12)

$$\xi_0 = X_0$$

$$\xi_1 = X_1$$

$$\xi_2 = X_2X_3^{0.8}$$

$$\xi_3 = X_3$$

Since $v_0 = 0$, we can directly apply Eq. (15.16), and since $\beta_2 = \alpha_3 = 0$, the result in S-system form is

$$\frac{d\xi_1}{d\xi_0} = 2\xi_1^{0.5}\xi_2 - 3\xi_1^2\xi_2^{2.5}$$

$$\frac{d\xi_2}{d\xi_0} = 2\xi_1\xi_2 - 0.8 \cdot 1.5\xi_1^{-2}\xi_2^2 \tag{15.19}$$

Setting the derivatives equal to zero, taking logarithms, and solving the set of algebraic equations yields the steady-state solution

$$\xi_{1S} = 0.82260$$

$$\xi_{2S} = 0.92772$$

The first value is the same as Z_{1S} in the original S-system, but the second value is different. In fact, the two S-systems (15.18) and (15.19) are different. How could that happen? The answer is that we arbitrarily defined the second invariant as $\xi_2 = X_2X_3^{0.8}$. As we discussed above, this choice is just one among many possible, and if instead we use as the second invariant $\xi_2^{2.5}$, which as a function of ξ_2 also is an invariant, the transformation to new coordinates leads back to the original S-system (15.18). As a matter of fact, for every power q of the invariant ξ_2 we obtain a slightly different S-system, and all these S-system can be transformed into each other via changes in coordinates of the type

$$\eta_2 = \xi_2^q$$

For instance, in such new coordinates with $q = 2.5$ Eq. (15.19) becomes Eq. (15.18).

The degree of freedom in choosing powers of invariants is related to the choice of powers in the recasting of GMA-systems as S-systems or of S-systems as Half-systems (cf. Chapter 12). It shows that different S-systems can be recast as the same Half-system. All S-systems leading to the same Half-system can be generated from each other by defining new variables as power functions of the old variables.

OTHER TYPES OF TRANSFORMATION GROUPS

The technique of recasting a differential equation as an S-system can be used to detect other types of groups. This will now be demonstrated with the prominent example of translation groups, which can be determined upon logarithmic coordinate transformation. For example, consider the differential equation

$$\frac{dY}{dX} = a(X - Y)^{-2}$$

[200:p. 9]. Define new variables as $X_0 = \exp(X)$, $X_1 = \exp(Y)$, which results in

$$\frac{dX_1}{dX_0} = \frac{X_1}{X_0} a\left(\ln(X_0) - \ln(X_1)\right)^{-2}$$

By defining $X_2 = \ln(X_0) - \ln(X_1)$, this equation is in Half-system form, and one obtains the additional S-system equation

$$\frac{dX_2}{dX_0} = X_0^{-1} - aX_0^{-1}X_2^{-2}$$

The symmetry matrix (C_{ij}) then is

$$(C_{ij}) = \begin{pmatrix} 0 & 0 & -2 \\ 0 & 0 & -1 \\ 0 & 0 & -3 \end{pmatrix}$$

Evidently, the rank is 1, and the solution corresponds to the transformation

$$\begin{aligned} Y_0 &= p^{v_0}X_0 \\ Y_1 &= p^{v_1}X_1 \\ Y_2 &= p^0 X_2 \end{aligned}$$

with real v_0 and v_1. However, the recasting constraint $X_2 = \ln(X_0) - \ln(X_1)$ must also be satisfied in the new coordinate system. Thus, in addition to the equation $Y_2 = X_2$, we have to require $Y_2 = \ln(Y_0) - \ln(Y_1)$ which yields

$$\ln(X_0 p^{-v_0}) - \ln(X_1 p^{-v_1}) = \ln(X_0) + \ln(p^{-v_0}) - \ln(X_1) - \ln(p^{-v_1})$$

which is only satisfied for $v_0 = v_1$. Hence, the S-system variables can be transformed as

$$\begin{aligned} Y_0 &= p^{v_0}X_0 \\ Y_1 &= p^{v_0}X_1 \\ Y_2 &= X_2 \end{aligned}$$

which corresponds to the translation

$$\tilde{Y} = Y + \epsilon$$

$$\tilde{X} = X + \epsilon$$

of the original equation (cf. [200: p. 9]), when we set $\epsilon = v_0 \ln p$.

In combination with recasting, other symmetries can be explored. For instance, it is possible to study scaling transformations in systems of higher-order differential equations.

OTHER FUNCTIONAL RELATIONSHIPS BETWEEN SYSTEM VARIABLES

While symmetry groups of scaling transformations can be computed in a straightforward manner (cf. Theorem 1), other symmetries are more difficult to detect. In this section, we discuss a general approach to characterizing functional relationships between system variables. This approach leads to systems of linear partial differential equations, and if these can be solved, at least partially, some of the original system variables can be eliminated, at the cost of losing the structural homogeneity of the GMA- or S-system equations. In many respects, this process provides a reversion of the recasting process.

Since we do not know which variables are expressed most easily as functions of some of the other variables, it is convenient to formulate the dependencies as

$$X_i = f_i(X_1, \ldots, X_n) \tag{15.20}$$

for $i = 1, \ldots, n$. Typically, f_i will be independent of X_i, but it is advantageous to allow dependence on X_i, since then (15.20) includes the identity mapping as a special case.

Differentiation of (15.20) with respect to the independent variable X_0 immediately yields

$$\frac{dX_i}{dX_0} = \sum_{j=1}^{n} \frac{\partial f_i}{\partial X_j} \frac{dX_j}{dX_0} \tag{15.21}$$

Using matrix notation, we can rewrite (15.21) as

$$(\boldsymbol{J} - \boldsymbol{I})\, \boldsymbol{DX} = \boldsymbol{o} \tag{15.22}$$

where $\boldsymbol{J}$ is the Jacobian matrix, $\boldsymbol{I}$ is the identity matrix, $\boldsymbol{DX}$ is the vector of derivatives, and $\boldsymbol{o}$ is the zero vector of dimension n. Obviously, (15.22) admits the trivial solution $\boldsymbol{J} = \boldsymbol{I}$ which corresponds to $\partial f_i / \partial X_j = \delta_{ij}$ and thus to the identity mapping $f_i(X_i) = X_i$. However, if we can find nontrivial solutions to (15.22), they immediately translate into functional relationships between the system variables. The process is best illustrated with an example.

Example:

$$\frac{dX_1}{dX_0} = X_2 X_3 \qquad X_1(0) = 1$$

$$\frac{dX_2}{dX_0} = 2X_1 X_2 X_3 \qquad X_2(0) = 2$$

$$\frac{dX_3}{dX_0} = 2X_1 X_2 X_3^2 - X_2 X_3^2 \quad X_3(0) = 1 \tag{15.23}$$

To test whether X_2 can easily be expressed in terms of the other variables, we study the second equation of the linear system (15.22):

$$X_2 X_3 \frac{\partial f_2}{\partial X_1} + 2X_1 X_2 X_3 \left(\frac{\partial f_2}{\partial X_2} - 1 \right) + (2X_1 X_2 X_3^2 - X_2 X_3^2) \frac{\partial f_2}{\partial X_3} = 0$$

We are not interested in the trivial solution $\partial f_2 / \partial X_1 = \partial f_2 / \partial X_3 = 0$, $\partial f_2 / \partial X_2 = 1$, and set $\partial f_2 / \partial X_2 = 0$. The previous equation then reduces to

$$\frac{\partial f_2}{\partial X_1} - 2X_1 + (2X_1 X_3 - X_3) \frac{\partial f_2}{\partial X_3} = 0$$

There is no obvious solution if both $\partial f_2 / \partial X_1$ and $\partial f_2 / \partial X_3$ are different from zero. For $\partial f_2 / \partial X_3 = 0$ the partial differential equation reduces to an ordinary differential equation, and with $f_2(X_1) = X_2(X_1)$ its solution is $X_2 = X_1^2 + c$. Employing the initial conditions, one identifies the integration constant c as unity.

Now we test whether X_3 also can be expressed in terms of the other variables; the third equation of the linear system (15.22) reads:

$$X_2 X_3 \frac{\partial f_3}{\partial X_1} + 2X_1 X_2 X_3 \frac{\partial f_3}{\partial X_2} + (2X_1 X_2 X_3^2 - X_2 X_3^2) \left(\frac{\partial f_3}{\partial X_3} - 1 \right) = 0$$

Setting $\partial f_3 / \partial X_2 = \partial f_3 / \partial X_3 = 0$ and $f_3(X_1) = X_3(X_1)$, this equation reduces to the ordinary equation

$$\frac{dX_3}{dX_1} - 2X_1X_3 + X_3 = 0$$

with the solution

$$X_3 = \exp(X_1^2 - X_1 + c')$$

where the integration constant vanishes because of the initial conditions. Thus, we have expressed X_2 and X_3 as functions of X_1, and the three-variables S-system (15.23) reduces to

$$\frac{dX_1}{dX_0} = (X_1^2 + 1)\exp(X_1^2 - X_1), \qquad X_1(0) = 1$$

The validity of this result is readily confirmed through recasting.

A COMPARISON OF METHODS THAT LOWER THE NUMBER OF EQUATIONS NECESSARY FOR SOLUTION

In Chapter 12, we illustrated recasting strategies with transformations of the single S-system equation

$$\frac{dX_1}{dX_0} = \alpha X_1^g - \beta X_1^h \tag{15.24}$$

It turned out that several recasting strategies resulted in the two-variable system

$$\frac{dP}{dX_0} = \beta(1 - h)P^2 - \alpha(1 - h)PQ$$

$$\frac{dQ}{dX_0} = \beta(1 - g)PQ - \alpha(1 - g)Q^2 \tag{15.25}$$

but it was not obvious how (15.24) could be retrieved from (15.25).

In this section, we use three different methods to derive (15.24) from (15.25). First we show that (15.25) admits a scaling symmetry that allows us to lower the number of simultaneous equations from two to one. Then, we solve a partial differential equation associated with (15.25), as was discussed in the previous section. And finally, we use the trajectorial solution dQ/dP to derive (15.24).

Symmetry Method

The symmetry matrix of (15.25) is

$$(C_{ij}) = \begin{pmatrix} 1 & 1 & 0 \\ 1 & 1 & 0 \\ 1 & 0 & 1 \\ 1 & 0 & 1 \end{pmatrix}$$

which admits the solution $v_0 = -v_2$, $v_1 = v_2$. Using as invariants

$$\xi_0 = X_0 Q^{-v_0/v_2} = X_0 Q$$
$$\xi_1 = PQ^{-1}$$
$$\xi_2 = Q$$

we obtain

$$\xi_2^{-1} \frac{d\xi_0}{dX_0} = 1 + \beta(1-g)\xi_0\xi_1 - \alpha(1-g)\xi_0$$

$$\begin{aligned}\xi_2^{-1} \frac{d\xi_1}{dX_0} &= \beta(1-h)\xi_1^2 - \alpha(1-h)\xi_1 - \beta(1-g)\xi_1^2 \\ &\quad + \alpha(1-g)\xi_1 \\ &= \beta(g-h)\xi_1^2 - \alpha(g-h)\xi_1 \end{aligned} \tag{15.26}$$

Division of the equations in (15.26) yields

$$\frac{d\xi_1}{d\xi_0} = \frac{(\beta\xi_1 - \alpha)(g-h)\xi_1}{1 + (\beta\xi_1 - \alpha)(1-g)\xi_0} \tag{15.27}$$

This nonautonomous first-order equation looks quite different from the original equation (15.24) but, in fact, is closely related to it. To see this, let's set $Q = X_1^q$, $P = X_1^p$ and, thus, according to the definitions of the invariants

$$\xi_0 = X_0 X_1^q$$
$$\xi_1 = X_1^{p-q} = X_1^r$$
$$\xi_2 = X_1^q$$

Then (15.27) in terms of X_0 and X_1 becomes

$$\begin{aligned}\frac{dX_1}{dX_0} &= \frac{dX_1}{d\xi_1}\frac{d\xi_1}{d\xi_0}\frac{d\xi_0}{dX_0} \\ &= r^{-1}\xi_1^{-1+1/r}\frac{(\beta\xi_1 - \alpha)(g - h)\xi_1}{1 + (\beta\xi_1 - \alpha)(1 - g)\xi_0} \\ &\quad \cdot \xi_2[1 + \beta(1 - g)\xi_0\xi_1 - \alpha(1 - g)\xi_0] \\ &= r^{-1}X_1^p\beta(X_1^{p-q} - \alpha)(g - h) \end{aligned} \tag{15.28}$$

which is identical to (15.24) upon setting $p = g$, $q = 2g - h$, and $r = h - g$.

Partial Differential Equation Method

Suppose that in (15.25) Q is a function of P, i.e., $Q = f(P)$. Then we have to find a nontrivial solution of the partial differential equation

$$\frac{\partial f}{\partial P}\frac{dP}{dX_0} + \left(\frac{\partial f}{\partial Q} - 1\right)\frac{dQ}{dX_0} = 0 \tag{15.29}$$

Since $Q = f(P)$, we have $\partial f/\partial Q = 0$ and $\partial f/\partial P = dQ/dP$. Substituting (15.25) in Eq. (15.29), Eq. (15.29) becomes

$$\begin{aligned}\beta(1 - h)P^2\frac{dQ}{dP} &- \alpha(1 - h)PQ\frac{dQ}{dP} \\ &- \beta(1 - g)PQ + \alpha(1 - g)Q^2 = 0\end{aligned} \tag{15.30}$$

When we identify the first with the third term and the second with the fourth term, we obtain in both cases the equation

$$(1 - h)P\frac{dQ}{dP} = \beta(1 - g)Q \tag{15.31}$$

with the solution $Q = cP^{(1-g)/(1-h)}$. This solution agrees with the recasting constraints introduced in Chapter 12. Substituting the solution in the first equation (15.25) and defining $X_1 = P^{1/(h-1)}$ yields (15.24).

When we identify the first term in (15.30) with the fourth term and the second with the third, we obtain the two differential equations

$$\frac{dQ}{dP}\beta(1 - h)P^2 = -\alpha(1 - g)Q^2 \tag{15.32}$$

with the solution

$$Q = \beta(1 - h)P[c_1 P - \alpha(1 - g)]^{-1} \tag{15.33}$$

and

$$\frac{dQ}{dP}\alpha(1 - h) = \beta(1 - g) \tag{15.34}$$

with the solution

$$Q = [\beta(1 - g)P + c_2][\alpha(1 - h)]^{-1} \tag{15.35}$$

where c_1 and c_2 are integration constants. Eqs. (15.33) and (15.34) are simultaneously satisfied only if P is a constant function. In this case, $dP/dX_0 = 0$ and $Q = \alpha^{-1}\beta P$. These constant solutions represent the family of steady states of (15.25). The unique steady state of (15.24), if existent, is retrieved when the family of steady states is constrained by the recasting condition introduced in the transformation of (15.24) into (15.25).

Trajectorial Solution

Division of the two equations in (15.25) yields the homogeneous equation

$$\frac{dP}{dQ} = \frac{(1 - h)}{(1 - g)} \frac{\beta P^2 - \alpha PQ}{\beta PQ - \alpha Q^2} \tag{15.36}$$

Introduction of the new coordinates $u = Q$, $v = PQ^{-1}$ yields

$$\begin{aligned}\frac{dv}{du} &= \frac{dP}{dQ}Q^{-1} - PQ^{-2} \\ &= \frac{(1 - h)}{(1 - g)} \frac{\beta u^2 v^2 - \alpha u^2 v}{\beta u^2 v - \alpha u^2} u^{-1} - vu^{-1} \\ &= \frac{(g - h)}{(1 - g)} \frac{v}{u}\end{aligned} \tag{15.37}$$

with the solution

$$v = cu^{(g-h)/(1-g)} \tag{15.38}$$

from which one immediately retrieves the same functional relationship between P and Q as previously seen (cf. Eq. [15.31] and the recasting constraint in Chapter 12):

$$Q = cP^{(1-g)/(1-h)} \tag{15.39}$$

Evaluation of the Three Methods

All three methods transform the recast S-system (15.25) back into the one-variable S-system (15.24). It may appear that the trajectorial method is the easiest. While this is true for the given example, the two other methods are by far more general. The trajectorial method is very useful for two-variable Half-systems (cf. Chapter 10) and for S-systems that happen to have favorable properties that allow us to solve the trajectorial solution, as was the case in the above example (see also [33]). The symmetry method is more general in that it can be applied to any S-system. However, it will be successful in lowering the number of equations only if the S-system admits a scaling symmetry. The partial differential equation method is the most general. In fact, if the partial differential equations associated with an S-system can be solved completely, we can solve the S-system. While generally the partial differential problem is more complicated than the original S-system, in many cases we can find partial solutions, and these already yield insight into relationships among the system variables. In particular, when we are interested in relationships between just two variables at a time, this method may be most helpful. Nevertheless, this method will almost always require some cleverness, intuition, or trial and error, whereas the other two methods are rather straightforward.

SUMMARY

This chapter discusses symmetries of Generalized Mass Action systems and their special cases. Symmetries correspond to conservation laws that remain invariant under the action of the differential equations. When symmetries exist, particular dynamical and steady-state solutions give rise to entire families of solutions. In general, the symmetry structure of a differential equation is difficult to reveal, but for GMA-systems, at least some simple symmetries can be detected with elementary algebraic means. Symmetries can be employed to decouple simultaneous GMA equations. In the case of S-system symmetries, this process often leads to GMA-systems that contain the same numbers of terms which, however, are distributed among fewer equations. The process can be considered the reverse of the process of recasting GMA-systems as S-systems. Other recasting steps, such as the introduction of new variables for functions of existing variables, can be reversed by deriving certain partial differential equations from the GMA-system or the S-system. Although general solutions for these partial differential equations usually are difficult to obtain, even particular solutions are very useful because they correspond directly to functional relationships between system variables as they come up as constraints during the recasting process.

ACKNOWLEDGMENTS

I wish to thank Donald C. Lewis for his valuable suggestions and criticisms, and to acknowledge the partial support of U.S. Public Health Service Grant GM 30054 from the National Institutes of Health.

APPENDIX

Proof of Theorem 1

1. Condition (15.3) is equivalent to the condition

$$v_h - v_0 = \sum_{j=0}^{n} v_j g_{hkj} \tag{15A.1}$$

for each term of the GMA-system (15.1). From (15.4) and (15A.1) we obtain the differential equations for Y_i as

$$\begin{aligned}\frac{dY_i}{dY_0} &= \frac{dY_i}{dX_i}\frac{dX_i}{dX_0}\frac{dX_0}{dY_0} = p^{v_i}\frac{dX_i}{dX_0}p^{-v_0} \\ &= p^{v_i - v_0}\sum_{k=1}^{m_i}\gamma_{ik}\prod_{j=0}^{n}(p^{-v_j}Y_j)^{g_{ikj}} \\ &= p^{v_i - v_0}\left(p^{-\Sigma_{j=0}^{n} v_j g_{ikj}}\sum_{k=1}^{m_i}\gamma_{ik}\prod_{j=0}^{n}Y_j^{g_{ikj}}\right) \\ &= \sum_{k=1}^{m_i}\gamma_{ik}\prod_{j=0}^{n}Y_j^{g_{ikj}}\end{aligned} \tag{15A.2}$$

Under composition, the transformations (15.4) form a one-parameter group with the identity transformation $Y_i = X_i$ when $p = 1$ and the inverse transformation $Y_i = (1/p)^{v_i}X_i$; closure and associativity are obvious.

2. If system (15.3) has two or more linearly independent solutions, then each solution v_h corresponds to a transformation

$$Y_{hj} = p^{v_{hj}}X_j \qquad j = 0, \ldots, n; \qquad h = 1, \ldots, n + 1 - \rho \tag{15A.3}$$

and the composition of these transformations generates transformations of type (15.6).

The transformations (15.6) leave the GMA-system (15.1) unaltered. As in part (1), we have for all $h \leq r$ and $i = 1, \ldots, n$

$$v_{hi} - v_{h0} = \sum_{j=0}^{n} v_{hj}\, g_{ikj} \tag{15A.4}$$

The differential equations for the new variables Y_i then are

$$\begin{aligned} \frac{dY_i}{dY_0} &= \prod_{h=1}^{r} p_h^{v_{hi}} \frac{dX_i}{dX_0} p_h^{-v_{h0}} \\ &= \sum_{k=1}^{m_i} \prod_{h=1}^{r} p_h^{v_{hi} - v_{h0}} \prod_{j=0}^{n} \left(\prod_{h=1}^{r} p_h^{-v_{hj}} \right)^{g_{ikj}} \gamma_{ik} \prod_{j=0}^{n} Y_j^{g_{ikj}} \\ &= \sum_{k=1}^{m_i} \gamma_{ik} \prod_{h=1}^{r} p^{v_{hi} - v_{h0} - \Sigma_{j=0}^{n} v_{hj} g_{ikj}} \prod_{j=1}^{n} Y_j^{g_{ikj}} \\ &= \sum_{k=1}^{m_i} \gamma_{ik} \prod_{j=0}^{n} Y_j^{g_{ikj}} \end{aligned} \tag{15A.5}$$

Thus, the r-parameter transformation (15.6) leaves the GMA-system (15.1) unaltered.

The linear independence of transformations of type (15A.3) is evident when one defines c_h such that $p^{c_h} = p_h$ and rewrites equation (15.6) as

$$Y_j = p^{\Sigma_{h=1}^{r} c_h v_{hj}} X_j \tag{15A.6}$$

where $r = n + 1 - \rho$. Thus, under composition, the collection of one-parameter groups (15A.3) generates an $n + 1 - \rho$ dimensional Lie group of transformations of type (15.6) with the identity transformation $Y_j = X_j$ when $p_h = 1$ for all h, and the inverse transformation $Y_j = \Pi_h (1/p_h)^{v_{hj}} X_j$; closure and associativity again are obvious.

Proof of Theorem 2

We prove (1) and (2) simultaneously. Without loss of generality, we can renumber variables and terms in such a way that the first variable in the first term of the i-th equation of (15.1) is replaced with a new variable; that is, we define $X_{n+1} = X_1^p (p \in R)$. The substitution affects two equations: the equation for X_i and the additional equation for X_{n+1}. The equation for X_i in the new system reads

$$\frac{dX_i}{dX_0} = \gamma_{i1} X_{n+1} X_1^{g_{i11}-p} \prod_{j=0, j\neq 1}^{n} X_j^{g_{i1j}} + \sum_{k=2}^{m_i} \gamma_{ik} \prod_{j=1}^{n} X_j^{g_{ikj}} \quad (15A.7)$$

and the equation for X_{n+1} reads

$$\begin{aligned}\frac{dX_{n+1}}{dX_0} &= pX_1^{p-1}\frac{dX_1}{dX_0}\\ &= pX_{n+1}X_1^{-1}\frac{dX_1}{dX_0}\end{aligned} \quad (15A.8)$$

In order for the first term in (15A.7) to satisfy Eq. (15A.1), the condition

$$v_i - v_0 = \sum_{j=0}^{n+1} v_j g'_{i1j} \quad (15A.9)$$

occurs when the g'_{i1j} are defined as

$$\begin{aligned} g'_{i1j} &= g_{i1j} \\ g'_{i11} &= g_{i11} - p \\ g'_{i1,n+1} &= 1 \end{aligned}$$

$(j = 0, \ldots, n; j \neq 1)$. Thus, Eq. (15A.9) can be rewritten as

$$\begin{aligned} v_i - v_0 &= \sum_{j=0, j\neq 1}^{n} v_j g_{ihj} + v_1(g_{ih1} - p) + v_{n+1} \\ &= \sum_{j=0}^{n} v_j g_{ihj} - pv_1 + v_{n+1} \end{aligned} \quad (15A.10)$$

Because Eq. (15A.1) is satisfied in the original system, the new system satisfies Eq. (15A.1) when we specify v_{n+1} as

$$v_{n+1} = pv_1 \quad (15A.11)$$

For Eq. (15A.8) we obtain in a similar fashion the condition

$$v_{n+1} - v_0 = \sum_{j=0}^{n+1} v_j g'_{n+1,k,j} \quad (15A.12)$$

which has to be satisfied for $k = 1, \ldots, m_{n+1}$, where

$$g'_{n+1,k,j} = g_{1kj}$$
$$g'_{n+1,k,1} = g_{1k1} - 1$$
$$g'_{n+1,k,n+1} = 1$$

$(j = 0, \ldots, n; j \neq 1)$. Therefore, Eq. (15A.12) becomes

$$v_{n+1} - v_0 = \sum_{j=0}^{n} v_j g_{1kj} - v_1 + v_{n+1} \tag{15A.13}$$

which is equivalent to Eq. (15A.1) for the first equation of the original system.

Proof of Theorem 3

Suppose X_i in the GMA-system (15.1) is replaced with the product of two new variables; i.e., $X_i = X_{n+1}X_{n+2}$. Differentiation yields

$$\frac{dX_i}{dX_0} = X_{n+1}\frac{dX_{n+2}}{dX_0} + X_{n+2}\frac{dX_{n+1}}{dX_0} \tag{15A.14}$$

Without loss of generality, we assign the first $\tilde{m}_i$ terms of the i-th equation of Eq. (15.1) to dX_{n+1}/dX_0 and the remaining $m_i - \tilde{m}_i$ terms to dX_{n+2}/dX_0 (cf. Chapter 12). The new differential equations then read

$$\frac{dX_{n+1}}{dX_0} = \sum_{k=1}^{\tilde{m}_i} \gamma_{ik} X_{n+2}^{-1} \prod_{j=0}^{n} X_j^{g_{ikj}}$$

and

$$\frac{dX_{n+2}}{dX_0} = \sum_{k=\tilde{m}_i+1}^{m_i} \gamma_{ik} X_{n+1}^{-1} \prod_{j=0}^{n} X_j^{g_{ikj}} \tag{15A.15}$$

Typically, the products of power-law functions contain X_i which is replaced by the product of X_{n+1} and X_{n+2}, and the result is

$$\frac{dX_{n+1}}{dX_0} = \sum_{k=1}^{\tilde{m}_i} \gamma_{ik} X_{n+1}^{g_{iki}} X_{n+2}^{g_{iki}-1} \prod_{j=0, j\neq 1}^{n} X_j^{g_{ikj}}$$

$$\frac{dX_{n+2}}{dX_0} = \sum_{k=\tilde{m}_i+1}^{m_i} \gamma_{ik} X_{n+1}^{g_{iki}-1} X_{n+2}^{g_{iki}} \prod_{j=0, j\neq 1}^{n} X_j^{g_{ikj}} \tag{15A.16}$$

Thus, the new parameters of the resulting GMA-system with variables X_j (j = 0, . . . , $n + 2; j \neq i$) are

$$
\begin{aligned}
\gamma_{n+1,k} &= \gamma_{i,k} \\
\gamma_{n+2,k} &= \gamma_{i,k} \\
g_{h,k,n+1} &= g_{hki} \\
g_{h,k,n+2} &= g_{hki} \\
g_{n+1,k,h} &= g_{ikh} \\
g_{n+2,k,h} &= g_{ikh} \\
g_{n+1,k,n+1} &= g_{iki} \\
g_{n+1,k,n+2} &= g_{iki} - 1 \\
g_{n+2,k,n+1} &= g_{iki} - 1 \\
g_{n+2,k,n+2} &= g_{iki}
\end{aligned}
\tag{15A.17}
$$

where $k = 1, \ldots, m_h$ and $h = 1, \ldots, n; h \neq i$. According to Eq. (15A.1), the original GMA-system (15.1) satisfies the conditions

$$v_h - v_0 = \sum_{j=0}^{n} v_j g_{hkj} \tag{15A.18}$$

In the new system with variables X_j these conditions become

$$
\begin{aligned}
v_h - v_0 &= \sum_{j=0, j\neq i}^{n+2} v_j g_{hkj} \\
&= \sum_{j=0}^{n} v_j g_{hkj} - v_i g_{hki} + v_{n+1} g_{h,k,n+1} + v_{n+2} g_{h,k,n+2}
\end{aligned}
\tag{15A.19}
$$

and hence, because of Eq. (15A.1)

$$v_i g_{hki} = v_{n+1} g_{h,k,n+1} + v_{n+2} g_{h,k,n+2}$$

and

$$v_i = v_{n+1} + v_{n+2} \tag{15A.20}$$

because of the definition of the new exponents g_{ikj} (cf. 15A.17).

For the equations of X_{n+1} and X_{n+2} in the new system, to satisfy Eq. (15A.1), the conditions

$$v_{n+1} - v_0 = \sum_{j=0, j \neq i}^{n+2} v_j g_{n+1,k,j} \quad k = 1, \ldots, \tilde{m}_i \tag{15A.21}$$

and

$$v_{n+2} - v_0 = \sum_{j=0, j \neq i}^{n+2} v_j g_{n+2,k,j} \quad k = \tilde{m}_i + 1, \ldots, m_i \tag{15A.22}$$

must hold. Expressed in terms of the old exponents (cf. [15A.17]), the condition (15A.21) becomes

$$\begin{aligned} v_{n+1} - v_0 &= \sum_{j=0}^{n} v_j g_{ikj} - v_i g_{iki} + v_{n+1} g_{iki} + v_{n+2}(g_{iki} - 1) \\ &= v_i - v_0 - v_{n+2} + g_{iki}(v_{n+1} + v_{n+2} - v_i) \end{aligned} \tag{15A.23}$$

which reduces to

$$0 = [v_i - v_{n+1} - v_{n+2}](1 - g_{iki}) \tag{15A.24}$$

and is equivalent to Eq. (15A.20). The same is true for the condition (15A.22).

Proof of Corollaries 1 and 2:

The corollaries follow immediately from Theorems 2 and 3.

16. A Qualitative Analysis of S-Systems: Hopf Bifurcations

Donald C. Lewis

S-systems of differential equations are endowed with extremely rich and complicated behavior. While the development of ESSYNS (Chapter 7) has greatly facilitated the numerical analysis of S-systems, when one is working with an S-system of large dimension, with unknown parameters, or with a sensitive dependence on parameters, then the direct interpretation of numerical solutions and how these solutions vary with parameters becomes a forbidding task. In this situation it is useful to identify and to study the qualitative properties of solutions. A knowledge of qualitative properties can guide both the synthesis of S-system models and the interpretation of numerical solutions.

In this chapter I will briefly describe what is meant by "qualitative property," present some key concepts from the qualitative theory of ordinary differential equations, and use these concepts to motivate a discussion of the Hopf bifurcation theorem. Two results for S-systems will be presented. The first is an easy application of the Hopf bifurcation theorem which gives a test for the existence of a Hopf bifurcation in an n-dimensional S-system. The second result concerns two-dimensional S-systems. In this case a more detailed analysis is possible, and a test for the existence of a Hopf bifurcation can be given explicitly in terms of the parameters of the S-system. A test for the stability of the limit cycle in a nondegenerate Hopf bifurcation is also presented. This test amounts to the evaluation of a "simple" expression involving the parameters of the general two-dimensional S-system.

QUALITATIVE PROPERTIES

Notions about qualitative properties are perhaps most easily grasped in the setting of geometry where we can rely on our rich (albeit trained) intuition about the shapes of figures and bodies. I think, for example, that most people would intuitively agree that a circle and an arbitrary ellipse have qualitatively the same shape; at least in contrast to a circle and an arbitrary hyperbola which seem to have qualitatively different shapes. Furthermore, there is a qualitative similarity between a circle and an ellipse which is independent of the scale or the orientation of these figures. This latter impression argues against thinking that a qualitative similarity is simply that of an approximate quantitative equality.

Suppose for the moment that we had a way to decide when two geometric figures were qualitatively the same. A qualitative property of a geometric object would then be a property that is shared by any objects that are qualitatively the same. That is, when two objects are qualitatively the same they have exactly the same qualitative properties. Our intuition might suggest, for example, that being a smooth simple closed curve is a qualitative property of figures in the plane. A qualitative analysis of a geometric object would then be an analysis of its qualitative properties. It is a remarkable state of affairs that many of the most widely applicable structures in mathematics have natural geometric interpretations. For example, the solution of a system of equations can be interpreted as a geometric figure in a space in which points are identified as all the possible values of the variables appearing in those equations. It is then natural to assert that systems of equations are qualitatively the same when their solutions, as geometric objects, are qualitatively the same. Using our circle and ellipse as examples, we can say that the equations $x^2 + y^2 = 1$ and $4x^2 + 9y^2 = 1$ are qualitatively the same and that both are qualitatively different from $x^2 - y^2 = 1$.

In the setting of differential equations (and S-systems in particular), solutions can be interpreted as t-parameterized curves in phase space. The family of all such solution curves is called a phase portrait. By a qualitative property of a system we mean therefore a qualitative property of its phase portrait. We say two systems are qualitatively the same whenever their phase portraits are qualitatively the same. For example, the S-systems

$$\frac{dx}{dt} = xy - x$$

$$\frac{dy}{dt} = xy - y$$

and

$$\frac{dx}{dt} = 3xy - x$$

$$\frac{dy}{dt} = 2xy - y$$

have similar phase portraits so it is natural to think of these systems as being qualitatively the same (see Fig. 16.1).

We are now faced with the task of finding a way to formalize the concept of qualitative similarity which is in accord with our intuition. In our example of the circle and the ellipse, a little algebra reveals that these figures are actually identical when they are compared in appropriate coordinate systems. When we change variables in the equation of the circle by the rule $x = 2\xi$, $y = 3\eta$ then the equation of the circle becomes the equation of the given ellipse. In contrast, there is no real continuous change of variables that will transform the equation of a circle into the equation of a hyperbola. In our S-system example the first system is identical to the second under the same change of variables $x = 2\xi$, $y = 3\eta$. In both of these examples the qualitative similarity exists because there are coordinate systems in which the objects are in fact identical; these qualitatively similar objects can be deformed into identical objects. In general, a qualitative property of an object will then be a property that does not depend on the choice of the coordinate system used to describe the object.

These examples suggest a way to study the qualitative properties of a system or a problem. State the problem in terms of coordinate geometry and then identify and investigate those properties that are invariant (unchanged) under all sufficiently smooth changes of coordinates. First, note that such properties are properties of the system *per se* since they don't depend upon the initial choice

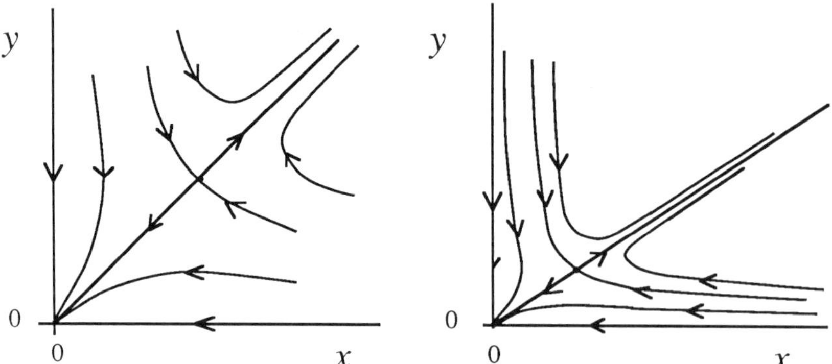

FIGURE 16.1. Two phase portraits that are qualitatively the same.

of coordinates we use to represent the problem. Second, such properties are not quantitative in that coordinate transformations may dramatically alter the quantitative relationships among the variables in our initial coordinate system. Third, such properties are compatible with our intuition about the qualitative similarity of shapes of figures and bodies in two or three dimensions.

In practice, we often try to determine whether or not a given system has a particular property that is known to be of interest. Since a qualitative property doesn't depend on our choice of coordinates, one way to made the determination is to find a convenient coordinate system in which the property is apparent. Often this amounts to placing the system in a canonical form, as when we diagonalize a symmetric matrix and display its eigenvalues on the main diagonal. In this scheme, the proof that a system has a particular property is accomplished by demonstrating the existence of a sufficiently smooth change of coordinates that places the system in a canonical form.

In the next section we will look at key concepts from the qualitative theory of ordinary differential equations. Our aim is to identify some useful qualitative properties of systems of equations. These properties will be properties of the phase portraits of solutions that are invariant under sufficiently smooth changes of coordinates.

SYSTEMS OF ORDINARY DIFFERENTIAL EQUATIONS

The starting point for the qualitative theory of systems of ordinary differential equations is the linear system

$$\frac{dx}{dt} = Ax$$

where

$$x = \begin{bmatrix} x_1 \\ x_2 \\ \vdots \\ x_n \end{bmatrix} \tag{16.1}$$

is a variable in R^n and A is an $n \times n$ real matrix.

For such a system, the generalized eigenspaces of A are invariant for solutions. That is, if a solution $x = \phi(t)$ is in an eigenspace, E, at some time, say $t = t_1$, then x is in E for all t. From the standpoint of the qualitative behavior of a linear system it is convenient to work with three other related invariant subspaces that are called the stable subspace, the unstable subspace, and the center subspace and are denoted E^s, E^u, and E^c respectively. These subspaces can be defined in the following way. Partition the set of all eigenvalues of A into three subsets: those with a negative real part, those with a positive real part, and those with a zero real part. For the subset of eigenvalues with a negative real part there is the associated collection of generalized eigenspaces. The vector subspace E^s is defined to be the span of the vectors in this entire collection of generalized eigenspaces. (Recall that the span of an empty subset is taken to be the trivial subspace of dimension 0.) For the subset of eigenvalues with a positive real part, collect the associated generalized eigenspaces and form E^u as the span of the vectors in this collection of eigenspaces. Similarly, E^c is the span of the generalized eigenspaces of A corresponding eigenvalues with a zero real part. As with the generalized eigenspaces, it is not difficult to see that E^s, E^u, and E^c are each invariant subspaces of Eq. 16.1 in that if a solution is in one of these subspaces at some time $t = t_1$ then the solution stays in that subspace for all t. Furthermore, for any A, $R^n = E^s \oplus E^u \oplus E^c$. That is, $E^s \cap E^u = E^u \cap E^c = E^c \cap E^s = 0$ and $E^s + E^u + E^c = R^n$ so that any x in R^n may be written uniquely as $x = x^s + x^u + x^c$ with x^s in E^s, x^u in E^u, and x^c in E^c (see Fig. 16.2).

These three subspaces characterize the dynamics of Eq. (16.1) in the following way. Any solution in $E^s - \{0\}$ converges to 0 as $t \to \infty$; solutions in the stable subspace move in to the origin. Any solution in $E^u - \{0\}$ diverges away from 0 as $t \to \infty$; nontrivial solutions in the unstable subspace move out from the origin. Solutions in E^c may be more complicated, as the following examples show.

Example 1

Let

$$\frac{dx}{dt} = \begin{bmatrix} 0 & 1 \\ 0 & 0 \end{bmatrix} x$$

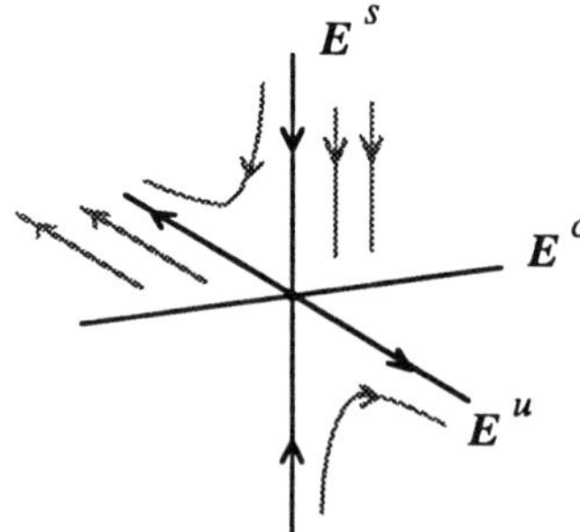

FIGURE 16.2. The stable, unstable, and center subspaces of a linear system $dx/dt = Ax$.

Here $E^c = R^2$. Particular solutions include

$$x(t) = \begin{bmatrix} 1 \\ 0 \end{bmatrix}$$

which is constant for all t and

$$x(t) = \begin{bmatrix} 0 \\ 1 \end{bmatrix} + t \begin{bmatrix} 1 \\ 0 \end{bmatrix}$$

which diverges as $t \to \infty$.

Example 2

Let

$$\frac{dx}{dt} = \begin{bmatrix} 0 & -1 \\ 1 & 0 \end{bmatrix} x$$

Here $E^c = R^2$ and for any constant $k \neq 0$

$$x(t) = k \begin{bmatrix} \cos(t) \\ \sin(t) \end{bmatrix}$$

is a bounded periodic solution.

Example 3

Let

$$\frac{dx}{dt} \begin{bmatrix} 0 & -1 & 0 & 0 \\ 1 & 0 & 0 & 0 \\ 0 & 0 & 0 & -\sqrt{2} \\ 0 & 0 & \sqrt{2} & 0 \end{bmatrix} x$$

Here $E^c = R^4$. The solution

$$x(t) = \begin{bmatrix} \sin(t) \\ \cos(t) \\ \sin(\sqrt{2}t) \\ \cos(\sqrt{2}t) \end{bmatrix}$$

is bounded but not periodic. This solution winds around a two-dimensional invariant torus, T, given by

$$T = \left\{ y \text{ in } R^4 \,\middle|\, y_1^2 + y_2^2 = 1 \text{ and } y_3^2 + y_4^2 = 1 \right\}$$

In fact, this solution curve is dense in T. (For any y in T there is a sequence $t_1 < t_2 < \ldots t_n < \ldots$ such that $\lim \|y - x(t_n)\| = 0$, for $n \to \infty$.)

Knowing about the dimensions of E^s, E^u, and E^c for a given linear system tells us quite a lot about the qualitative nature of its solutions, particularly when the dimension of E^c is 0. In this case a solution is *hyperbolic* in that it has a component, x^s, in E^s that converges to 0, and if it has a nonzero component, x^u, in E^u, then that component diverges. On the other hand, when the dimension of E^c is greater than 0 then any qualitatively interesting (complicated) behavior happens in E^c. In fact, any solution either diverges or converges to E^c as $t \to \infty$, so the solutions in E^c are characteristic of the long-term or *asymptotic behavior* of the system.

If we have a linear system, say $dx/dt = Ax$, with the stable, unstable, and center subspaces identified, then by introducing a nonlinear change of coordinates we would obtain a nonlinear system of equations and the invariant subspaces E^s, E^u, and E^c would map to curved surfaces or *manifolds* of identical dimensions. Let's call these manifolds the stable manifold, the unstable manifold, and the center manifold and denote them by W^s, W^u, and W^c. These manifolds inherit the qualitative properties of E^s, E^u, and E^c; namely, W^s, W^u, and W^c are invariant, solutions in W^s move into the equilibrium, solutions in W^u move away from the equilibrium, and solutions in W^c may have a more complicated behavior. There are no surprises here since W^s, W^u, and W^c are just E^s, E^u, and E^c presented in a new coordinate system. On the other hand it is a remarkable fact that in a neighborhood of an equilibrium of any (smooth) system of differential equations we can find *local* manifolds W^s_{loc}, W^u_{loc}, and W^c_{loc} that have these same qualitative properties. Furthermore, these local manifolds can be extended to global invariant manifolds, denoted W^s, W^u, and W^c, in such a way that $W^s \supset W^s_{loc}$, $W^u \supset W^u_{loc}$, and $W^c \supset W^c_{loc}$. The global manifolds W^s, W^u, and W^c, of a nonlinear system of differential equations may support complicated global dynamic behavior that cannot exist in a linear system (or a system that is linear in appropriate coordinates). It can happen, for example, that W^s and W^u intersect at some point other than the equilibrium, so that some solution that leaves a neighborhood of the equilibrium on W^u may eventually return to that neighborhood on W^s. In this chapter, we will focus on the behavior of solutions near the equilibrium, and when necessary, explicitly restrict our attention to the local manifolds W^s_{loc}, W^u_{loc}, and W^c_{loc}, which "look like" E^s, E^u, and E^c near an equilibrium. The existence of W^s, W^u, and W^c for a general system of first-order differential equations is insured by the *center manifold theorem*.

THE CENTER MANIFOLD THEOREM

The center manifold theorem describes a way in which the qualitative behavior of solutions of a general nonlinear system are inherited from a related linear system of equations. For clarifying discussions see [105, 205, 208]. For a proof see [207].

The following is an informal statement of the center manifold theorem. Suppose

$$\frac{dx}{dt} = f(x)$$

represents a system of n simultaneous first-order differential equations so that

$$x = \begin{bmatrix} x_1 \\ x_2 \\ \vdots \\ x_n \end{bmatrix} \text{ is in } R^n$$

and

$$f\colon R^n \to R^n \text{ with } f(x) = \begin{bmatrix} f_1(x) \\ f_2(x) \\ \vdots \\ f_n(x) \end{bmatrix}$$

Suppose f has $k > 0$ continuous derivatives and that $f(x_0) = 0$ so that $x = x_0$ is an equilibrium. Let A be the $n \times n$ matrix given by $A = Df(x_0)$ where

$$Df(x_0) = \begin{bmatrix} \dfrac{\partial f_1}{\partial x_1} & \dfrac{\partial f_1}{\partial x_2} & \cdots & \dfrac{\partial f_1}{\partial x_n} \\ \dfrac{\partial f_2}{\partial x_1} & & \cdots & \\ \vdots & & & \\ \dfrac{\partial f_n}{\partial x_1} & & \cdots & \dfrac{\partial f_n}{\partial x_n} \end{bmatrix}_{x = x_0}$$

The linear system $dx/dt = Ax$ is called the *linearization of* $dx/dt = f(x)$ *about* x_0. Let E^s, E^u, and E^c be the stable, unstable, and center subspaces of $dx/dt = Ax$. Suppose the dimensions of E^s, E^u, and E^c are p, q, and r, respectively. Then there exist stable, unstable, and center manifolds, denoted W^s, W^u, and W^c, such that

1. W^s, W^u, and W^c are each invariant for the solutions of $dx/dt = f(x)$,
2. The dimensions of W^s, W^u, and W^c are p, q, and r, respectively,
3. W^s, W^u, and W^c intersect at $x = x_0$ and are tangent to E^s, E^u, and E^c at $x = x_0$,
4. W^s and W^u are unique but W^c need not be unique.

The center manifold theorem tells us quite a lot about the qualitative nature of solutions near $x = x_0$ (see Fig. 16.3). If we restrict the differential equation to W^s, we see that $x = x_0$ is an equilibrium and that all the eigenvalues of the linearization of the restricted system have strictly negative real parts. Invoking the results of Lyapunov or Hartman [105] we find that any solution in W^s which is near $x = x_0$ at $t = 0$ converges to x_0 as $t \to \infty$. Similarly, if we restrict the differential equation to W^u, we see that all the eigenvalues of the linearization of the restricted system have strictly positive real parts. Therefore, a solution in W^u which is close to x_0 at $t = 0$ either leaves a neighborhood of x_0 or is the constant solutions $x = x_0$. On the other hand, if we restrict the differential equation to W^c, more complicated things may happen near x_0. In particular, the linearization of the restricted system tells us nothing about the stability of x_0 in W^c.

Let us now consider what happens to solutions in the original unrestricted system. Because of the continuity of the system, the trajectory of a point that

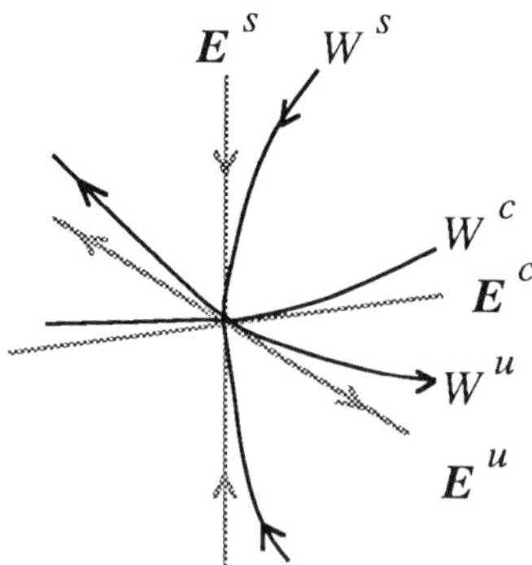

FIGURE 16.3. The stable, unstable, and center manifolds of an equilibrium of a nonlinear system are tangent to the stable, unstable, and center subspaces of the linearized system.

starts near x_0 is steered by the flow in any nearby invariant manifold. So let N be a small neighborhood of x_0, and suppose x is in N at $t = 0$. If x is close enough to W^u at $t = 0$, then x leaves N at some $t > 0$ point. The flow in W^u "pulls" x out of N. If x is close enough to W^s at $t = 0$, then W^s "pushes" x toward W^c until either x converges on W^c or until x gets close enough to W^u to be pulled out of N. If x is close to W^c at $t = 0$, then either x is pushed onto W^c (because of W^s) or pulled out of N (because x gets close enough to W^u). This means that solutions that stay in N for a long time must be approaching solutions in W^c, so that the interesting (nonhyperbolic) *large-time* behavior near x_0 is characterized by the behavior of solutions in W^c. The special case when the dimension of W^u equals 0 is particularly significant because then all solutions in N approach W^c for the $t > 0$ point.

The center manifold theorem guides our analysis of the qualitative behavior of solutions near an equilibrium. Since all solutions near an equilibrium are either approaching the center manifold or leaving a neighborhood of the equilibrium, our primary task is to analyze the behavior of solutions in the center manifold. Knowledge of the behavior of solutions in W^c together with what we already know about the hyperbolic behavior in W^s and W^u allows us to describe the behavior of all solutions near an equilibrium. Knowing the behavior in W^c "fills in" the phase portrait.

A stunning practical consequence of the center manifold theorem is that, by restricting our attention to solutions in W^c, we can reduce the dimension of the system we are to analyze to the dimension of W^c. For example, suppose that our original system comprises 100 simultaneous equations but that the dimension of W^c is 2. We know that all the interesting behavior (near the equilibrium) is taking place in W^c. To analyze this behavior, we restrict our system of 100 equations to its action on W^c which, in the appropriate coordinates, will be given by a system of two first-order equations; the two dependent variables will be coordinates for points in W^c.

A DIRECT NOTATION FOR S-SYSTEMS

In order to ease the presentation of the material in upcoming sections, I wish to introduce the reader to a convenient *direct* notation for S-systems. This notation facilitates many routine computations with S-systems and frequently produces a clean and "digestible" display of results. I call this notation "direct" by analogy with the terminology of tensor analysis where the notation using polyadic expressions allows one to express relations among tensors directly, instead of expressing the relations indirectly, using the components of the tensors in a particular coordinate system. For S-systems, the idea is to have a notation in which the explicit display of indices of the variables and exponents can be suppressed, but which still admits the expressive and computational power afforded by those indices. This introduction to the notation will be as informal and as brief as possible. For that reason it may require some initiative on the part of the reader. I will present the notation in small pieces and each piece will be followed by one or more observations. I suggest that the reader verify the observations and thereby acquire the notation. Once the notation is described I will use it to define *regular* S-systems and the $p + q$ *expansion* of an S-system. As an illustration in the use of the notation, I will solve for all the equilibria of a regular S-system and describe two canonical forms for any regular S-system that has an equilibrium.

Basic Definitions

Let $x = \begin{bmatrix} x_1 \\ \vdots \\ x_n \end{bmatrix}$ and $y = \begin{bmatrix} y_1 \\ \vdots \\ y_n \end{bmatrix}$ be in R^n. The notations $x > 0$, $x \geq 0$, $x < 0$, $x \leq 0$, and $x \neg 0$ mean, respectively, $x_i > 0$, $x_i \geq 0$, $x_i < 0$, $x_i \leq 0$, and $x_i \neq 0$ for each $i = 1, \ldots, n$. The notation $x \neq 0$ means $x_i \neq 0$ for some $i = 1, \ldots, n$.

Let $A = \begin{bmatrix} a_{11} & a_{12} & \ldots & a_{1n} \\ a_{21} & a_{22} & \ldots & a_{2n} \\ \vdots & & & \\ a_{m1} & a_{m2} & \ldots & a_{mn} \end{bmatrix}$ be an $m \times n$ matrix with real components.

If B is an $m \times n$ matrix with real components, we'll say that B is in $R^{m \times n}$. If B is in $R^{n \times p}$, then AB is the usual matrix product of A and B so that AB is in $R^{m \times p}$.

For any x in R^n, let $[x]$ denote the $n \times n$ diagonal matrix that has the components of x along its main diagonal. That is,

$$[x] = \begin{bmatrix} x_1 & 0 & \ldots & 0 \\ 0 & x_2 & \ldots & 0 \\ \vdots & & & \\ 0 & 0 & \ldots & x_n \end{bmatrix}$$

so that $[x]$ is in $R^{n \times n}$. For any x in R^n and y in R^n, define xy in R^n by the rule $[xy] = [x]\,[y]$, where $[x]\,[y]$ is the usual matrix product of $[x]$ and $[y]$. We will call xy the *S-product* of x and y. If $y \neg 0$, define x/y by the rule $[x/y] = [x][y]^{-1}$ and define y^{-1} by the rule $[y^{-1}] = [y]^{-1}$. Observe that the S-product is associative and commutative and that if $y \neg 0$ then $x/y = x(y^{-1}) = (y^{-1})x$.

Let the S-product take precedence over the *matrix by vector* product so if B is in $R^{m \times n}$ by $Bxy = B(xy)$ by definition. Observe that in general $B(xy) \neq (Bx)y$. However, if z is in R^n and $B = [z]$, then $B(xy) = [z](xy) = z(xy) = (zx)y = ([z]x)y = (Bx)y$.

Next, we define for $x > 0$ in R^n and A in $R^{m \times n}$ the *S-exponential* of x by A as a vector of power-law terms in which the i-th component consists of a product of all X_j, raised to a_{ij}:

$$X^A = \begin{bmatrix} X_1^{a_{11}} \cdot X_2^{a_{12}} \cdot \ldots \cdot X_n^{a_{1n}} \\ X_1^{a_{21}} \cdot X_2^{a_{22}} \cdot \ldots \cdot X_n^{a_{2n}} \\ \vdots \\ X_1^{a_{m1}} \cdot X_2^{a_{m2}} \cdot \ldots \cdot X_n^{a_{mn}} \end{bmatrix}$$

The S-exponential in many ways behaves like a scalar exponential. For instance, if B is in $R^{m \times n}$, then $x^{(A+B)} = x^A x^B$. However, because matrix multiplication in general is not commutative, the order of matrix exponents cannot be ignored. In particular, if B is in $R^{p \times m}$, then $(x^A)^B = x^{(BA)}$ which is in R^p. S-exponentials take precedence over S-products so that $yx^A = y(x^A)$ by definition. In the absence of parentheses, one evaluates the exponent before exponentiation: $x^{BA} = x^{(BA)} \neq (x^B)^A$. Observe that in general $y(x^A) \neq (yx)^A$, but that $(xy)^A = (x^A)(y^A) = x^A y^A$. If I is the identity matrix in $R^{n \times n}$, then $x^I = x$, $x^{-I} = x^{-1}$, and $x^{-A} = (x^A)^{-1}$. If G is in $R^{n \times n}$ then $x^{-1} x^G = x^{G-I}$.

Suppose G is in $R^{n \times n}$, G^{-1} exists, $x > 0$, and $y > 0$. If $x^G = y$, then $y = x^{G^{-1}}$, since $(x^G)^{G^{-1}} = y^{G^{-1}}$, $x^{G^{-1}G} = y^{G^{-1}}$, $x^I = y^{G^{-1}}$, and $x = y^{G^{-1}}$.

Finally, we define functions of the diagonal matrix $[x]$ in terms of functions of the vector x. Let R^+ be defined as $R^+ = \{t \text{ in } R \mid t > 0\}$. If f: $R^+ \to R$ or f: $R \to R$, extend the definition of f to R^n by letting f act on components. That is,

$$f(x) = \begin{bmatrix} f(x_1) \\ f(x_2) \\ \vdots \\ f(x_n) \end{bmatrix}$$

when x is in R^n and each x_i is in the domain of f. Observe that $\ln(\exp(x)) = x$ and $\exp(\ln(y)) = y$ when $y > 0$. Let e be Euler's number. If e is interpreted as an element of R^1 and x in R^n is interpreted as an element of $R^{n \times 1}$, then $e^x = \exp(x)$ for all x in R^n. If G is in $R^{n \times n}$, $x > 0$, and $y > 0$, then $\ln(x^G) = G \ln(x)$ and $e^{G \ln x} = x^G$. Also, $x^G = y$ implies $G \ln(x) = \ln(y)$.

If u is a real variable and a is real number, then f: $R^+ \to R^+$, defined by $f(u) = u^a$ when $u > 0$, extends to real exponentiation of a vector by the rule

$$f(x) = x^a = \begin{bmatrix} x_1^a \\ x_2^a \\ \vdots \\ x_n^a \end{bmatrix}$$

We then have $x^a = x^{aI}$, for $x > 0$, and $xx = x^2$, $xxx = x^3$, . . . , for all x in R^n.

The definition of $f(x)$ extends directly to $f([x])$ via the rule $f([x]) = [f(x)]$. As an important example, we write the derivative of the S-exponential, $f(x) = x^A$, as $Df(x) = [x^A]\, A\, [x]^{-1}$.

To indicate the usefulness and power of the direct notation, let's describe S-systems and demonstrate why the S-product and S-exponentiation are natural for S-systems. First of all, we can represent an S-system entirely index free, without losing information:

An *n-dimensional S-system* is a system of the form

$$\frac{dx}{dt} = \alpha x^G - \beta x^H, \ x \geq 0$$

where x, α, and β are in R^n with $\alpha \geq 0$ and $\beta \geq 0$, and where G, H are in $R^{n \times n}$. It is convenient to distinguish the special case in which $x > 0$; we call such an S-system *regular*:

A *regular* n-dimensional S-system is a system of the form

$$\frac{dx}{dt} = \alpha x^G - \beta x^H, \ x > 0$$

The *domain* of the general S-system and the *domain* of a regular S-system are defined to be $\{x \text{ in } R^n \,|\, x \geq 0\}$ and $\{x \text{ in } R^n \,|\, x > 0\}$, respectively.

It is advantageous to distinguish these two cases because regular S-systems are analytic at each point in their domain while, in general, S-systems need not even be continuous in their domain. These definitions do *not* say that the solutions of a regular S-system remain in $\{x \text{ in } R^n/x > 0\}$ for all time or that the solutions of a general S-system remain in $\{x \text{ in } R^n/x \geq 0\}$ for all time. They say that we only consider a solution of an S-system to exist during the time interval for which it remains in its domain. Computations proceed more smoothly in regular S-systems. For example, ln (x) and x^{-1} always exist in regular S-systems. Since ESSYNS uses the Taylor series (cf. Chapter 7), ESSYNS solves regular S-systems.

To illustrate the use of the direct notation in a computation, let's consider the problem of determining the equilibria of a regular S-system. We want to find all x such that $\alpha x^G - \beta x^H = 0$ and $x > 0$. We can easily solve for x when $\alpha > 0$ as follows:

$$\begin{aligned} \alpha x^G &= \beta x^H \\ x^G x^{-H} &= (\beta/\alpha) \\ x^{G-H} &= (\beta/\alpha) \\ x &= (\beta/\alpha)^{(G-H)^{-1}}, \text{ provided det } (G - H) \neq 0, \text{ or} \\ (G - H) \ln (x) &= \ln (\beta/\alpha), \text{ in any case.} \end{aligned}$$

When det $(G - H) = 0$, the theory of linear algebra tells us that this last equation either has an infinity of solutions for ln (x) or else no solution for ln (x), depending on whether or not ln (β/α) is in the range of $(G - H)$. So suppose det $(G - H) = 0$ and ln (β/α) is in the range of $(G - H)$. Let r be the rank of $(G - H)$ and let $k = n - r$ be the corank of $(G - H)$. Then by linear algebra we know that there is a vector v_0 in R^n and a linearly independent set of k vectors in R^n, $\{v_1, \ldots, v_k\}$, such that the general solution for ln (x) is given by

$$\ln (x) = v_0 + c_1 v_1 + c_2 v_2 + \ldots + c_k v_k$$

with k arbitrary real parameters, $c_1, c_2, \ldots, c_k$.

Exponentiating we find

$$\begin{aligned} x &= e^{v_0 + c_1 v_1 + \ldots c_k v_k} \\ x &= e^{v_0} e^{c_1 v_1} e^{c_2 v_2} \ldots e^{c_k v_k} \\ x &= u_0 u_1^{c_1} u_2^{c_2} \ldots u_k^{c_k} \end{aligned}$$

where

$$u_i = e^{v_i} \qquad \text{for } i = 0, 1, \ldots, k$$

This shows that the set of equilibria is a k parameter surface in R^n which passes through u_0 when $c_i = 0$ for $i = 1, 2, \ldots, k$. When we redefine terms we see that the set of equilibria may be written as

$$x = u_0 \sigma_1^{v_1} \sigma_2^{v_2} \ldots \sigma_k^{v_k}, \text{ where } \sigma_i = e^{c_i} \text{ and } u_0 = e^{v_0}$$

Now using direct notation for the σ_i and v_i where

$$\sigma = \begin{bmatrix} \sigma_1 \\ \sigma_2 \\ \vdots \\ \sigma_k \end{bmatrix} \text{ is in } R^k \text{ and } V = [v_1 v_2 \ldots v_k] \text{ is in } R^{n \times k}$$

we can condense the expression for the set of equilibria into the simple formula

$$x = u_0 \sigma^V$$

In summary, for a regular S-system with $\alpha > 0$, let k be the corank of $(G - H)$. If $k = 0$, then there is a unique equilibrium given by

$$x = (\beta/\alpha)^{(G-H)^{-1}}$$

If $k > 0$, then either there is no equilibrium or there is a k-dimensional surface of equilibria that is parameterized by σ in R^k, $\sigma > 0$, and given by

$$x = u_0 \sigma^V$$

Here u_0 is an element of R^n such that $(G - H) \ln (u_0) = \ln (\beta/\alpha)$, and V is any matrix in $R^{n \times k}$ such that V has rank k and $(G - H)V = 0$.

The Decomposition of S-Systems

Any S-system where $n > 1$ may be *expanded* into two coupled S-systems of dimensions p and q where $p + q = n$, $p > 0$, and $q > 0$. Any coupled pair of S-systems may be *condensed* into a single S-system. Here is how. Assume $n > 1$, and let $n = p + q$ with $p > 0$ and $q > 0$. First, let's expand a vector. Let v be in R^n, let v_1 be in R^p, and let v_2 be in R^q. Define the statement

$$v = \begin{bmatrix} v_1 \\ v_2 \end{bmatrix}$$

to mean that v_1 is the vector of the first p components of v, and v_2 is the vector of the last q components of v. Next, let's expand a square matrix. Let A be in $R^{n \times n}$ with real components a_{ij}. Let A_{11} be in $R^{p \times p}$, let A_{12} be in $R^{p \times q}$, let A_{21} be in $R^{q \times p}$, and let A_{22} be in $R^{q \times q}$. Define the statement

$$A = \begin{bmatrix} A_{11} & A_{12} \\ A_{21} & A_{22} \end{bmatrix}$$

to mean that

$$A_{11} = \begin{bmatrix} a_{11} & \cdots & a_{1p} \\ \vdots & & \\ a_{p1} & \cdots & a_{pp} \end{bmatrix},$$

$$A_{12} = \begin{bmatrix} a_{1p+1} & \cdots & a_{1n} \\ \vdots & & \\ a_{pp+1} & \cdots & a_{pn} \end{bmatrix}$$

$$A_{21} = \begin{bmatrix} a_{p+11} & \cdots & a_{p+1p} \\ \vdots & & \\ a_{n1} & \cdots & a_{np} \end{bmatrix},$$

$$A_{22} = \begin{bmatrix} a_{p+1p+1} & \cdots & a_{p+1n} \\ \vdots & & \\ a_{np+1} & \cdots & a_{nn} \end{bmatrix}$$

A remark is in order: We may no longer assume that a subscripted object is a real number; it may also be either a matrix or a vectór. In any context, the dimension and shape of a subscripted object will be determined either explicitly or implicitly. For example, if x_1 is in R^4 and x_2 is in R^3, then $\begin{bmatrix} x_1 \\ x_2 \end{bmatrix}$ is in R^7 and not in R^2.

With these definitions we have all the necessary tools to expand an S-system. Let

$$\frac{dx}{dt} = \alpha x^G - \beta x^H, \ x \geq 0$$

be an n-dimensional S-system where $n > 1$. Let x_1, α_1, and β_1 be in R^p, let x_2, α_2, and β_2 be in R^q, let G_{11} and H_{11} be in $R^{p\times p}$, let G_{12} and H_{12} be in $R^{p\times q}$, let G_{21} and H_{21} be in $R^{q\times p}$, and let G_{22} and H_{22} be in $R^{q\times q}$, where $p > 0$, $q > 0$, and $p + q = n$. Letting

$$x = \begin{bmatrix} x_1 \\ x_2 \end{bmatrix}, \begin{bmatrix} x_1 \\ x_2 \end{bmatrix} \geq 0$$

and

$$\alpha = \begin{bmatrix} \alpha_1 \\ \alpha_2 \end{bmatrix}, \beta = \begin{bmatrix} \beta_1 \\ \beta_2 \end{bmatrix}, G = \begin{bmatrix} G_{11} & G_{12} \\ G_{21} & G_{22} \end{bmatrix}, H = \begin{bmatrix} H_{11} & H_{12} \\ H_{21} & H_{22} \end{bmatrix}$$

we can write the S-system as

$$\frac{d}{dt}\begin{bmatrix} x_1 \\ x_2 \end{bmatrix} = \begin{bmatrix} \alpha_1 \\ \alpha_2 \end{bmatrix}\begin{bmatrix} x_1 \\ x_2 \end{bmatrix}^{\begin{bmatrix} G_{11} & G_{12} \\ G_{21} & G_{22} \end{bmatrix}} - \begin{bmatrix} \beta_1 \\ \beta_2 \end{bmatrix}\begin{bmatrix} x_1 \\ x_2 \end{bmatrix}^{\begin{bmatrix} H_{11} & H_{12} \\ H_{21} & H_{22} \end{bmatrix}}$$

Grinding out the definitions of S-product and S-exponentiation in terms of the p components of x_1 and the q components of x_2, we find that this is identical to the following coupled pair of S-systems:

$$\begin{aligned} \frac{dx_1}{dt_2} &= \alpha_1 x_1^{G_{11}} x_2^{G_{12}} - \beta_1 x_1^{H_{11}} x_2^{H_{12}} \\ & \qquad\qquad\qquad\qquad , x_1 \geq 0, x_2 \geq 0 \\ \frac{dx_2}{dt} &= \alpha_2 x_1^{G_{21}} x_2^{G_{22}} - \beta_2 x_1^{H_{21}} x_2^{H_{22}} \end{aligned}$$

We call this coupled system the $p + q$ *expansion* of the original system. The $p + q$ expansion has the same form as the scalar equations of a two-dimensional S-system, which again indicates that the direct notation is natural for S-systems. One can think of any $p + q$ expansion as a type of generalized S-system in which the α_i, β_i, and x_i are vectors and the G_{ij} and H_{ij} are matrices of the appropriate sizes. It is not too difficult to see that if $n > 2$ then there is a natural $p + q + r$ *expansion* which is possible for $p > 0$, $q > 0$, $r > 0$, and $p + q + r = n$, and so on. The $1 + 1 + \ldots + 1$ expansion of an S-system is identical to its scalar form. Reversing the expansion process, we can *condense* the coupled differential equations of a $p + q$ expansion into the single equation of a $(p + q)$-dimensional S-system.

The $p + q$ expansion allows us to compute the equilibria of a regular

S-system where $\alpha \geq 0$. Without loss of generality assume that the first p components of α are strictly positive and that the remaining $q = n - p$ are all 0. (If not, just renumber the equations.) If $p = 0$, then the system has no equilibria unless $\beta = 0$, in which case each point in the domain of the S-system is an equilibrium. If $p = n$, then $\alpha > 0$, which is the case we just analyzed. So assume $p > 0$, $q > 0$, and $p + q = n$. Performing a $p + q$ expansion of the system we see that $\alpha_2 = 0$, so the expansion has the form

$$\begin{aligned} \frac{dx_1}{dt} &= \alpha_1 x_1^{G_{11}} x_2^{G_{12}} - \beta_1 x_1^{H_{11}} x_2^{H_{12}} \\ & \qquad\qquad\qquad\qquad , x_1 > 0,\ x_2 > 0 \\ \frac{dx_2}{dt} &= \qquad\qquad\quad - \beta_2 x_1^{H_{21}} x_2^{H_{22}} \end{aligned}$$

This regular system has no equilibrium unless $\beta_2 = 0$. Assume $\beta_2 = 0$, so that the second equation in the expansion is just $dx_2/dt = 0$. Then, a $p + q$ expansion with identical solutions is given by

$$\begin{aligned} \frac{dx_1}{dt} &= \alpha_1 x_1^{G_{11}} x_2^{G_{12}} - \beta_1 x_1^{H_{11}} x_2^{H_{12}} \\ & \qquad\qquad\qquad\qquad , x_1 > 0,\ x_2 > 0 \\ \frac{dx_2}{dt} &= \gamma x_1^0 x_2^0 \qquad - \gamma x_1^0 x_2^0 \end{aligned}$$

where γ is in R^q and $\gamma > 0$. So condense this $p + q$ expansion into an n-dimensional S-system

$$\frac{dx}{dt} = \tilde{\alpha} x^{\tilde{G}} - \tilde{\beta} x^{\tilde{H}},\ x > 0$$

whose solutions are identical to the solutions of the original systems. In particular, they have the same equilibria. Looking for the equilibria of this equivalent system, we find that $\tilde{\alpha} > 0$, and that is a case we have already analyzed.

Focusing on the geometry, let's summarize what we've found. In a regular n dimensional S-system exactly one of the following possibilities occurs:

1. There are no equilibria.
2. There is a unique isolated equilibrium.
3. There is a k-parameter surface of equilibria $0 < k \leq n$.

A Canonical Form for S-Systems

It is obvious that changes in coordinates affect the parameters of an S-system. Using direct notation, we now introduce changes in coordinates that simplify the S-system structure. In contrast to the recasting of S-systems as Half-systems or Binary systems (cf. Chapter 12), we only allow transformations that do not increase the dimension of the system.

Suppose that u is in R^n and that $x = u$ is an equilibrium of a regular n-dimensional S-system:

$$\frac{dx}{dt} = \alpha x^G - \beta x^H,\ x > 0$$

Define y in R^n by the equation $x = uy$. Then y satisfies the S-system

$$\frac{dy}{dt} = \gamma(y^G - y^H),\ y > 0$$

where $\gamma = \alpha u^{(G-I)}$, γ in R^n. To see this, just substitute uy for x in the original system and simplify as follows:

$$\frac{duy}{dt} = \alpha(uy)^G - \beta(uy)^H$$

$$u\frac{dy}{dt} = \alpha u^G y^G - \beta u^H y^H$$

But u is an equilibrium, so $\alpha u^G = \beta u^H$, and we can factor out this common term on the right-hand side to obtain

$$u\frac{dy}{dt} = \alpha u^G(y^G - y^H)$$

Now because $u^{-1} = u^{-I}$ exists (u is an equilibrium of a regular S-system so $u > 0$) and because the S-product commutes, we have

$$\frac{dy}{dt} = \alpha u^G u^{-1}(y^G - y^H)$$

The addition rule for S-exponents now yields the desired form:

$$\frac{dy}{dt} = \alpha u^{G-I}(y^G - y^H)$$

Note that $\gamma > 0$ whenever $\alpha > 0$. In algebraic terms, this change of coordinates eliminates n parameters from the original S-system and places the S-system in a simpler form where $\alpha = \beta$. In geometric terms, this change of coordinates rescales the system and places the equilibrium at the location

$$y = \begin{bmatrix} 1 \\ 1 \\ \vdots \\ 1 \end{bmatrix}$$

in the new coordinate system.

Now suppose $\gamma > 0$. Define z in R^n by the equation

$$y = z^{[\gamma]}$$

Then z satisfies the S-system

$$\frac{dz}{dt} = z^{\hat{G}} - z^{\hat{H}}, \; z > 0$$

where

$$\hat{G} = (G - I)[\gamma] + I \text{ and } \hat{H} = (H - I)[\gamma] + I$$

To see this, substitute and simplify as follows:

$$\frac{dz^{[\gamma]}}{dt} = \gamma\left((z^{[\gamma]})^G - (z^{[\gamma]})^H\right)$$

Using our rule for differentiating S-exponentials, the chain rule, and our rule for multiplication of S-exponents, we obtain:

$$[z^{[\gamma]}][\gamma][z]^{-1}\frac{dz}{dt} = \gamma(z^{G[\gamma]} - z^{H[\gamma]})$$

In a regular S-system with $\gamma > 0$, each matrix on the left-hand side exists and has an inverse. So move each to the right-hand side and simplify:

$$[\gamma][z]^{-1}\frac{dz}{dt} = [z^{[\gamma]}]^{-1}\gamma(z^{G[\gamma]} - z^{H[\gamma]})$$

$$[\gamma][z]^{-1}\frac{dz}{dt} = z^{-[\gamma]}\gamma(z^{G[\gamma]} - z^{H[\gamma]})$$

$$[\gamma][z]^{-1}\frac{dz}{dt} = \gamma(z^{(G-I)[\gamma]} - z^{(H-I)[\gamma]})$$

Now use the fact that

$$[\gamma]^{-1}\gamma = \begin{bmatrix} 1 \\ 1 \\ \vdots \\ 1 \end{bmatrix}$$

to get

$$[z]^{-1}\frac{dz}{dt} = z^{(G-I)[\gamma]} - z^{(H-I)[\gamma]}$$

Finally, multiply both sides by $[z]$ and obtain

$$\frac{dz}{dt} = z^{(G-I)[\gamma]+I} - z^{(H-I)[\gamma]+I}$$

or

$$\frac{dz}{dt} = z^{\hat{G}} - z^{\hat{H}}$$

Even when $\gamma \geq 0$ one can show that a regular S-system with an equilibrium is equivalent to a system of this form. Suppose $\gamma \geq 0$. Without loss of generality, assume that the first p components of γ are all positive and that the last $q = n - p$ components are identical with 0. (If not, just renumber the equations.) If $p = n$ then $\gamma > 0$, which is the case we just treated. If $p = 0$ the system is equivalent to the system

$$\frac{dy}{dt} = 0$$

which is equivalent to

$$\frac{dy}{dt} = y^0 - y^0$$

so if $y = z$, $\hat{G} = 0$, and $\hat{H} = 0$ the system has the form

$$\frac{dz}{dt} = z^{\hat{G}} - z^{\hat{H}}$$

If $0 < p < n$, then let $q = n - p$. If we perform a $p + q$ expansion, we see that $\gamma_2 = 0$ so that this expansion has the form

$$\frac{dy_1}{dt} = \gamma_1 y_1^{G_{11}} y_2^{G_{12}} - \gamma_1 y_1^{H_{11}} y_2^{H_{12}}$$

$$\frac{dy_2}{dt} = 0$$

But this system has the same solutions as the $p + q$ expansion

$$\frac{dy_1}{dt} = \gamma_1 y_1^{G_{11}} y_2^{G_{12}} - \gamma_1 y_1^{H_{11}} y_2^{H_{12}}$$

$$\frac{dy_2}{dt} = \rho_2 y_1^0 y_2^0 - \rho_2 y_1^0 y_2^0$$

where

$$\rho_2 = \begin{bmatrix} 1 \\ 1 \\ \vdots \\ 1 \end{bmatrix}$$

is in R^q. Condensing this new system we obtain

$$\frac{dy}{dt} = \tilde{\gamma}(y^{\tilde{G}} - y^{\tilde{H}})$$

where

$$\tilde{\gamma} = \begin{bmatrix} \gamma_1 \\ \rho_2 \end{bmatrix}, \ \tilde{G} = \begin{bmatrix} G_{11} & G_{12} \\ 0 & 0 \end{bmatrix}, \text{ and } \tilde{H} = \begin{bmatrix} H_{11} & H_{12} \\ 0 & 0 \end{bmatrix}$$

Now letting

$$y = z^{[\tilde{\gamma}]},\ \hat{G} = (\tilde{G} - I)[\tilde{\gamma}] + I, \text{ and } \hat{H} = (\tilde{G} - I)[\tilde{\gamma}] + I$$

and using the fact that $\tilde{\gamma} > 0$, we obtain

$$\frac{dz}{dt} = z^{\hat{G}} - z^{\hat{H}}$$

In summary, any regular n-dimensional S-system that has an equilibrium is equivalent, under an appropriate change of coordinates, to a regular n-dimensional S-system of the form

$$\frac{dx}{dt} = x^{G} - x^{H}$$

APPLICATIONS OF THE CENTER MANIFOLD THEOREM TO S-SYSTEMS

In this section we discuss the center manifold theorem in the context of any regular S-system that has an equilibrium. Since we are studying qualitative properties, we are free to choose any convenient coordinate system. Given any particular S-system, we may as well choose the "z" coordinates given at the end of the previous section and study the equivalent S-system of the form

$$\frac{dx}{dt} = x^{G} - x^{H},\ x > 0$$

where the equilibrium in question is located at

$$x = \begin{bmatrix} 1 \\ 1 \\ \vdots \\ 1 \end{bmatrix}$$

Linearizing this equation, we find

$$Df(x) = D(x^{G} - x^{H}) = [x^{G}]G[x]^{-1} - [x^{H}]H[x]^{-1}$$

so that at the equilibrium

$$A = Df(x_0) = G - H$$

To characterize the center manifold of the equilibrium, we now consider three cases. We will discuss a fourth case in a later section.

1. *$G - H$ has no eigenvalue whose real part is zero*. In this case, (1) the equilibrium is the only equilibrium of the S-system, and (2) it is hyperbolic—that is, dim $(W^c) = 0$.

This is easy to see. Since $G - H$ has no eigenvalue whose real part is 0, it has no eigenvalue that is equal to zero. Therefore det $(G - H) \neq 0$. Therefore, $(G - H)^{-1}$ exists. Therefore the equilibrium is unique (as described in the previous section). But $G - H$ is also the linearization of the system about this isolated equilibrium. Since no eigenvalue of the linearization has a real part equal to zero, the center manifold theorem says that dim $(W^c) = r = 0$.

2. *$G - H$ has exactly one zero eigenvalue and no other eigenvalue whose real part is zero*. In this case, (1) there is a one dimensional curve of equilibria passing through the given equilibrium, and (2) this curve is a center manifold of the given equilibrium.

Since $G - H$ has exactly one zero eigenvalue, the kernel of $G - H$ is one-dimensional. Let v be in R^n and $v \neq 0$ be in the kernel of $G - H$. From our work in the previous section, we know that all equilibria of the system are given by the μ-parameterized curve $x = \mu^v$ where μ is in R and $\mu > 0$. When $\mu = 1$, we have the given equilibrium. Let $W = \{x \text{ in } R^n/x = \mu^v, \mu \text{ is in } R^+\}$ be the one-dimensional curve of equilibria described in (1). To see that $W = W^c$, observe that W intersects the given equilibrium and that (since every point of W is invariant) W is invariant. So we only need to check that dim $(W) =$ dim (E^c) and that W is tangent to E^c at the equilibrium. Since $G - H$ is the linearization of the system about the given equilibrium and since it has exactly one zero eigenvalue and no other eigenvalue whose real part is zero, it must be that E^c is the one-dimensional kernel of $G - H$. Therefore dim $(W) =$ dim (E^c). Furthermore, the tangents to W at any point in W are all the real scalar multiples of

$$\frac{dx}{d\mu} = \frac{d\mu^v}{d\mu} = \frac{\mu^v}{\mu} v$$

so that at the given equilibrium, where $\mu = 1$, the tangents to W are all the scalar multiples of v. But v is in the kernel of $G - H$, so these tangents are the kernel of $G - H$. Therefore these tangents are in E^c. Therefore W is tangent to E^c at the given equilibrium. Therefore, as we see in (2), $W = W^c$.

It is natural to wonder whether all the equilibria that are in W have W as a center manifold. The answer is no. It is not too difficult to construct examples where some equilibria in W have a higher dimensional center manifold that contains W. However, all the equilibria in W that are close enough to the given equilibrium have W as a center manifold.

It is not difficult to see that if an equilibrium of a regular S-system has a one-dimensional center manifold, then it has a center manifold that is a curve of equilibria.

3. *$G - H$ has a zero eigenvalue that is repeated exactly k times and it has no other eigenvalue whose real part is zero.* Suppose that k is the corank of $G - H$ (so there are k linearly independent eigenvectors for the zero eigenvalue). Then, (1) there is a k-dimensional "surface" or manifold of equilibria that intersects the given equilibrium, and (2) this manifold is a center manifold of the given equilibrium.

Comment: This can be justified by the same line of argument given for the previous fact, which is really the special case when $k = 1$. If the corank of $G - H$ is less than k, then there will still be a k-dimensional center manifold, but this manifold will not consist entirely of equilibria.

In the case when $G - H$ has a single pair of purely imaginary eigenvalues and no other eigenvalues whose real parts are zero, there is a two-dimensional center manifold. We will treat this case when we discuss the Hopf bifurcation.

PARAMETERIZED SYSTEMS AND BIFURCATIONS

One of the main tasks of the qualitative theory of differential equations is the analysis of systems that contain parameters. The problem is to determine if changes in the parameters of the system affect any of the qualitative properties of its solutions. Such changes in qualitative properties are called *bifurcations*. For technical reasons, a value of the parameters is called a bifurcation value whenever the system is not *structurally stable* for that value of the parameters, that is, whenever some arbitrarily small perturbation of the system yields a system that is qualitatively different from the unperturbed system. This definition admits the possibility that the system has a bifurcation at some value of the parameters even though there is no qualitative change in the phase portrait of the solutions at that value of the parameters. On the other hand, structural stability itself is a qualitative property of any system of differential equations and so the definition is natural.

The simplest cases involve a single real parameter, μ, and we can write the system as

$$\frac{dx}{dt} = f_\mu(x)$$

where μ is in R and x is in R^n. Among the bifurcations we may encounter when we vary μ are those involving changes in the number or stability of equilibria and those involving the creation or annihilation of more complicated behavior, such as periodic or chaotic motion. In an S-system, μ might be one of the exponents, say g_{ij}. The problem would be to determine how the structure of solutions depends on the value of g_{ij}. For instance, if the S-system is of the form

$$\frac{dx}{dt} = x^G - x^H, \; x > 0$$

and has an isolated equilibrium at

$$x = \begin{bmatrix} 1 \\ 1 \\ \vdots \\ 1 \end{bmatrix}$$

for some value g_{ij}, say $g_{ij} = g^*$, then we would like to know at least what happens to the equilibrium and its invariant manifolds for the nearby values of g_{ij}. In this case we know, since det $(G - H)$ is a continuous function of g_{ij} and det $(G - H) \neq 0$ at $g_{ij} = g^*$, that det $(G - H) \neq 0$ for all values of g_{ij} near g^*. This means that

$$x = \begin{bmatrix} 1 \\ 1 \\ \vdots \\ 1 \end{bmatrix}$$

is an isolated equilibrium for all values of g_{ij} near g^*. A similar continuity argument shows that if none of the eigenvalues of $G - H$ has a zero real part when $g_{ij} = g^*$, then none of the eigenvalues has a zero real part when g_{ij} is near g^*. This means that a hyperbolic equilibrium remains a hyperbolic equilibrium for all g_{ij} near g^*. In other words, if dim $(W^c) = 0$ when $g_{ij} = g^*$, then dim $(W^c) = 0$ for all g_{ij} near g^*. In contrast, if the equilibrium for $g_{ij} =$

g^* is hyperbolic but g_{ij} is not near g^*, then it can happen that either det $(G - H) = 0$ or $G - H$ has a pair of pure imaginary eigenvalues so that dim $(W^c) \neq 0$ for some value of g_{ij}. In this case a bifurcation must occur in the parameterized system since the dimensions of either W^s or W^u will change as we vary the parameter, and this is definitely a qualitative change in the behavior of solutions near the equilibrium.

The center manifold theorem is a powerful tool in the analysis of bifurcations. One reason for this is that we can sometimes arrange things so that the bifurcations occur in the center manifold of an equilibrium, and then the analysis of a bifurcation in a system of large dimension reduces to the analysis of solutions in a center manifold of low dimension. As a consequence, bifurcation results are often, in a sense, independent of the dimension of the system of equations under consideration.

A standard way to use the center manifold theorem to study a parameterized system is to treat μ as an independent variable and to study the related system

$$\frac{dx}{dt} = f_\mu(x)$$

$$\frac{d\mu}{dt} = 0$$

By this means the parameterized system is suspended in a system of higher dimension. If the original system has an equilibrium at $x = x(\mu)$ and a bifurcation at $\mu = \mu_0$, then the suspended system has an equilibrium at $x = x(\mu_0)$, $\mu = \mu_0$ and a nontrivial center manifold passing through that equilibrium. If the bifurcation of the original system involves a change in the dimension of the center manifold, then we suspect that the center manifold of the suspended system contains the interesting qualitative dynamics associated with that bifurcation. If this is the case, then to study the (local) bifurcation we only need to find a convenient coordinate system for the center manifold of the suspended system, and to study the solutions restricted to this center manifold. The procedure for finding such convenient coordinates is described in Carr and in Guckenheimer and Holmes; it won't be described in this chapter. Suffice it to say that, using this procedure, we can sometimes reduce the analysis of a bifurcation in a system of large dimension to an analysis of a system of several dimensions. The remainder of this chapter will be devoted to the analysis of the Hopf bifurcation in S-systems.

HOPF BIFURCATIONS

A common and particularly well-understood bifurcation is the Hopf bifurcation [206]. It describes a situation where the vector field in a neighborhood of an equilibrium undergoes a bifurcation that gives birth to a family of periodic orbits in a neighborhood of the equilibrium. We'll say a system has a Hopf bifurcation at $\mu = 0$ when there is a coordinate system for the suspended system such that the third-order Taylor expansion for the system about the equilibrium (and restricted to the center manifold) is given by

$$\frac{dx}{dt} = \left(d\mu + a(x^2 + y^2)\right)x - \left(\omega + c\mu + b(x^2 + y^2)\right)y$$
$$\frac{dy}{dt} = \left(\omega + c\mu + b(x^2 + y^2)\right)x + \left(d\mu + a(x^2 + y^2)\right)y$$
$$\frac{d\mu}{dt} = 0 \qquad (16.\text{HBR})$$

which has the polar form, where $r^2 = x^2 + y^2$ and $\tan(\theta) = y/x$,

$$\frac{dr}{dt} = (d\mu + ar^2)r$$
$$\frac{d\theta}{dt} = (\omega + c\mu + br^2)$$
$$\frac{d\mu}{dt} = 0 \qquad (16.\text{HBP})$$

It is assumed that $d \neq 0$ and that $\omega \neq 0$.

For fixed values of b, c, and ω we can sketch various phase portraits of this Taylor expansion in the $x - y$ plane (ignoring higher-order terms) as follows:

For $d > 0$, $a > 0$

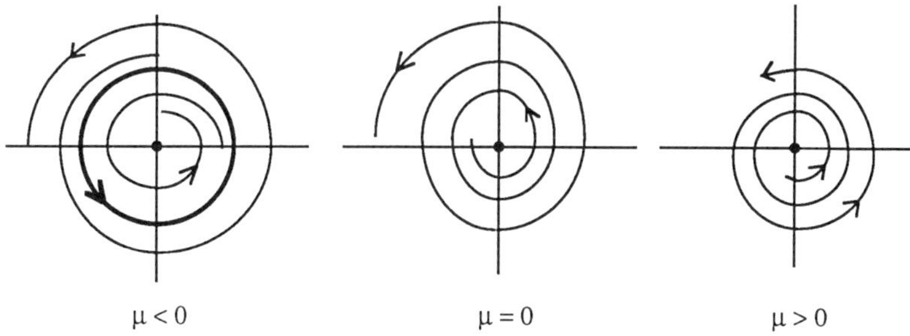

For $d > 0$, $a = 0$

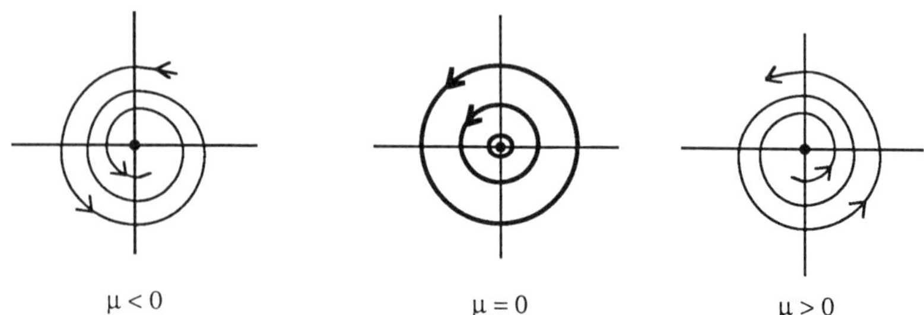

For $d > 0$, $a < 0$

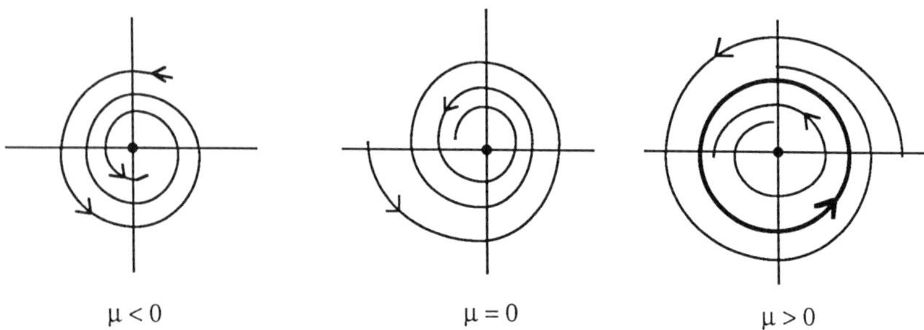

The phase portraits for the cases when $d < 0$ look the same as those when $d > 0$ except that the direction of the inequality for μ is reversed. (That is, for $d < 0$, relabel the phase portraits shown above by replacing $\mu < 0$ with $\mu > 0$, $\mu > 0$ with $\mu < 0$, and $\mu = 0$ with $\mu = 0$). Note that when $a \neq 0$, there is a periodic orbit (a limit cycle) whose radius is proportional to $\sqrt{|\mu|}$.

When $a > 0$, then the orbit is unstable (or repelling), but when $a < 0$ then the orbit is stable (or attracting). When $a = 0$, there is an entire family of periodic orbits centered about the equilibrium for the value $\mu = 0$. In every case, there is a bifurcation at $\mu = 0$, since there is a qualitatively different phase portrait for some nearby value of μ.

The Hopf bifurcation theorem (1) gives conditions that tell us when there is a coordinate system that expresses the Taylor expansion in the center manifold in the form given above, and (2) asserts, when $a \neq 0$, that the third-order Taylor expansion actually determines the qualitative behavior of the system (that is, we don't need to look at higher-order terms in the Taylor expansion to determine the qualitative behavior of the system). The following informal statement of the Hopf bifurcation theorem is taken from [105]. For a proof see [208].

Consider the following smooth system of differential equations:

$$\frac{dx}{dt} = f_\mu(x)$$

$$\frac{d\mu}{dt} = 0 \tag{16.2}$$

where x is in R^n and μ is in R. Assume this system has an equilibrium at $x = x_0$ and $\mu = \mu_0$. Let

$$\frac{dx}{dt} = Ax$$

be the linearization of

$$\frac{dx}{dt} = f_{\mu_0}(x)$$

at the point where $x = x_0$. Suppose A has a simple (nonrepeated) pair of pure imaginary eigenvalues whose values are $+i\omega$ and $-i\omega$ ($\omega \neq 0$), and it has no other eigenvalues with zero real part. Under these conditions, the Hopf bifurcation theorem asserts:

1. The system given in Eq. (16.2) has a smooth curve of equilibria parameterized by μ, $(x(\mu), \mu)$, which passes through (x_0, μ_0).
2. There are two smooth complex-valued functions that depend on μ, $\lambda(\mu)$, and $\lambda^*(\mu)$, which give eigenvalues of the linearization of

$$\frac{dx}{dt} = f_\mu(x)$$

   at the equilibrium where $x = x(\mu)$ such that $\lambda(\mu_0) = +i\omega$ and $\lambda^*(\mu_0) = -i\omega$.
3. In addition, if

$$\frac{d}{d\mu} \operatorname{Re} \{\lambda(\mu)\}\bigg|_{\mu = \mu_0} = d \neq 0$$

   then the system given in Eq. (16.2) has a unique three-dimensional center manifold passing through (x_0, μ_0) in $R^n \times R$, and there is a system of coordinates in which its third-order Taylor expansion, restricted to this center manifold, is given by the system displayed in Eq. (16.HBR).

4. If $a \neq 0$ in Eq. (16.HBR), then the solutions to Eq. (16.HBR) are qualitatively the same as the solutions near (x_0, μ_0) in the center manifold of the system given in Eq. (16.2). In particular, there is a surface of periodic solutions in the center manifold of the system given in Eq. (16.2) that is approximated to the second order by the paraboloid

$$\mu = -\frac{a}{d}(x^2 + y^2)$$

5. If $a < 0$, then these periodic solutions are stable (attracting) limit cycles. The situation in the center manifold may be visualized as shown in Fig. 16.4.
6. If $a > 0$, then these periodic solutions are unstable (repelling) limit cycles.

If $a = 0$, then, in appropriate coordinates, the system given in Eq. (16.2) still has a third-order Taylor expansion in its center manifold given by Eq. (16.HBR); but it can happen that the solutions in the center manifold of Eq. (16.2) don't look like the solutions to Eq. (16.HBR). When $a = 0$, we need a higher order Taylor expansion to determine the qualitative properties of the

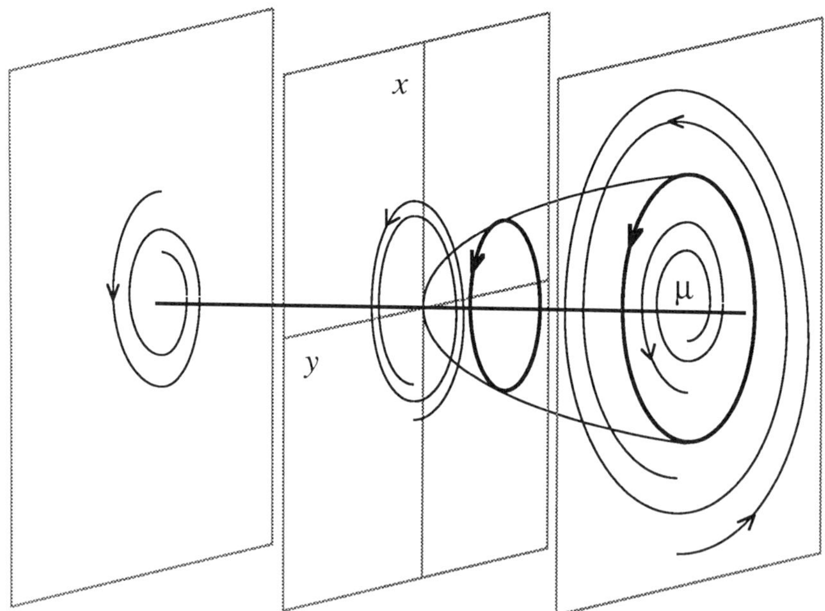

FIGURE 16.4. When $a < 0$, there are periodic orbits associated with a Hopf bifurcation that lie on a paraboloid in the three-dimensional center manifold of the extended system given in Eq. (16.2).

solutions to Eq. (16.2). On the other hand, even when $a = 0$, the solutions in the center manifold of Eq. (16.2) may look like the solutions to Eq. (16.HBR). In that case, the entire family of periodic orbits appears at the single parameter value $\mu = \mu_0$.

A Test for Hopf Bifurcations in S-Systems

It is easy to describe a test for the Hopf bifurcation in a regular n-dimensional S-system. Since the system must have an equilibrium if it is to have a Hopf bifurcation, we may assume, without loss of generality, that the μ-parameterized S-system is written in the form

$$\frac{dx}{dt} = \gamma(x^G - x^H)$$

where γ, G, and H may depend (smoothly) upon μ, and $\gamma \geq 0$. For convenience, let Q be in $R^{n \times n}$ and define Q by the rule $Q = G - H$. For such a system, the conditions of the Hopf bifurcation theory can be stated as follows.

If, when $\mu = \mu_0$, $[\gamma]Q$ has a simple (nonrepeated) pair of pure imaginary eigenvalues, $\lambda = +i\omega$ and $\lambda^* = -i\omega$, (with $\omega \neq 0$), but all other eigenvalues have nonzero real parts, and if

$$\frac{d}{d\mu} = \text{Re}\,\{\lambda(\mu)\}\bigg|_{\mu = \mu_0} \neq 0$$

then the equilibrium at

$$x = \begin{bmatrix} 1 \\ 1 \\ \vdots \\ 1 \end{bmatrix}$$

has a Hopf bifurcation at $\mu = \mu_0$.

Remarks: Since the matrix of the linearization of the parameterized S-system is just $[\gamma]Q$, this test follows directly from the statement of the Hopf bifurcation theorem. Note that for a parameterized S-system written in this form, the equilibrium does not move as μ is varied. The sign of a, which determines the stability of the periodic orbits associated with the bifurcation, may or may not be easy to determine by looking at numerical solutions of the system for $\mu = \mu_0$, particularly if $|a|$ is small.

To illustrate the application of this theorem we'll look at a example involving a regular three-dimensional S-system.

Example 4

Let

$$\frac{dx}{dt} = \gamma(x^G - x^H)$$

where

$$G = \begin{bmatrix} 1 & 1 & 0 \\ 1 & 0 & 2 \\ 2 & 1 & 1 \end{bmatrix}, H = \begin{bmatrix} 0 & 2 & 1 \\ 1 & 0 & 1 \\ 0 & 2 & 3 \end{bmatrix}, \gamma = \begin{bmatrix} 1 + \mu \\ 1 \\ 1 \end{bmatrix}$$

This system has an isolated equilibrium at

$$x = \begin{bmatrix} 1 \\ 1 \\ 1 \end{bmatrix}$$

for all $\mu > -1$ and a linearization about this equilibrium whose matrix, $[\gamma]Q$, is given by

$$[\gamma]Q = \begin{bmatrix} (1 + \mu) & -(1 + \mu) & -(1 + \mu) \\ 0 & 0 & 1 \\ 2 & -1 & -2 \end{bmatrix}$$

The characteristic equation of $[\gamma]Q$, defined by $\det([\gamma]Q - \lambda I) = 0$, is simply $(\lambda + 1)(\lambda^2 - \mu\lambda + (\mu + 1)) = 0$, so the eigenvalues of $[\gamma]Q$ comprise the set $\{-1, (\mu \pm \sqrt{\mu^2 - 4(\mu + 1)})/2\}$. When $\mu = 0$, the eigenvalues are $\{-1, i, -i\}$, so that $[\gamma]Q$ has a simple pair of pure imaginary eigenvalues ($\omega = 1 \neq 0$) and no other eigenvalue whose real part is zero. Defining $\lambda(\mu)$ by $\lambda(\mu) = (\mu + \sqrt{\mu^2 - 4(\mu + 1)})/2$, we find that

$$\frac{d}{d\mu} \operatorname{Re}\{\lambda(\mu)\}\bigg|_{\mu=0} = \frac{d}{d\mu}\left(\frac{\mu}{2}\right)\bigg|_{\mu=0} = \frac{1}{2}$$

so that $d = 1/2 \neq 0$. Our test now shows that this S-system has a Hopf bifurcation when $\mu = 0$ and we expect to see a family of periodic orbits associated with the equilibrium at

$$x = \begin{bmatrix} 1 \\ 1 \\ 1 \end{bmatrix}$$

To find out more about the qualitative properties of this family of periodic orbits we need to determine the value, or at least the sign, of a. In general, it is quite tedious to compute a—for to do so, in effect, requires us to find the Taylor approximation of a change of coordinates which places our original S-system in the form given by Eq. (16.HBR). (Actually, if we don't care about the values of b and c in Eq. (16.HBR), we can simplify the computation quite a lot, but it may still be a daunting task, since we first need to find a convenient coordinate system for the center manifold of the original S-system.) Fortunately, we can sometimes easily determine the sign of a by a numerical analysis. Inspecting Eq. (16.HBP) we see that when $\mu = 0$, if $a < 0$, then the equilibrium is stable for solutions in the center manifold of Eq. (16.2), and if $a > 0$, then the equilibrium is unstable for solutions in the center manifold of Eq. (16.2). In our present example, dim $(W^u) = 0$ when $\mu = 0$, so all solutions near the equilibrium approach W^c as $t \to \infty$. This means that when $\mu = 0$, the stability of the equilibrium when this system is restricted to its center manifold is the same as the stability of the equilibrium in the original, unrestricted system. In our example (assuming $a \neq 0$), it follows that the sign of a is determined by the stability of the equilibrium when $\mu = 0$. If the equilibrium is stable, then $a < 0$. If the equilibrium is unstable, then $a > 0$. A practical consequence of this is that a numerical analysis can be used to determine the stability of the equilibrium when $\mu = 0$, and the results can be used to determine the sign of a. ESSYNS was used to determine the stability of the equilibrium for this example (see Fig. 16.5). We'll conclude that $a > 0$. Since $d = 1/2 > 0$, it follows that there is a family of unstable periodic orbits for $\mu < 0$ (see the ESSYNS plot in Fig. 16.6). In the next section, I'll present a formula for the computation of a for any regular two-dimensional S-system that has a Hopf bifurcation.

Tests for Hopf Bifurcations and Stability in Regular Two-Dimensional S-Systems

This section comprises two results. The first result is really just a special case of the test described in the last section; however, in the two-dimensional case,

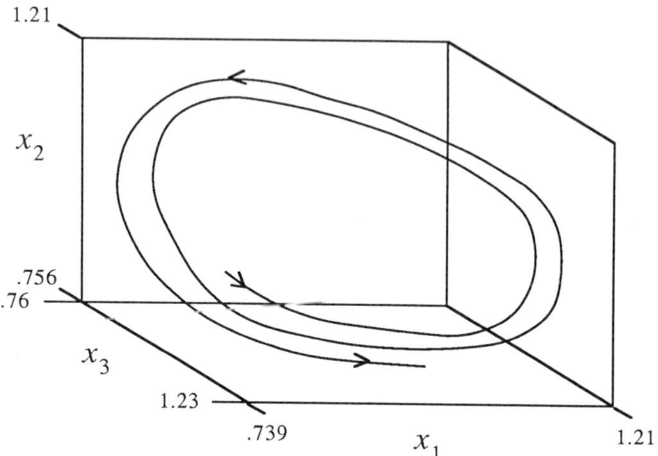

FIGURE 16.5. Sketch of an ESSYNS plot for Example 4 when $\mu = 0$.

the general test can be restated in a particularly convenient form. The second result is a test for the stability of the periodic orbits of Hopf bifurcations in regular two-dimensional S-systems—a formula is presented which explicitly gives the value of a, the coefficient of r^3 in Eq. (16.HBP), expressed as a modest polynomial in the coefficients and exponents of the S-system.

Consider a μ-parameterized, regular two-dimensional S-system that has an equilibrium. If necessary, rescale the system and place it in the form

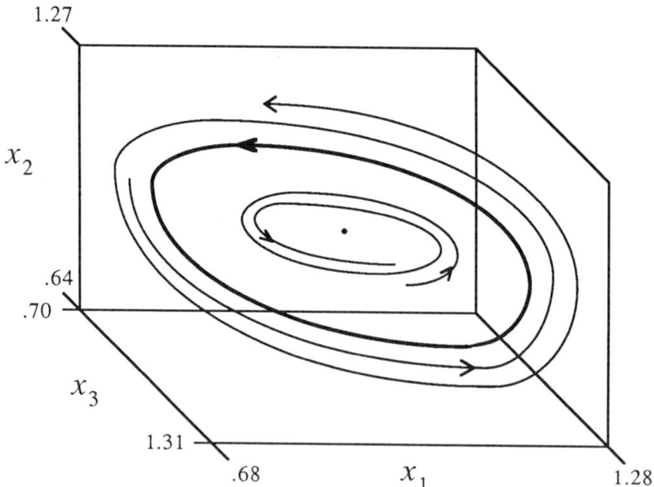

FIGURE 16.6. Sketch of an ESSYNS plot for Example 4 when $\mu = .05$.

$$\frac{dx}{dt} = \gamma(x^G - x^H)$$

where $\gamma \geq 0$ in R^2, $G = R^{2 \times 2}$, and H in $R^{2 \times 2}$ may depend smoothly upon the parameter μ. For convenience, let $Q = G - H$.

Suppose that trace $([\gamma]Q) = 0$ and det $([\gamma]Q) > 0$ when $\mu = \mu_0$. If

$$\frac{d}{d\mu}(\text{trace}\,([\gamma]Q))\bigg|_{\mu=\mu_0} \neq 0$$

then the two-dimensional S-system has a Hopf bifurcation at $\mu = \mu_0$.

This test follows directly from the previous test. In the two-dimensional case, however, we can state the test directly in terms of the coefficients and exponents of the S-system without having to mention any eigenvalues. This is possible because the eigenvalues of $[\gamma]Q$ are given by

$$\lambda = \frac{\text{trace}\,([\gamma]Q) \pm \sqrt{\text{trace}^2\,([\gamma]Q) - 4\,\det\,([\gamma]Q)}}{2}$$

We see that $[\gamma]Q$ has a nonzero, simple pair of pure imaginary eigenvalues if and only if trace $([\gamma]Q) = 0$ and det $([\gamma]Q) > 0$. In this case

$$\lambda = i\sqrt{\det\,([\gamma]Q)},\ \lambda^* = -i\sqrt{\det\,([\gamma]Q)}$$

and there are no other eigenvalues with a zero real part. Since $[\gamma]Q$ depends smoothly on μ, trace$^2([\gamma]Q) - 4\det([\gamma]Q) < 0$ for all μ near μ_0, provided trace $([\gamma]Q) = 0$ and det $([\gamma]Q) > 0$ when $\mu = \mu_0$. Letting

$$\lambda(\mu) = \frac{\text{trace}\,([\gamma]Q) + \sqrt{\text{trace}^2\,([\gamma]Q) - 4\,\det\,([\gamma]Q)}}{2}$$

it follows that Re $\{\lambda(\mu)\} = $ trace $([\gamma]Q)/2$, and therefore

$$\frac{d}{d\mu}(\text{trace}\,([\gamma]Q))\bigg|_{\mu=\mu_0} \neq 0$$

if and only if

$$\frac{d}{d\mu}\text{Re}\,\{\lambda(\mu)\}\bigg|_{\mu=\mu_0} \neq 0$$

That establishes the test for the existence of a Hopf bifurcation in a regular two-dimensional S-system.

We now turn to the question of the stability of the periodic orbits associated with a Hopf bifurcation. For convenience, let $P = G + H - 2I$ (where I is the 2×2 identity matrix). Suppose that the regular two-dimensional S-system given above has a Hopf bifurcation at $\mu = \mu_0$. Here is a stability test.

Let

$$a = \frac{-\gamma_1\gamma_2 q_{12} q_{22}}{64}\left((\gamma_1 q_{21} + \gamma_2 q_{12}) \det(Q) + \gamma_2 q_{21}(p_{12} - p_{22})^2 + \gamma_1 q_{12}(p_{11} - p_{21})^2\right)$$

If $a < 0$, then there is a family of stable limit cycles. If $a > 0$, then there is a family of unstable limit cycles. If

$$\left.\frac{d}{d\mu}\left(\text{trace}\,([\gamma]Q)\right)\right|_{\mu=\mu_0} > 0$$

then the stable limit cycles occur for $\mu > \mu_0$ and the unstable limit cycles occur for $\mu < \mu_0$. If

$$\left.\frac{d}{d\mu}\left(\text{trace}\,([\gamma]Q)\right)\right|_{\mu=\mu_0} < 0$$

then the stable limit cycles occur for $\mu < \mu_0$, and the unstable limit cycles occur for $\mu > \mu_0$.

The formula for a gives the coefficient r^3 in the right-hand side of Eq. (16.HBP). For any regular two-dimensional S-system with a Hopf bifurcation, the value of a was found by a standard procedure described in Guckenheimer and Holmes. The procedure involves approximating the S-system by its third-order Taylor polynomial, changing coordinates, and evaluating several complicated higher derivatives. The formula for a given here was obtained in a computer-assisted computation using MACSYMA. MACSYMA was able to simplify the value of a to an expression several pages long. This result was further simplified and factored by hand to yield our expression for a.

Here are three examples that illustrate the use of this test.

Example 5

Let

$$\gamma = \begin{bmatrix} (1+\mu) \\ 1 \end{bmatrix}, \quad G = \begin{bmatrix} 4 & 0 \\ 3 & 0 \end{bmatrix}, \quad H = \begin{bmatrix} 3 & 2 \\ 2 & 1 \end{bmatrix}$$

so that the scalar form of the system may be written as

$$\frac{dx_1}{dt} = (1 + \mu)(x_1^4 - x_1^3 x_2^2)$$

$$\frac{dx_2}{dt} = x_1^3 - x_1^2 x_2$$

Here

$$Q = \begin{bmatrix} 1 & -2 \\ 1 & 1 \end{bmatrix} \text{ and } P = \begin{bmatrix} 5 & 0 \\ 3 & -1 \end{bmatrix}$$

Computing the trace and determinant of $[\gamma]Q$ we find that trace $([\gamma]Q) = \mu$ and det $([\gamma]Q) = (1 + \mu)$. Observe that, if $\mu = 0$, then trace $([\gamma]Q) = 0$ and det $([\gamma]Q) > 0$. Computing the derivative of the trace, we find that

$$\frac{d \text{ trace } ([\gamma]Q)}{d\mu} = \frac{d\mu}{d\mu} = 1 > 0$$

Our test for existence now asserts that this S-system has a Hopf bifurcation when $\mu = 0$. Grinding out the formula for a, we find that $a = -.25 < 0$. Our stability test now asserts that this S-system has a family of stable limit cycles for $\mu > 0$. For an ESSYNS plot of this system, see Fig. 16.7.

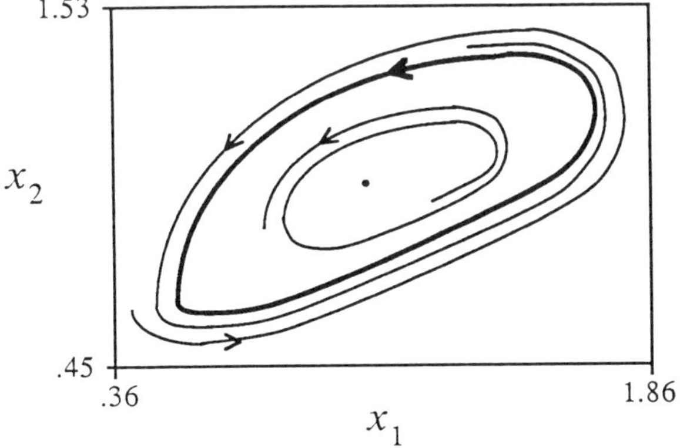

FIGURE 16.7. Sketch of an ESSYNS plot for Example 5 when $\mu = .05$.

Example 6

Let

$$\gamma = \begin{bmatrix} (1+\mu) \\ 1 \end{bmatrix}, \; G = \begin{bmatrix} 3 & 1 \\ 1 & 2 \end{bmatrix}, \; H = \begin{bmatrix} 2 & 3 \\ 0 & 3 \end{bmatrix}$$

Here $[\gamma]Q$ is identical to the $[\gamma]Q$ of Example 5, so this system also has a Hopf bifurcation when $\mu = 0$. Using

$$P = \begin{bmatrix} 3 & 2 \\ -1 & 3 \end{bmatrix}$$

in the stability formula, we find that $a = .25 > 0$. In this case, our stability test asserts that this system has unstable limit cycles for $\mu < 0$. For an ESSYNS plot of this system, see Fig. 16.8.

Example 7

Let

$$\gamma = \begin{bmatrix} (1+\mu) \\ 1 \end{bmatrix}, \; G = \begin{bmatrix} 2 & 1 \\ 2 & 1 \end{bmatrix}, \; H = \begin{bmatrix} 1 & 3 \\ 1 & 2 \end{bmatrix}$$

Here $[\gamma]Q$ is identical to the $[\gamma]Q$ of Example 5, so the system in this example also has a Hopf bifurcation when $\mu = 0$. Using

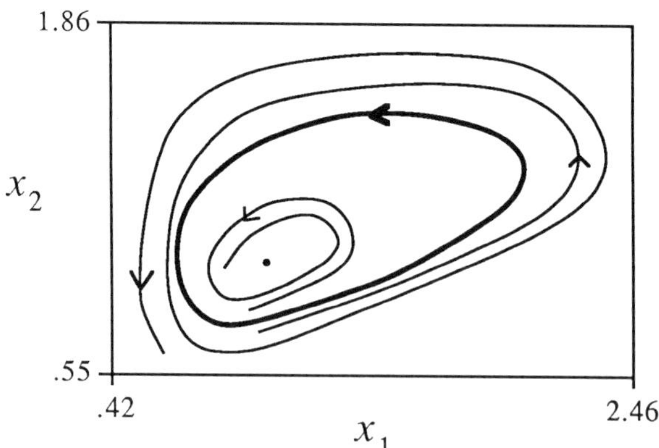

FIGURE 16.8. Sketch of an ESSYNS plot for Example 6 when $\mu = -.05$.

$$P = \begin{bmatrix} 1 & 2 \\ 1 & 1 \end{bmatrix}$$

in the stability formula, we find that $a = 0$. In this case, our stability test fails to provide any information about the stability of the periodic motions associated with the Hopf bifurcation. Some further analysis is required. As it turns out, we can actually solve the S-system of this example when $\mu = 0$. Setting $\mu = 0$, the scalar form of the system is

$$\frac{dx_1}{dt} = x_1^2 x_2 - x_1 x_2^3$$

$$\frac{dx_2}{dt} = x_1^2 x_2 - x_1 x_2^2$$

so that, eliminating t, we obtain a first-order differential equation for the solution curves in the phase plane

$$(x_1^2 x_2 - x_1 x_2^2)\, dx_1 + (x_1 x_2^3 - x_1^2 x_2)\, dx_2 = 0$$

This equation has an integrating factor, $(x_1 x_2)^{-1}$. Multiplying through by this integrating factor we obtain an equivalent exact equation,

$$(x_1 - x_2)\, dx_1 + (x_2^2 - x_1)\, dx_2 = 0$$

whose solutions are given by

$$3x_1^2 - 6x_1 x_2 + 2x_2^3 = c$$

A little calculus shows that these solutions include an entire family of closed orbits about the equilibrium at $x_1 = 1$, $x_2 = 1$. We conclude that the family of periodic orbits associated with the Hopf bifurcation occurs ''all at once'' when $\mu = 0$. In this example, our S-system has solutions that look like the solutions to Eq. (16.HBR) with $a = 0$. For an ESSYNS plot of this system, see Fig. 16.9.

SUMMARY

This chapter presents some key concepts from the qualitative theory of ordinary differential equations and discusses the Hopf bifurcation theorem. Two results for S-systems are presented. The first is an application of the Hopf bifurcation theorem that gives a test for the existence of a Hopf bifurcation in an n-dimen-

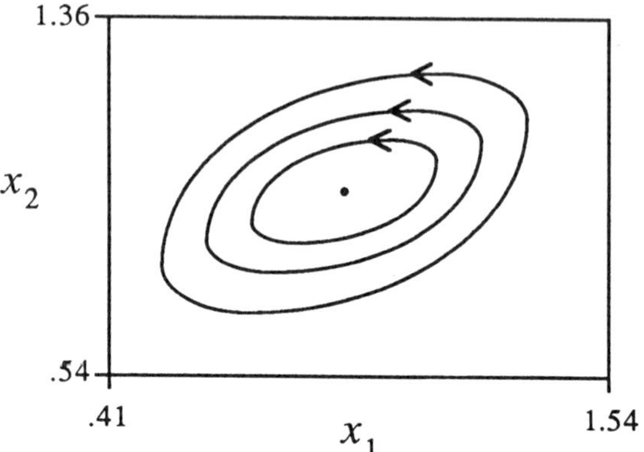

FIGURE 16.9. Sketch of an ESSYNS plot for Example 7 when $\mu = 0$.

sional S-system. The second result concerns two-dimensional S-systems. In this case the test for the existence of a Hopf bifurcation is given explicitly in terms of the parameters of the S-system. A test for the stability of the limit cycle in a nondegenerate Hopf bifurcation amounts to the evaluation of a simple expression involving the parameters of the general two-dimensional S-system.

References

The following list of references is divided into four groups. The first three, *Books and Reviews, Modeling,* and *Recasting,* contain work specifically addressing S-systems and the power-law formalism. These lists contain articles published in referred journals by spring 1990. Within each group, the articles are arranged in temporal order. The fourth group contains other references, arranged alphabetically for each chapter.

BOOKS AND REVIEWS

[1] Savageau, M. A. 1972. The behavior of intact biochemical control systems. *Curr. Top. Cell Reg. 6:*63–130.

[2] Savageau, M. A. 1976. *Biochemical Systems Analysis: A Study of Function and Design in Molecular Biology.* Reading, Mass.: Addison-Wesley.

[3] Savageau, M. A., and E. O. Voit. 1982. Power-law approach to modeling biological systems; I. Theory. *J. Ferment. Technol. 60*(3):221–228.

[4] Savageau, M. A. 1985. Mathematics of organizationally complex systems. *Biomed. Biochim. Acta 44:*839–844.

[5] Savageau, M. A. 1985. A theory of alternative designs for biochemical control systems. *Biomed. Biochim. Acta 44:*875–880.

[6] Savageau, M. A. 1987. Introduction to S-systems and the underlying power-law formalism. *Mathem. Comput. Modelling 11:*546–551.

[7] Voit, E. O., D. H. Irvine, and M. A. Savageau. 1989. *The User's Guide to ESSYNS.* Charleston, S.C.: Medical University of South Carolina Press.

MODELING

[8] Savageau, M. A. 1969. Biochemical systems analysis, I. Some mathematical properties of the rate law for the component enzymatic reactions. *J. Theor. Biol. 25:*365–369.

[9] Savageau, M. A. 1969. Biochemical systems analysis, II. The steady-state solutions for an n-pool system using a power-law approximation. *J. Theor. Biol. 25:*370–379.

[10] Savageau, M. A. 1970. Biochemical systems analysis, III. Dynamic solutions using a power-law approximation. *J. Theor. Biol. 26:*215–226.

[11] Savageau, M. A. 1971. Parameter sensitivity as a criterion for evaluating and comparing the performance of biochemical systems. *Nature 229:*542–544.

[12] Savageau, M. A. 1971. Concepts relating the behavior of biochemical systems to their underlying molecular properties. *Arch. Biochem. Biophys. 145:*612–621.

[13] Savageau, M. A. 1974. Comparison of the classical and autogenous system of regulation in inducible operons. *Nature 252:*546–549.

[14] Savageau, M. A. 1974. Optimal design of feedback control by inhibition: Steady-state considerations. *J. Mol. Evol. 4:*139–156.

[15] Smith, D. F. 1974. Quantitative analysis of the functional relationships existing between ecosystem components; II. Analysis of non-linear relationships. *Oecologia 16:*107–117.

[16] Savageau, M. A. 1975. Optimal design of feedback control by inhibition: Dynamic considerations. *J. Mol. Evol. 5:*199–222.

[17] Savageau, M. A. 1975. Significance of autogenously regulated and constitutive synthesis of regulatory proteins in repressible biosynthetic systems. *Nature 258:*208–214.

[18] Smith, D. F. 1975. Quantitative analysis of the functional relationships existing between ecosystem components; III. Analysis of ecosystem stability. *Oecologia 21:*17–29.

[19] Savageau, M. A. 1979. Growth of complex systems can be related to the properties of their underlying determinants. *Proc. Nat. Acad. Sci. 76:*5413–5417.

[20] Savageau, M. A. 1979. Allometric morphogenesis of complex systems: Derivation of the basic equations from first principles. *Proc. Nat. Acad. Sci. 76:*6023–6025.

[21] Savageau, M. A., and G. Jacknow. 1979. Feedforward inhibition in biosynthetic pathways: Inhibition of the aminoacyl-t RNA synthetase by intermediates of the pathway. *J. Theor. Biol. 77:*405–425.

[22] Voit, E. O., and M. A. Savageau. 1982. Power-law approach to modeling biological systems, II. Application to ethanol production. *J. Ferment. Technol. 60:*229–232.

[23] Voit, E. O., and M. A. Savageau. 1982. Power-law approach to modeling biological systems, III. Methods of analysis. *J. Ferment. Technol. 60:*233–241.

[24] Voit, E. O., and M. A. Savageau. 1984. Analytical solutions to a generalized growth equation. *J. Mathem. Anal. Appl. 103:*380–386.

[25] Irvine, D. H., and M. A. Savageau. 1985. Network regulation of the immune response: Alternative control points for suppressor modulation of effector lymphocytes. *J. Immunol. 134:*2100-2116.

[26] Irvine, D. H., and M. A. Savageau. 1985. Network regulation of the immune response: Modulation of suppressor lymphocytes by alternative signals including contrasuppression. *J. Immunol. 134:*2117-2130.

[27] Savageau, M. A., E. O. Voit, and D. H. Irvine. 1987. Biochemical systems theory and metabolic control theory, 1. Fundamental similarities and differences. *Mathem. Biosci. 86:*127–145.

[28] Savageau, M. A., E. O. Voit, and D. H. Irvine. 1987. Biochemical systems theory and metabolic control theory, 2. The role of summation and connectivity relationships. *Mathem. Biosci. 86:*147–169.

[29] Voit, E. O., and M. A. Savageau. 1987. Accuracy of alternative representations for integrated biochemical systems. *Biochemistry 26:*6869–6880.

[30] Irvine, D. H. 1988. Efficient solution of nonlinear models expressed in S-system canonical form. *Mathem. Comput. Modelling 11:*123–128.

[31] Johnson, T. 1988. Estimation and simulation of S-systems. *Mathem. Comput. Modelling* 11:134–139.

[32] Sorribas, A. 1988. Alternative S-system representations for reversible biochemical pathways. *Mathem. Comput. Modelling 11:*129–133.

[33] Voit, E. O. 1988. Dynamics of self-thinning plant stands. *Ann. Bot. 62:*67–78.

[34] Savageau, M. A., and A. Sorribas. 1989. Constraints among molecular and systemic properties: Implications for physiological genetics. *J. Theor. Biol. 141:*93–115.

[35] Sorribas, A., and M. A. Savageau. 1989. A comparison of variant theories of intact biochemical systems, 1. Enzyme-enzyme interactions and biochemical systems theory. *Mathem. Biosci. 94:*161–193.

[36] Sorribas, A., and M. A. Savageau. 1989. A comparison of variant theories of intact biochemical systems, 2. Flux oriented and metabolic control theories. *Mathem. Biosci. 94:*195–238.

[37] Sorribas, A., and M. A. Savageau. 1989. Strategies for representing metabolic pathways within biochemical systems theory: Reversible pathways. *Mathem. Biosci. 94:*239–269.

[38] Irvine, D. H., and M. A. Savageau. 1990. Efficient solution of nonlinear ordinary differential equations expressed in S-system canonical form. *SIAM J. Num. Anal. 27:*704–735.

[39] Voit, E. O. 1990. S-system analysis of endemic infections. *Comp. and Mathem. with Applications. 20:*161–173.

[40] Voit, E. O., and P. N. Yi. 1990. Comparison of isoeffect relationships in radiotherapy. *Bull. Mathem. Biol. 52:*657–675.

RECASTING

[41] Savageau, M. A. 1980. Growth equations: A general equation and a survey of special cases. *Mathem. Biosci. 48:*267–278.
[42] Savageau, M. A. 1982. A suprasystem of probability distributions. *Biometrical J. 24:*323–330.
[43] Voit, E. O. 1985. Cell cycles and growth laws. The CCC-model. *J. Theor. Biol. 114:*589–599.
[44] Voit, E. O., and H. J. Anton. 1985. Derivation of the frequency distribution of cycle durations from continuous-labeling curves. *J. Theor. Biol. 112:*575–588.
[45] Voit, E. O., and M. A. Savageau. 1986. Equivalence between S-systems and Volterra-systems. *Mathem. Biosci. 78:*47–55.
[46] Savageau, M. A., and E. O. Voit. 1987. Recasting nonlinear differential equations as S-systems: A canonical nonlinear form. *Mathem. Biosci. 87:*83–115.
[47] Voit, E. O. 1988. Recasting nonlinear models as S-systems. *Mathem. Comput. Modelling 11:*140–145.
[48] Rust, P. F., and E. O. Voit. 1990. Statistical densities, cumulatives, quantiles, and power obtained by S-system differential equations. *JASA 85:*572–578.
[49] Voit, E. O., and P. F. Rust. 1990. Evaluation of the noncentral t distribution with S-systems. *Biometrical J. 32.* In press.
[50] Chaudhuri, K., and T. Johnson. 1990. Bioeconomic dynamics of a fishery modeled as an S-system. *Mathem. Biosci. 99:*231–249.

See also [3], [18], [24], [33], [38–39].

OTHER REFERENCES

Chapter 1

[51] von Bertalanffy, L. 1968. *General Systems Theory.* New York: George Braziller.
[52] Black, H. S. 1924. U.S. Patent 1,686,792.
[53] Bode, H. W. 1945. *Network Analysis and Feedback Amplifier Design.* Princeton, N.J.: Van Nostrand.
[54] Broad, W., and N. Wade. 1982. *Betrayers of the Truth.* New York: Simon & Schuster.
[55] Cascante, M., R. Franco, and E. I. Canela. 1989. Use of implicit methods from general sensitivity theory to develop a systematic approach to metabolic control, 1. Unbranched chains. *Mathem. Biosci. 94:*271–288.
[56] Cascante, M., R. Franco, and E. I. Canela. 1989. Use of implicit meth-

ods from general sensitivity theory to develop a systematic approach to metabolic control, 2. Complex systems. *Mathem. Biosci. 94:*289–309.

[57] Casti, J. L. 1989. *Paradigms Lost.* New York: Morrow.

[58] Chua, L. O. 1969. *Introduction to Nonlinear Network Theory.* New York: McGraw-Hill.

[59] Cruz, J. B., ed. 1973. *System Sensitivity Analysis.* Stroudsburg, PA: Dowden, Hutchinson and Ross.

[60] Davidson, M. 1983. *Uncommon Sense.* Los Angeles: J. P. Tarcher.

[61] Davis, B. D. 1961. The teleonomic significance of biosynthetic control mechanisms. *Cold Spring Harbor Symp. Quant. Biol. 26:*1–10.

[62] Goodwin, B. 1963. *Temporal Organization in Cells.* New York: Academic Press.

[63] Heinrich, R., and T. M. Rapoport. 1974. A linear steady-state treatment of enzymatic chains: General properties control and effector strength. *Eur. J. Biochem. 42:*89–95.

[64] Jacob, F., and J. Monod. 1961. Genetic regulatory mechanisms in the synthesis of proteins. *J. Mol. Biol. 3:*318–356.

[65] Jurgen, R. K. 1972. Feedforward correction—late-blooming design. *IEEE Spectrum.* (April):42–43.

[66] Kacser, H., and J. A. Burns. 1973. The control of flux. *Symp. Soc. Exp. Biol. 27:*65–104.

[67] Mason, S. J. 1953. Feedback theory—some properties of signal flow graphs. *Proc. I.R.E. 41:*1144–1156.

[68] Mead, T. British Patent (Old Series) No. 1628 (1787). [Noted in: O. Mayr. *The Origins of Feedback Control.* Cambridge: The M.I.T. Press, 1970.]

[69] Monod, J. 1971. *Chance and Necessity.* New York: Knopf.

[70] Monod, J., J. Wyman, and J. P. Changeux. 1965. On the nature of allosteric transitions: A plausible model. *J. Mol. Biol. 12:*88–118.

[71] Okamoto, M., and M. A. Savageau. 1984. Integrated function of a kinetic proofreading mechanism: Steady-state analysis testing internal consistency of data obtained *in vivo* and *in vitro* and predicting parameter values. *Biochemistry 23:*1701–1709.

[72] Okamoto, M., and M. A. Savageau. 1984. Integrated function of a kinetic proofreading mechanism: Dynamic analysis separating the effects of speed and substrate competition on accuracy. *Biochemistry 23:*1710–1715.

[73] Peschel, M., and W. Mende 1986. *The Predator-Prey Model: Do we Live in a Volterra World?* Berlin: Akademie-Verlag.

[74] Root-Bernstein, R. S. 1989. *Discovering.* Cambridge: Harvard University Press.

[75] Rosen, R. 1978. Dynamical similarity and the theory of biological transformations. *Bull. Math. Biophys. 40:*549–579.

[76] Savageau, M. A. 1968. Biochemical amplification and control in enzyme-proenzyme reactions. *Proc. 21st Annual Conference on Engineering in Medicine and Biology.* Houston: 29.8.

[77] Savageau, M. A. 1974. Genetic regulatory mechanisms and the ecological niche of *Escherichia coli. Proc. Nat. Acad. Sci. USA 71:*2453–2455.

[78] Savageau, M. A. 1974. Comparison of classical and autogenous systems of regulation in inducible operons. *Nature 252:*546–549.

[79] Savageau, M. A. 1989. Are there rules governing patterns of gene regulation? In: *Theoretical Biology—Epigenetic and Evolutionary Order.* ed. B. C. Goodwin and P. T. Saunders, 42–66. Edinburgh: Edinburgh University Press.

[80] Savageau, M. A. 1990. A critique of the enzymologist's test tube. In *Foundations of Medical Cell Biology,* ed. W. Bittman. Greenwich, CT: JAI Press Inc.

[81] Singer, P. A., M. Levinthal, and L. S. Williams. 1984. Synthesis of the isoleucyl- and valyl-tRNA synthetases and the isoleucine-valine biosynthetic enzymes in a threonine deaminase regulatory mutant of *Escherichia coli* K-12. *J. Mol. Biol. 175:*39–55.

[82] Smith, G. R., and B. Magasanik. 1971. Nature and self-regulated synthesis of the repressor of the hut operons in Salmonella typhimurium. *Proc. Nat. Acad. Sci. USA 68:*1493–1497.

[83] Truxal, J. G. 1955. *Automatic Feedback Control Systems Synthesis.* New York: McGraw-Hill.

[84] Van Valkenburg, M. E. 1958. *Network Analysis.* Englewood Cliffs, N.J.: Prentice-Hall.

Also cited in this Chapter: [1–2], [4], [8–9], [11–12], [14], [16], [19], [21–30], [33], [35–39], [41–42], [46–49].

Chapter 2

[85] Cotton, R. H. G., and F. Gibson. 1965. The biosynthesis of phenylalanine and tyrosine; enzymes converting chorismic acid into prephenic acid and their relationships to prephenate dehydratase and prephenate dehydrogenase. *Biochim. Biophys. Acta 100:*76–88.

[86] Patte, J. C., P. Truffa-Bachi, and G. N. Cohen. 1966. The threonine-sensitive homoserine dehydrogenase and aspartokinase activities of *Escherichia coli*, I. Evidence that the two activities are carried by a single protein. *Biochim. Biophys. Acta 128:*426–439.

[87] Roels, J. A. 1983. *Energetics and Kinetics in Biotechnology.* New York: Elsevier.

[88] Stadtman, E. R. 1963. Symposium on multiple forms of enzymes and

control mechanisms, II. Enzyme multiplicity and function in the regulation of divergent metabolic pathways. *Bacteriol. Rev. 27:*170–181.

[89] Routh, E. J. 1930. *Advanced Part of Dynamics of a System of Rigid Bodies.* 6th ed. Vol. 2, London: Macmillan.

[90] Dorf, R. C. 1974. *Modern Control Systems.* 2nd ed. Reading, Mass.: Addison-Wesley.

Also cited in this Chapter: [1–2], [4], [8–9], [11–12], [14], [16], [19], [21–30], [33–39], [41–42], [46–49].

Chapter 3

[91] Garfinkel, D., L. Garfinkel, M. Pring, S. B. Green, and B. Chance. 1970. Computer applications to biochemical kinetics. *Ann. Rev. Biochem. 39:*473–498.

[92] Savageau, M. A. 1985. Coupled circuits of gene regulation. In *Sequence Specificity in Transcription and Translation,* ed. R. Calendar and L. Gold, 633–642. New York: Alan R. Liss.

[93] Wright, B. E., and P. J. Kelly. 1981. Kinetic models of metabolism in intact cells, tissues, and organisms. *Curr. Tops. Cell. Reg. 19:*103–158.

Also cited in this Chapter: [1–3], [8–9], [11], [14], [16], [25–29], [35–37].

Chapter 4

[94] Frank, P. M. 1978. *Introduction to System Sensitivity Theory.* New York: Academic Press.

[95] Rabitz, H. K., M. Kramer, and D. Dacol. 1983. Sensitivity analysis in chemical kinetics. *Ann. Rev. Phys. Chem. 34:*1419–1461.

[96] Šiljak, D. D. 1969. *Nonlinear Systems. The Parameter Analysis and Design.* New York: J. Wiley & Sons, Inc.

Also cited in this Chapter: [1–2], [11], [27–29], [34–37], [55–56], [59], [63], [66].

Chapter 5

[97] Al-Adra, A. R., and L. M. Pilarski. 1978. Antigen-specific suppression of cytotoxic T-cell responses in mice, I. Suppressor T cells are not cytotoxic cells. *Eur. J. Immunol. 8:*504–511.

[98] Bellanti, J. A. 1985. *Immunology III.* Philadelphia: W. B. Saunders.

[99] Benacerraf, B., and R. N. Germain. 1979. Specific suppressor responses to antigen under I region control. *Fed. Proc. 38:*2053–2057.

[100] Gorczynski, R. M., and S. McRae. 1979. Suppression of cytotoxic re-

sponse to histoincompatible cells, I. Evidence for two types of T lymphocyte-derived suppressors acting at different stages in the induction of a cytotoxic response. *J. Immunol. 122:*737–746.

[101] Milburn, G. L., R. G. Hoover, and R. G. Lynch. 1982. Immuno-regulatory cell interactions that govern the growth and differentiation of murine myeloma cells. In: *B and T Cell Tumors,* ed. E. S. Vitetta, 335. New York: Academic Press.

[102] Savageau, M. A. 1990. Evolution of regulation examined with a novel strategy for comparative optimization. In *Evolution Theory and Combinatorial Optimization,* ed. M. Peschel. Berlin: Akademie-Verlag. In press.

[103] Susskind, B. M., V. J. Merluzzi, R. B. Faanes, M. A. Pallandio, and Y. S. Choi. 1983. Regulatory mechanisms in cytotoxic T lymphocyte development, I. A suppressor T cell subset that regulates the proliferative stage of CTL development. *J. Immunol. 130:*527–532.

[104] Takemori, T., and T. Tada. 1975. Properties of antigen-suppressive T cell factor in regulation of antibody response of the mouse, I. *In vitro* activity and immunochemical characterizations. *J. Exp. Med. 142:*1241–1253.

Also cited in this Chapter: [1–2], [5], [25–26].

Chapter 6

[105] Guckenheimer, J., and P. Holmes. 1986. *Nonlinear Oscillations, Dynamical Systems, and Bifurcations of Vector Fields.* New York: Springer-Verlag.

[106] Henson, M. C. 1987. *Elements of Functional Languages.* Computer Science Texts. Oxford: Blackwell Scientific Publications.

[107] Kemper, J. T. 1978. The effects of asymptomatic attacks on the spread of infectious disease. *Bull. Mathem. Biol. 40:*707–718.

[108] Rand, R. H. 1984. *Computer Algebra in Applied Mathematics: An Introduction to MACSYMA.* Boston: Pitman Advanced Publishing Program.

[109] Schwarz, F. 1989. *Computer-Algebra.* Schloß Birlinghoven, D-5205 Sankt Augustin 1, Germany: GMD, Institut F1.

[110] Symbolics, Inc. 1985. *VAX UNIX MACSYMA™ Reference Manual.* Cambridge, Mass.: Symbolics, Inc.

[111] Voit, E. O. and M. A. Savageau. 1985. S-system analysis of biological systems. In: *Mathematics in Biology and Medicine. Lecture Notes in Biomathematics* 57, ed. V. Capasso, E. Grosso, and S. L. Paveri-Fontana, 517–524. Berlin: Springer-Verlag.

Also cited in this Chapter: [2], [14], [16], [21], [24–26], [30], [38–39].

Chapter 7:

Cited in this Chapter: [2], [7–9], [21], [24–26], [38].

Chapter 8:

[112] Bergy. Personal communication.

[113] Blundell, M., and D. Kennell. 1974. Evidence for endonucleolytic attack in decay of *lac* messenger RNA in *Escherichia coli*. *J. Mol. Biol. 83:*143–161.

[114] Bourgeois, S. 1971. The *lac* repressor. *Curr. Top. Cellul. Reg. 4:*39–75.

[115] Casadaban, M. J. 1976. Regulation of the regulatory gene for the arabinose pathway *araC*. *J. Mol. Biol. 104:*557–566.

[116] Debarbouille, M., and M. Schwartz. 1979. The use of gene fusions to study the expression of *malT*, the positive regulator gene of the maltose regulon. *J. Mol. Biol. 132:*521–534.

[117] Englesberg, E., and G. Wilcox. 1974. Regulation: positive control. *Ann. Rev. Genet. 8:*219–242.

[118] Englesberg, E., J. Irr, J. Power, and N. Lee. 1965. Positive control of enzyme synthesis by gene C in the L-arabinose system. *J. Bacteriol. 90:*946–957.

[119] Goldberger, R. F. 1974. Autogenous regulation of gene expression. *Science 183:*810–816.

[120] Huber, R. E., K. Wallenfels, and G. Kurz. 1975. Action of β-galactosidase on allolactose. *Can. J. Biochem. 53:*1035–1038.

[121] Irvine, D. H., and M. A. Savageau. Using the special structure of S-systems for efficient solution of nonlinear ordinary differential equations with Taylor series. SIAM *J. Num. Anal.* Submitted.

[122] Jacob, F., and J. Monod. 1961. Genetic regulatory mechanisms in the synthesis of proteins. *J. Mol. Biol. 3:*318–356.

[123] Koch, A. L., and H. R. Levy. 1955. Protein turnover in growing cultures of *Escherichia coli*. *J. Biol. Chem. 217:*947–957.

[124] Lehnigk, S. H. 1966. *Stability Theorems for Linear Motions, with an Introduction to Liapunov's Direct Method.* Englewood Cliffs, N.J.: Prentice-Hall.

[125] Leive, L., and V. Kollin. 1967. Synthesis, utilization and degradation of lactose operon mRNA in *Escherichia coli*. *J. Mol. Biol. 24:*247–259.

[126] McFall, E., and M. C. Heincz. 1983. Identification and control of synthesis of the *dsdC* activator protein. *J. Bacteriol. 153:*872–877.

[127] Miller, J. H., and W. S. Reznikoff, eds. 1980. *The Operon.* New York: Cold Spring Harbor Laboratory.

[128] Müller-Hill, B., and J. Kania. 1974. *Lac* repressor can be fused to β-galactosidase. *Nature 249:*561–563.

[129] Ni'Bhriain, N. N., S. Silver, and T. Foster. 1983. Tn5 insertion mutations in the mercuric ion resistance genes derived from plasmid R100. *J. Bacteriol. 155:*690–703.

[130] Raibaud, O., and M. Schwartz. 1984. Positive control of transcription initiation in bacteria. *Ann. Rev. Genet. 18:*173–206.

[131] Ritzenthaler, P., and M. Mata-Gilsinger. 1982. Use of *in vitro* gene fusions to study the *uxuR* regulatory gene in *Escherichia coli* K-12: Direction of transcription and regulation of its expression. *J. Bacteriol. 150:*1040–1047.

[132] Rotman, B., and S. Speigelman. 1954. On the origin of the carbon in the induced synthesis β-galactosidase in *Escherichia coli. J. Bacteriol. 68:*419–429.

[133] Savageau, M. A. 1977. Design of molecular control mechanisms and the demand for gene expression. *Proc. Natl. Acad. Sci. USA 74:*5647–5651.

[134] Savageau, M. A. 1979. Autogenous and classical regulation of gene expression: A general theory and experimental evidence. In *Biological Regulation and Development*, ed. R. F. Goldberger, 57–108. Vol. 1, New York: Plenum.

[135] Savageau, M. A. 1983. Models of gene function: General methods of kinetic analysis and specific ecological correlates. In: *Foundations of Biochemical Engineering: Kinetics and Thermodynamics in Biological Systems,* ed. H. W. Blanch, E. T. Papoutsakis, and G. Stephanopoulos, 3–25. Washington, D.C.: American Chemical Society.

[136] Savageau, M. A. 1983. Regulation of differentiated cell-specific functions. *Proc. Natl. Acad. Sci. USA 80:*1411–1415.

[137] Savageau, M. A. 1983. *Escherichia coli* habitats, cell types, and molecular mechanisms of gene control. *Am. Nat. 122:*732–744.

[138] Savageau, M. A. 1990. Molecular modes of control in coupled gene circuits: Implications of demand theory. Submitted.

[139] Smith, G. R., and B. Magasanik. 1971. Nature and self-regulated synthesis of the repressor of the *hut* operons in *Salmonella typhimurium. Proc. Natl. Acad. Sci. USA 68:*1493–1497.

Also cited in this Chapter: [1–5], [7], [9], [12–13], [17], [19], [25], [38], [92].

Chapter 9

[140] Dix, D. B. 1983. Accuracy in protein biosynthesis: a kinetic analysis of the interaction of Phe-tRNA and Leu-$tRNA_2$ with poly(U)-programmed ribosomes. Ph.D. Diss., Temple University.

[141] Eccleston, J. F., D. B. Dix, and R. C. Thompson. 1985. The rate of

cleavage of GTP on the binding of Phe-tRNA · elongation factor Tu · GTP to poly(U)-programmed ribosomes of *Escherichia coli*. *J. Biol. Chem. 260:*16237–16241.

[142] Edelmann, P., and J. Gallant. 1977. Mistranslation in *E. coli*. *Cell 10:*131–137.

[143] Grosjean, H. J., S. DeHenau, and D. M. Crothers. 1978. On the physical basis for ambiguity in genetic coding interactions. *Proc. Natl. Acad. Sci. USA 75:*610–614.

[144] Hopfield, J. J. 1974. Kinetic proofreading: A new mechanism for reducing errors in biosynthetic processes requiring high specificity. *Proc. Natl. Acad. Sci. USA 71:*4135–4139.

[145] Loftfield, R. B., and D. Vanderjagt. 1972. The frequency of errors in protein biosynthesis. *Biochem. J. 128:*1353–1356.

[146] Okamoto, M., and M. A. Savageau. 1986. Integrated function of a kinetic proofreading mechanism: Double-stage proofreading by isoleucyl-tRNA synthetase. *Biochemistry 25:*1969–1975.

[147] Okamoto, M., J. Yoshii, and K. Hayashi. Kinetic proofreading model of the reaction of poly(U)-programmed ribosomes with a leucyl-tRNA2·elongation factor Tu·GTP complex: Steady-state analysis and predicting parameter values. In preparation.

[148] Thompson, R. C., and D. B. Dix. 1982. Accuracy of protein biosynthesis: A kinetic study of the reaction of poly(U)-programmed ribosomes with a leucyl-tRNA$_2$-elongation factor Tu-GTP complex. *J. Biol. Chem. 257:*6677–6682.

[149] Thompson, R. C., D. B. Dix, and J. F. Eccleston. 1979. Single turnover kinetic studies of guanosine triphosphate hydrolysis and peptide formation in the elongation factor Tu-dependent binding of aminoacyl-tRNA to *Escherichia coli* ribosomes. *J. Biol. Chem. 255:*11088–11090.

[150] Thompson, R. C., and A. M. Karim. 1982. The accuracy of protein biosynthesis is limited by its speed: High fidelity selection by ribosomes of aminoacyl-tRNA ternary complexes containing GTP [γS]. *Proc. Natl. Acad. Sci. USA 79:*4922–4926.

[151] Thompson, R. C., and P. J. Stone. 1977. Proofreading of the codon-anticodon interaction on ribosomes. *Proc. Natl. Acad. Sci. USA 74:*198–202.

Also cited in this Chapter: [2–3], [29], [71–72].

Chapter 10

[152] Lloyd, F. T., and W. R. Harms. 1986. An individual stand growth model for mean plant size based on the rule of self-thinning. *Ann. Bot. 57:*681–688.

Also cited in this Chapter: [2], [19], [22], [31], [33], [38], [46].

Chapter 11

[153] Arnold, L. 1974. *Stochastic Differential Equations, Theory and Application.* New York: John Wiley & Sons.

[154] Banks, H. T., and G. M. Groome Jr. 1973. Convergence theorems for parameter estimation by quasilinearization. *J. Mathem. Anal. and Appl. 42:*91–109.

[155] Erickson, R. V. 1971. Constant coefficient linear differential equations driven by white noise. *Ann. Mathem. Stat. 42:*820–823.

[156] Garcia, O. 1983. A stochastic differential equation model for the height growth of forest stands. *Biometrics 39:*1059–1072.

[157] Golub, G. H., and C. F. Van Loan. 1983. *Matrix Computations.* Baltimore: The Johns Hopkins University Press.

[158] Jazwinski, A. H. 1970. *Stochastic Processes and Filtering Theory.* New York: Academic Press.

[159] Johnson, T. 1986. Effects of specification errors in continuous dynamic systems. Econometric Society Meetings.

[160] Kalaba, R. E., and K. Spingarn. 1982. *Control, Identification and Input Optimization.* New York: Plenum Press.

[161] McGarty, T. P. 1974. *Stochastic Systems and State Estimation.* New York: John Wiley & Sons.

[162] Smith, W. D. 1988. *Modeling Sweep of Loblolly Pines.* Ph.D. Diss., Department of Forestry, North Carolina State University.

[163] Van Loan, C. F. 1978. Computing integrals involving the matrix exponential. *IEEE Trans. Auto. Control* AC-23 *3:*(June).

Also cited in this Chapter: [24], [30–31], [38], [46].

Chapter 12

[164] Arneodo, A., P. Coullet, and C. Tresser. 1980. Occurrence of strange attractors in three-dimensional Volterra equations. *Phys. Lett. 79A:*259–263.

[165] Chen, C. T. 1984. *Linear System Theory and Design.* New York: Holt, Rinehart, & Winston.

[166] Ebeling, W., and M. Peschel, eds. 1985. *Lotka-Volterra-Approach to Cooperation and Competition in Dynamic Systems.* Berlin: Akademie-Verlag.

[167] Lotka, A. J. 1956. *Elements of Physical Biology.* Williams & Wilkins, Baltimore (1924): reprinted as *Elements of Mathematical Biology.* New York: Dover.

[168] May, R. M. 1973. *Stability and Complexity in Model Ecosystems.* Princeton, N.J.: Princeton Univ. Press.

[169] Mende, W., and M. Peschel. 1981. Strukturentwurf für instationäre und nichtlineare Systeme. *Messen-Steuern-Regeln 24:*581–583.

[170] Peschel, M., and W. Mende. 1982. Volterra-Modelle für nichtlineare und instationäre Systeme und deren Äquivalenztransformationen. *Messen-Steuern-Regeln 25:*102–104.

[171] Peschel, M., and W. Mende. 1983. Riccati-Darstellung nichtlinearer Systeme und einige meßtechnische Konsequenzen. *Messen-Steuern-Regeln 25:*67–69.

[172] Peschel, M., and W. Mende. 1984. A unified modelling concept for nonlinear systems with Lotka-Volterra equations. *Syst. Anal. Model. Simul. 1:*17–26.

[173] Redheffer, R., and W. Walter. 1984. Solution to the stability problem for a class of generalized Volterra prey-predator systems. *J. Differential Equations 52(2):*245–263.

[174] Redheffer, R., and Z. Zhiming. 1981. Global asymptotic stability for a class of many-variable Volterra prey-predator systems. *Nonlinear Anal. Theor. Meth. Appl. 5(12):*1309–1329.

[175] Richter, O. 1985. *Simulation des Verhaltens ökologischer Systeme: Mathematische Methoden und Modelle.* Weinheim, Germany: Verlag Chemie.

[176] Rössler, O. E. 1979. Continuous chaos—four prototype equations. In *Bifurcation Theory and Applications in Scientific Disciplines,* ed. O. Gurel and O. E. Rössler, *Ann. NY Acad. Sci. 316:*376–392.

[177] Saaty, T. L., and J. Bram. 1964. *Nonlinear Mathematics.* New York: Dover Publications, Inc. 214.

[178] Volterra, V. 1931. *Leçons sur la Théorie Mathématique de la Lutte pour la Vie.* Paris: Gauthier-Villars.

[179] Wright, R. 1987. The razor's edge. *The Sciences 4:*2–5.

Also cited in this Chapter: [2–3], [7], [9], [19], [24], [27–30], [35–39], [41–50], [73], [111].

Chapter 13

[180] Abramowitz, M., and I. A. Stegun, eds. 1972. *Handbook of Mathematical Functions.* 9th printing. New York: Dover Publications.

[181] Carlson, B. C. 1977. *Special Functions of Applied Mathematics.* New York: Academic Press.

[182] Rust, P. F. 1986. Programming common distributions for precise P-values. *Computational Statistics & Data Analysis 4:*205–209.

Also cited in this Chapter: [42], [49].

Chapter 14

[183] Brooks, F. P. 1985. Physiology of the stomach. In *Bockus Gastroenterology,* ed. J. E. Berk. Volume 2, p. 874. W. B. Saunders.

[184] Diamant, N. E., P. K. Rose, and E. J. Davison. 1970. Computer simulation of intestinal slow-wave frequency gradient. *Am. J. Physiol. 219:*1684–1690.

[185] Glandsdorff, P. and I. Prigogine. 1971. *Thermodynamic Theory of Structure, Stability and Fluctuations.* New York: Wiley & Sons.

[186] Larter, R., H. Rabitz, and M. Kramer. 1984. Sensitivity analysis of limit cycles with application to the Brusselator. *J. Chem. Phys. 80:*4120–4128.

[187] Nicolis, G. 1971. Stability and dissipative structures in open systems far from equilibrium. *Adv. Chem. Phys. 19:*209–324.

[188] Nicolis, G., and I. Prigogine. 1977. *Self-Organization in Non-Equilibrium Systems. From Dissipative Structures to Order Through Fluctuations.* New York: Wiley & Sons.

[189] Plant, R. E., and M. Kim. 1976. Mathematical description of a bursting pacemaker neuron by a modification of the Hodgkin-Huxley equations. *Biophys. J. 16:*227–244.

[190] Prigogine, I., and I. Stengers. 1984. *Order Out of Chaos: Man's New Dialogue With Nature.* London: New Science Library.

[191] Ritger, P., and N. Rose. 1968. *Differential Equations with Applications,* New York: McGraw-Hill. 325.

[192] Robertson-Dunn, and D. A. Linkens. 1974. A mathematical model of the slow-wave electrical activity of the human small intestine. *Med. Bio. Eng. 12:*750–757.

[193] Scratcherd, T., and D. Grundy. 1984. The physiology of intestinal motility and secretion. *Br. J. Anaesth. 56:*3–18.

[194] Szurszewski, J. H. 1987. Electrical basis for gastrointestinal motility. In *Physiology of the Gastrointestinal Tract,* ed. L. R. Johnson, 2nd ed., Vol. 1, 383. New York: Raven Press.

[195] Tyson, J. J. 1973. Some further studies of nonlinear oscillations in chemical systems. *J. Chem. Phys. 58:*3919–3930.

[196] van der Pol, B. 1927. Forced oscillations in a circuit with non-linear resistance (reception with reactive triode). *Phil. Mag. (7)3:*65–80.

[197] van der Pol, B. 1928. The heart beat considered as a relaxation oscillation, and an electrical model of the heart. *Phil. Mag. (7)6:*763–775.

[198] Wever, R., and R. Lund. 1973. Free and forced internal desynchronisation of circadian rhythms. *21st International Congress of Aviation and Space Medicine,* Munich.

Also cited in this Chapter: [35].

Chapter 15

[199] Arnold, V. I. 1983. *Geometrical Methods in the Theory of Ordinary Differential Equations.* New York: Springer-Verlag.

[200] Hill, J. M. 1982. *Solution of Differential Equations by Means of One-Parameter Groups.* Boston: Pitman Advanced Publishing Program.

[201] Lie, S. (1891) 1967. *Vorlesungen über Differentialgleichungen mit bekannten infinitesimalen Transformationen,* ed. G. Scheffers. Leipzig: Teubner. Reprint: New York: Chelsea.

[202] Noether, E. 1918. Invariante Variationsprobleme. Nachr. Königl. Gesell. Wissensch. Göttingen. *Mathem.-Phys. Kl.:*235–257.

[203] Olver, P. J. 1986. *Applications of Lie Groups to Differential Equations.* New York: Springer-Verlag.

[204] Schwarz, F. 1988. Symmetries of differential equations: From Sophus Lie to computer algebra. *SIAM Review 30:*450–481.

Also cited in this Chapter: [33], [42].

Chapter 16

[205] Carr, J. 1981. *Applications of Center Manifold Theory.* New York: Springer-Verlag.

[206] Hopf, E. 1942. Abzweigung einer periodischen Lösung von einer stationären Lösung eines Differential-Systems. *Ber. Mathem.-Phys. Kl. Sächs. Acad. Wiss. Leipzig 94:*1–22.

[207] Kelley, A. 1967. The stable, center stable, center, center unstable and unstable manifolds. *J. Diff. Eqns. 3:*546–570.

[208] Marsden, J. E., and M. McCracken. 1976. *The Hopf Bifurcation and Its Applications.* New York: Springer-Verlag.

Also cited in this Chapter: [105].

Index

Accumulation, 48, 51-52, 94. *See also* Production *and* Synthesis
Accuracy, 8, 18, 28, 36, 40-41, 50, 63-65, 70, 72-74, 173, 176-179, 235-236, 245, 254-255, 258-261
Activation, 22, 51
Adams method, 134, 136
Aggregation, 17, 41, 53, 72-74, 84-89, 227, 230, 233
Algebraic equation, 57, 173, 189-190
Allometric, 32-33
Amino acid, 26, 29, 34, 70, 166-180
Amphibolic, 42, 73-74
Amplification, 167. *See also* Logarithmic gain
Antibody, 92-93
Anticodon, 166
Antigen, 91-109
Approximation, 4, 10, 23, 34, 40-41, 68-69, 134, 183-184, 225, 227
  linear, 4, 183-184
  power-law, 23, 48-50, 53, 86, 172, 178, 213, 232, 236, 258-264, 276
  stochastic processes, 202
  successive, 12
Arctangent transformation, 224, 255-256
Arrow diagram, 51, 53
Autogenous, 146-164

von Bertalanffy, 3, 9, 43
Bessel function, 65, 214, 224, 242-243
Beta function, 249-251
Bifurcation, 77, 123, 232, 235, 279, 304, 328-344
Binary half-system, 222, 226-227, 230, 282
Binary system, 39, 222, 224, 226-227, 230-232, 282, 322
Biochemical systems theory, 72, 78-79, 82-89. *See also* BST
Black box, 258, 267, 276
B lymphocyte, 92
BMDP, 57, 188-189, 191, 195, 198
Bode, 5-7, 11-12, 23, 48
Brusselator, 261-266, 274, 276
BST, 79, 82-89. *See also* Biochemical systems theory

Canonical forms, 38-40, 213-237
Carrying capacity, 185
Cascade, 24, 27
Cauchy distribution, 255
Chain rule, 34, 39, 215, 228, 245-246, 251
Channeling, 14, 63
Chaos, 23, 217-219, 228, 329
Characteristic equation, 59, 61, 158
Characteristic polynomial, 119, 128
Chi-square, 240-246, 254-255
Chyme, 273-274
Circuit, 10, 22, 27, 31-32, 37, 145-164
Classification, 65, 214, 230, 233-234
Clearance, 48, 51-52, 91, 95-96
Closed system, 53
Cognate, 167, 176, 178
Competition, 228
Complex, 9-10, 14, 22-23, 27, 166
Computer algebra, 110-132
Conservation
  flow, 52
  law, 279-285

Constraint, 33, 108, 129–130, 152, 155–157, 161–162, 184–185, 264, 290
among parameters, 26–27, 53, 56–57, 80–81, 184–185, 188, 213
among variables, 80, 173
recasting, 232, 295
Control
coefficient, 49, 84
group, 182
theory, 21, 28. *See also* MCT
Controlled comparison, 25, 29–31, 36–37, 42, 73, 90–109, 148, 152–153
Cooperation, 228
Correlation, 31, 188, 205–206
Cumulative distribution, 34, 240–243, 247, 251
Cycle
cell, 42
limit, 16, 77, 131, 183–184, 259, 262, 264–265
substrate, 115

Degradation, 17, 26, 37, 52, 72, 74–75, 79–80, 82, 89, 91, 149–151, 156–157, 232, 261, 263–264. *See also* Clearance
Delay differential equation, 217
Demand, 31, 164, 176, 178
Differential equation
delay, 217
implicit, 217
integro, 217
order, 203, 217, 242, 244
stochastic, 200–206
Discrimination, 168, 174–178
Distribution
beta, 249–251
Cauchy, 255
chi-square, 240–246, 254–255
exponential, 239
F, 250
gamma, 224, 249–250
Gompertz, 286
normal, 202, 239–240
Pearson family, 238
t, 236, 250, 255
Durbin-Watson statistic, 205

E. coli, 22, 24, 31, 35-36, 145–147, 150–151, 167–168
Ecology, 31, 228, 232, 236
Editing, 168
Effector, 69
cells, 92–109
Efficiency, 21, 37, 39–40, 43, 65, 133, 214, 221, 236
Eigenspace, 307–308
Eigenvalue, 111, 119, 121, 124, 137, 307–339
Einstein, 4, 23
Elasticity, 84, 88
Elongation factor, 166–180
Enzyme, 14, 22–23, 27, 35, 41, 62–63, 69–70, 79, 84, 150, 154–155
activity, 56
complexes, 14, 35, 63
key, 79
kinetic, 11
Enzymology, 68
Equilibrium, 42, 74, 310–343. *See also* Steady state
Equivalence, 4, 39, 223–224, 229–230, 233–234
internal, 25, 152–154
external, 25, 29, 152–157, 162
Error, 40, 168
computational, 131
experimental, 11, 36
random, 204–205
Escherichia coli, 22, 24, 31, 35–36, 145–147, 150–151, 167–168
ESSYNS, 39–40, 42–43, 63, 65, 133–141, 152, 198, 214, 231, 235–236, 239–240, 246, 249–250, 278, 304, 341–343
Euler, 4
Expansion of S-system, 314, 319–321, 325

F distribution, 250
F-factor, 59–60, 120, 154, 158–159, 172–173
F-test, 206
Feedback, 10, 25–29, 34–35, 64, 68, 70–71, 73, 121–122, 126
Feedforward, 26, 34–35
Filter, 203–205
Flow, 52, 176, 203, 313
Flux, 5, 14, 17–18, 20, 26–27, 41–42, 53, 56, 59, 73–74, 76–89, 104, 106, 110, 119–120, 167–168, 173–174, 177–179
Forestry, 181–199, 201
Functional effectiveness, 91–109, 148, 151–152, 158, 162–163

Gain, 20, 24–25, 56–57, 61–62, 78, 83, 88, 96, 98, 100–101, 103, 105, 107, 111,

116, 121-123, 128, 134, 137, 152-153, 156, 188, 206
Galileo, 2-4, 8
Gamma function, 224, 249
Gauss, 202, 208
Gene circuit, 31-32, 37, 145-164
Gene expression, 10, 27, 30
Generalized mass action, 14-15, 38, 40, 49-50, 64, 71, 216, 225-226, 265, 276, 279. *See also* GMA *and* Mass action
Gene regulation, 11, 24, 31, 68, 145-164
Genetic control, 73
GMA, 40-42, 71-72, 74-75, 216-217, 225-227, 229, 232, 279-302. *See also* Generalized mass action
Graphics, 139
Growth, 3, 26-27, 32-35, 41, 91, 95-96, 112, 150, 181-199, 201, 203, 214, 230, 234-235
Group, 41, 232, 235, 278-303
GTP, 166-180

Half-system, 16, 37-39, 50, 74, 190-191, 196-198, 222-235, 278, 282, 287, 297, 322
Herodotus, 1
Hill, 10-11, 55
Hodgin-Huxley, 274
Hopf, 131, 232, 235, 304, 328, 330-344
Horner scheme, 220
Hydrolysis, 167, 178

Idiotype, 92
Immune network, 36-37, 63, 68, 73, 91-109
Immune response, 63, 92-109
Immune system, 22, 36, 92-109
Implicit
  differential equation, 217
  representation, 84
  solution, 113
Induction characteristic, 150, 153, 155, 159, 161-162
Infection, 115
Inhibition, 10, 25-29, 34, 51, 63, 121-122
Integro-differential equation, 217
Invariant, 280-289
Inversion of S-system, 245
Irreversible, 167

Jacobian matrix, 85, 116, 119

Kalman filter, 203, 205
Kalman gain, 206
Kinetic, 3, 11, 13-14, 23, 29, 53, 167
  order, 13, 20, 24, 28, 49, 55-56, 69, 82-84, 88, 95, 150-151, 154, 156-157, 172
Kirchhoff, 4-5, 14, 16-18

Lac operon, 31, 146, 150, 158, 161, 163-164
Laplace, 23
Least squares, 202, 204-205
Lie, 278-281, 299
Lienard transformation, 270
Lifetime
  biochemical complexes, 166
Limit cycle, 16, 77, 131, 183-184, 259, 262, 264-265
Linear
  dependence, 20, 121
  independence, 117, 281, 198-199, 317, 328
  piecewise, 7
Linearization, 12, 25, 47, 121, 183, 201, 203, 312, 326-327, 333, 335
LISP, 111, 126-127, 130-131
Loblolly Pine, 181-182, 189-190
Logarithmic gain, 20, 24-25, 56-57, 61-62, 78, 83, 88, 105, 107, 111, 116, 121-123, 128, 134, 137, 152-153, 156, 188
Logarithmic coordinates, 13, 20, 33
Logarithmic space, 13, 48, 64
Log-log plot, 5-8, 12, 20, 33
Lotka, 214, 222, 226-228, 233
Lymphocyte, 63, 92-108

MACSYMA, 110-132, 340
Manifold, 121, 221, 310-313, 326-330, 333-335, 337
Martingale, 200
Mass action, 14. *See also* Generalized mass action
Matrix
  gain, 83, 206
  inverse, 19-20, 58, 287
  Jacobian, 85, 116, 119, 291
  regular, 58
  sensitivity, 85-89
MCT, 79, 84-85, 87-89
Mean, 247
Metabolic
  control, 42, 49, 79, 84-85
  control theory. *See* MCT
  pathway, 70-71, 74
  physiology, 3
  rate, 8
  system, 79-80

Michaelis-Menten, 15–16, 23, 27, 33, 36, 55, 59, 68–70, 260-261
Misreading probability, 166
Mode, 239
Modulation, 51–52, 64
Moment, 247–248
Moving average, 206
Multinomial system, 216, 225

Network theory, 4–5, 7, 11, 20, 23, 25, 68
Newton, 23, 190, 208, 210
Node, 5, 14, 16–18
Normal distribution, 202, 239–240

Open system, 53
Operating point, 13, 54–56, 65, 70, 73–75, 172–174, 213, 259-261, 263-264
Optimal
  recasting, 215, 221, 226-227, 229-233
  representation, 73–74
Orbit, 331-343
Oscillations, 15, 23, 47, 77, 228, 236, 259, 262, 266-276

Pathway, 25-35, 53, 70-75, 79, 81, 88-89, 260, 262
Parameter
  estimation, 12-13, 20, 41, 50, 53-57, 77-81, 168, 178, 181, 186-198, 200-206
  group, 232, 235, 278-303
  kinetic, 133, 167-180
  meaning, 49, 232-233
  sensitivity, 20, 25, 47, 56-57, 61-62, 76-89, 111, 122, 124-128, 154
Percentile, 256. *See also* Quantile
Periodic, 309-310, 329-343
Perturbation, 24, 70, 74, 78, 93, 152, 162, 174, 176, 213, 269, 328
Phase plane, 139
Pine, 181-182, 189-190, 201, 203
PLF *See* Power-law formalism
van der Pol, 266, 269-277
Polar coordinates, 224
Poles, 232
Polynomial, 34, 59-61, 119, 128, 219-221, 338, 340
Population, 182
Power (of test), 246, 253-254
Power-law formalism, 2, 11-12, 16-41, 48-50, 63, 68-70, 149, 161, 172, 214, 236
Power-law function, 3, 8, 11-14, 17-18, 33-34, 49-50, 64, 69-71, 149, 172, 185, 190, 215-216, 225, 227, 280, 282-284
Precursor, 26, 94, 96
Predator-prey, 15, 22, 214, 227
Predictions, 62-63, 167
Probability distribution, 34, 42, 214-215, 238-257
Production, 51, 104, 167-168, 228, 232. *See also* Synthesis *and* Accumulation
Proofreading, 36, 166-180

Quantile, 245-246
Quasi homogeneous, 282
Quasi linearization, 201
Quasi steady state, 33, 173, 184

Rank, 130, 280-281, 288, 290, 317-318
Rate, 6, 8, 10, 12-13, 20, 71-73, 104-105, 149-150, 233
  constant, 13, 20, 24, 49, 53-54, 69, 94, 104, 112, 115, 150, 154, 156, 158, 168, 172, 260, 263-264
  law, 6, 11, 14, 16-17, 33, 53, 56, 64, 67-71, 79-84, 87-89, 236, 259-261. *See also* Michaelis-Menten *and* Hill
  limiting step, 79
Recasting, 38-43, 65, 131, 204, 213-277, 282, 287-297
Recycling, 174, 177
Regulation, 10, 21, 24, 27-32, 68, 73, 90-94, 145-164. *See also* Gene regulation
Regular S-system, 314-328, 335-340
Regulatory network, 50
Regression
  linear, 13, 20, 53, 56, 186, 188, 192, 195
  nonlinear, 57, 186, 188-190, 200
Relaxation time, 173
Response function, 201
Ribosome, 149, 166-169
RNA, 149-151, 153-154, 157, 167-169, 173-179
Robustness, 47, 73, 188
Rotation, 283
Routh-Hurwitz Criterion, 37, 60-61, 119-121, 152
Runge-Kutta Method, 134, 136, 188-189, 198

Saturable, 49
Saturation, 11, 23, 176, 183
SCoP, 189-190
Self-replication, 22

Self-starvation, 26
Self-thinning, 185
Sensitivity, 20, 24, 49, 56–57, 61–62, 76–89, 97, 101, 103, 107, 122, 124–128, 134, 137, 154, 188, 278. *See also* Parameter sensitivity
  analysis, 36, 56–57, 76–89
  coefficient, 78, 83–89
  theory, 21, 25, 42
S-exponential, 315–316, 323
Similarity, 2–3, 8
Simulation, 57, 136, 184–185
Smoothing, 190
Solution
  analytical, 15, 35, 37, 47, 63, 110–132, 201, 214, 235
  explicit, 18, 20, 154
  numerical, 11, 18, 37, 40, 134, 136–138, 150, 161, 172, 178, 214, 231
  speed, 214, 229, 231, 235–236, 245
  symbolic, 57, 134, 153
  trajectorial, 193–196, 293, 296–297
S-product, 315–316, 320, 322
SSE, 204–206
Stability, 28–29, 35, 37, 47, 50, 59–61, 110–111, 121–122, 152, 157–160, 162, 269, 278, 304, 308–312, 328, 334, 337–343
  analytical computation, 47, 59–61, 110
  biochemical complexes, 166
  marginal, 121
Statistical mechanics, 32
Steady state, 9, 11–24, 28–29, 33, 38, 47, 50, 57–59, 73–74, 77–89, 97–108, 110–111, 114–126, 130–131, 150–153, 156–161, 167, 185, 221, 230–231, 235, 278, 287–289, 296–297. *See also* Equilibrium
Stiff, 36, 172–173, 178, 231
Stochastic
  differential equation, 200–202
  process, 200–202, 206
Strategy
  modeling, 50, 53, 67–75, 236
  recasting, 215–236, 240–242, 293
Stretching group, 282
Structure design principle, 228–229, 232
Suppression, 63, 93–94
Sweep, 203–204
Symbolic
  analysis, 10, 18, 62, 108, 110–132, 236
  computation, 56, 61, 110–132
  computer analysis, 110–132, 236
Symmetry, 278–303
Synergism, 11, 23, 47, 183
Synergistic, 18, 49, 149, 152, 232
Synthesis, 4–5, 17, 20, 26, 37, 44, 51, 72, 74–75, 79–80, 82, 89, 149, 154, 156–157, 260, 263–264. *See also* Production *and* Accumulation

Tail, 251–256
T-cell receptor, 92
Test
  chi-square, 246
  F, 206
  power, 246, 253–254
T lymphocytes, 92
Taylor, 4, 12, 16, 33, 48, 54–56, 64–65, 78, 134–137, 140, 201, 317, 331–334, 337, 340
Telescopic property, 33
Temporally dominant, 29, 33, 149
Thermodynamic equilibrium, 74
Threshold, 96, 152, 155, 162, 266–268, 272
Tolerance, 92
Transfer function, 5–7, 11, 217, 221
Trajectory, 37–38, 185–186, 193–196, 198, 259, 293, 296–297
Tree, 181–198, 201–206
Turnover, 59

User function, 111–112, 116–117, 124–130

Variable
  aggregate, 33
  ancillary, 217, 232–234, 261
  auxiliary, 184, 218, 240, 245–246, 261
  dependent, 14, 19, 20, 24, 79–80, 83, 88, 117–118, 136–137, 149, 156–157, 192, 246
  external, 79, 82–83, 86–87, 258, 263
  independent, 19–20, 24, 48, 79, 81, 88, 136–137, 148, 153, 156, 183, 192, 245–247, 255, 280, 291
  unobservable, 203, 205
Variance, 247
Volterra, 39, 214, 222, 226–228

Waste energy, 168, 174
Wiener process, 202–203

Yield, 181, 203